Parasitology: A Conceptual Approach

Parasitology: A Conceptual Approach

Edited by
Grayson Barker

Larsen & Keller
www.larsen-keller.com

Parasitology: A Conceptual Approach
Edited by Grayson Barker
ISBN: 978-1-63549-213-2 (Hardback)

⊟ Larsen & Keller

Published by Larsen and Keller Education,
5 Penn Plaza,
19th Floor,
New York, NY 10001, USA

Cataloging-in-Publication Data

Parasitology : a conceptual approach / edited by Grayson Barker.
    p. cm.
Includes bibliographical references and index.
ISBN 978-1-63549-213-2
1. Parasitology. 2. Parasites. I. Barker, Grayson.
QL757 .P37 2017
591.785 7--dc23

The publisher's policy is to use permanent paper from mills that operate a sustainable forestry policy. Furthermore, the publisher ensures that the text paper and cover boards used have met acceptable environmental accreditation standards.

Printed and bound in the United States of America.

For more information regarding Larsen and Keller Education and its products, please visit the publisher's website www.larsen-keller.com

Table of Contents

Preface

The book aims to shed light on some of the unexplored aspects of parasitology. It talks about the different studies performed and techniques used in detail in this field. Parsitology refers to the study of parasites, hosts and their relationship and interactions with each other. Most of the topics introduced in this text cover new techniques and the applications of this field. It explores all the important aspects of the subject in the present day scenario. This text will serve as a reference to a broad spectrum of readers. It will have immensely beneficial to professionals and students involved in the area at various levels.

A detailed account of the significant topics covered in this book is provided below:

Chapter 1- The study of parasites is known as parasitology. It studies the relationship between parasites and their hosts; human beings are hosts to nearly 300 species of parasites, some of them lethal and others, beneficial. This chapter focuses on the subject of parasitology.

Chapter 2- The major areas related to parasitology have been discussed in the following text. Some of these areas are human parasites, veterinary parasitogy, quantitative parasitology and structural parasitology. Human parasites include parasites that cause parasitic diseases. They can be divided into either endoparasites or ectoparasites. Endoparasites cause infections inside the body whereas ectoparasites cause infections rapidly within the skin. The section closely examines the major areas related to parasitology.

Chapter 3- Protozoa is a single-celled organism eukaryotic organism. Some of the topics listed in this chapter are plasmodium, plasmodium falciparum, plasmodium vivax, plasmodium knowlesi and trypanosome cruzi. This chapter is written in a manner that will introduce briefly all the significant aspects of protozoa.

Chapter 4- Tapeworms are parasitic and live in the digestive systems of adults. Tapeworms are hermaphroditic; they have both male and female reproductive systems in their bodies. Some of the types of tapeworms are diphyllobothrium, echinococcus granulosus, hymenolepis nana and taenia solium. In order to understand tapeworms, it is necessary to understand all the types of tapeworms.

Chapter 5- Flukes have numerous classifications; some of these are liver flukes, paragonimus and schistosoma. Adult flukes that are found in the liver of mammals are known as liver flukes. They feed on blood and also reproduce into the intestine. This section helps the reader in understanding flukes and their characteristics.

Chapter 6- Roundworms species are almost one million in number. The types of roundworms explained within this chapter are anisakis, brugia malayi, dracunculus medinensis, loa loa filariasis and thelzia callipaeda. This text will not only provide an overview, it will also delve into the topics related to it.

Chapter 7- Parasites are usually species that benefit at the expense of the host. Head louse cause head lice infestation. Head lice are insects that spend their entire life on the human scalp and feed on human blood. Scabies is another kind of parasite that causes skin infections, and its symptoms are itchiness and rashes. The aspects elucidated in this chapter are of vital importance, and provide a better understanding of parasites.

I would like to make a special mention of my publisher who considered me worthy of this opportunity and also supported me throughout the process. I would also like to thank the editing team at the back-end who extended their help whenever required.

Editor

Introduction to Parasitology

The study of parasites is known as parasitology. It studies the relationship between parasites and their hosts; human beings are hosts to nearly 300 species of parasites, some of them lethal and others, beneficial. This chapter focuses on the subject of parasitology.

Parasitology

Parasitology is the study of parasites, their hosts, and the relationship between them. As a biological discipline, the scope of parasitology is not determined by the organism or environment in question, but by their way of life. This means it forms a synthesis of other disciplines, and draws on techniques from fields such as cell biology, bioinformatics, biochemistry, molecular biology, immunology, genetics, evolution and ecology.

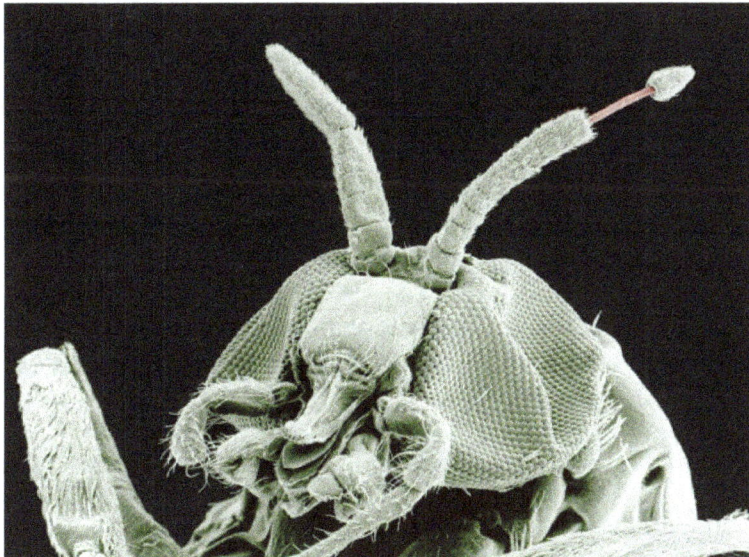

Adult black fly (*Simulium yahense*) with (*Onchocerca volvulus*) emerging from the insect's antenna. The parasite is responsible for the disease known as river blindness in Africa. Sample was chemically fixed and critical point dried, then observed using conventional scanning electron microscopy. Magnified 100×.

Fields

The study of these diverse organisms means that the subject is often broken up into simpler, more focused units, which use common techniques, even if they are not studying the same organisms or diseases. Much research in parasitology falls somewhere between two or more of these definitions. In general, the study of prokaryotes falls under the field of bacteriology rather than parasitology.

Medical Parasitology

"Humans are hosts to nearly 300 species of parasitic worms and over 70 species of protozoa, some derived from our primate ancestors and some acquired from the animals we have domesticated or come in contact with during our relatively short history on Earth".

The Italian Francesco Redi, considered to be the father of modern parasitology, he was the first to recognize and correctly describe details of many important parasites.

One of the largest fields in parasitology, medical parasitology is the subject which deals with the parasites that infect humans, the diseases caused by them, clinical picture and the response generated by humans against them. It is also concerned with the various methods of their diagnosis, treatment and finally their prevention & control. A parasite is an organism that live on or within another organism called the host . These include organisms such as:

- *Plasmodium* spp., the protozoan parasite which causes malaria. The four species of malaria parasites infective to humans are *Plasmodium falciparum*,*Plasmodium malariae*, *Plasmodium vivax* & *Plasmodium ovale*.

- *Leishmania donovani*, the unicellular organism which causes leishmaniasis

- *Entamoeba* and *Giardia*, which cause intestinal infections (dysentery and diarrhoea)

- Multicellular organisms and intestinal worms (helminths) such as *Schistosoma* spp., *Wuchereria bancrofti*, *Necator americanus* (hookworm) and *Taenia* spp. (tapeworm)

- Ectoparasites such as ticks, scabies and lice

Medical parasitology can involve drug development, epidemiological studies and study of zoonoses.

Veterinary Parasitology

The study of parasites that cause economic losses in agriculture or aquaculture operations, or which infect companion animals. Examples of species studied are:

- *Lucilia sericata*, a blowfly, which lays eggs on the skins of farm animals. The maggots hatch and burrow into the flesh, distressing the animal and causing economic loss to the farmer

- *Otodectes cynotis*, the cat ear mite, responsible for Canker.

- *Gyrodactylus salaris*, a monogenean parasite of salmon, which can wipe out populations which are not resistant.

Structural Parasitology

This is the study of structures of proteins from parasites. Determination of parasitic protein structures may help to better understand how these proteins function differently from homologous proteins in humans. In addition, protein structures may inform the process of drug discovery.

Quantitative Parasitology

Parasites exhibit an aggregated distribution among host individuals, thus the majority of parasites live in the minority of hosts. This feature forces parasitologists to use advanced biostatistical methodologies.

Parasite Ecology

Parasites can provide information about host population ecology. In fisheries biology, for example, parasite communities can be used to distinguish distinct populations of the same fish species co-inhabiting a region. Additionally, parasites possess a variety of specialized traits and life-history strategies that enable them to colonize hosts. Understanding these aspects of parasite ecology, of interest in their own right, can illuminate parasite-avoidance strategies employed by hosts.

Conservation Biology of Parasites

Conservation biology is concerned with the protection and preservation of vulnerable species, including parasites. A large proportion of parasite species are threatened by extinction, partly due to efforts to eradicate parasites which infect humans or domestic animals, or damage human economy, but also caused by the decline or fragmentation of host populations and the extinction of host species.

Taxonomy and Phylogenetics

The huge diversity between parasitic organisms creates a challenge for biologists who wish to describe and catalogue them. Recent developments in using DNA to identify separate species and to investigate the relationship between groups at various taxonomic scales has been enormously useful to parasitologists, as many parasites are highly degenerate, disguising relationships between species.

History

[10]"Our knowledge of parasitic infections extends into antiquity, and descriptions of parasites and parasitic infections are found in the earliest writings and have been confirmed by the finding of parasites in archaeological material".

Parasitism

In biology/ecology, parasitism is a non-mutual symbiotic relationship between species, where one species, the parasite, benefits at the expense of the other, the host. Traditionally *parasite* (in biological usage) referred primarily to organisms visible to the naked eye, or macroparasites (such as helminths). Parasites can be microparasites, which are typically smaller, such as protozoa, viruses, and bacteria. Examples of parasites include the plants mistletoe and cuscuta, and animals such as hookworms.

Brood parasitism is a form of parasitism

A *Tetragnatha montana* spider parasitized by an *Acrodactyla quadrisculpta* larva.

A *Lithognathus* fish parasitized by a *Cymothoa exigua* parasite.

Unlike predators, parasites typically do not kill their host, are generally much smaller than their host, and will often live in or on their host for an extended period. Both are special cases of consumer-resource interactions. Parasites show a high degree of specialization, and reproduce at a faster rate than their hosts. Classic examples of parasitism include interactions between vertebrate hosts and tapeworms, flukes, the *Plasmodium* species, and fleas. Parasitism differs from the parasitoid relationship in that parasitoids generally kill their hosts.

Parasites reduce host biological fitness by general or specialized pathology, such as parasitic castration and impairment of secondary sex characteristics, to the modification of host behavior. Parasites increase their own fitness by exploiting hosts for resources necessary for their survival, e.g.

food, water, heat, habitat, and transmission. Although parasitism applies unambiguously to many cases, it is part of a continuum of types of interactions between species, rather than an exclusive category. In many cases, it is difficult to demonstrate harm to the host. In others, there may be no apparent specialization on the part of the parasite, or the interaction between the organisms may remain short-lived.

Etymology

First used in English 1539, the word *parasite* comes from the Medieval French *parasite*, from the Latin *parasitus*, the latinisation of the Greek (*parasitos*), "one who eats at the table of another" and that from (*para*), "beside, by" + (*sitos*), "wheat". Coined in English in 1611, the word *parasitism* comes from the Greek (*para*) + (*sitismos*) "feeding, fattening". In its original sense, it was not strictly pejorative in nature; being a *parasitos* was an accepted lifestyle, whereby a person could live off the hospitality of others, and in return provide "flattery, simple services, and a willingness to endure humiliation".

Types

Parasites are classified based on their interactions with their hosts and on their life cycles. An obligate parasite is totally dependent on the host to complete its life cycle, while a facultative parasite is not. A direct parasite has only one host while an indirect parasite has multiple hosts. For indirect parasites, there will always be a definitive host and an intermediate host.

Human head lice (*Pediculus humanus capitis*) are ectoparasites.

Parasites that live on the outside of the host, either on the skin or the outgrowths of the skin, are called ectoparasites (e.g. lice, fleas, and some mites).

Those that live inside the host are called endoparasites (including all parasitic worms). Endoparasites can exist in one of two forms: intercellular parasites (inhabiting spaces in the host's body) or intracellular parasites (inhabiting cells in the host's body). Intracellular parasites, such as protozoa, bacteria or viruses, tend to rely on a third organism, which is generally known as the carrier or vector. The vector does the job of transmitting them to the host. An example of this interaction is the transmission of malaria, caused by a protozoan of the genus *Plasmodium*, to humans by the bite of an anopheline mosquito.

Those parasites living in an intermediate position, being half-ectoparasites and half-endoparasites, are called mesoparasites.

An epiparasite is one that feeds on another parasite. This relationship is also sometimes referred to as *hyperparasitism,* exemplified by a protozoan (the hyperparasite) living in the digestive tract of a flea living on a dog.

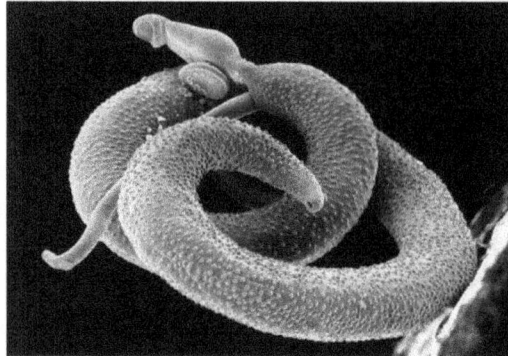

Schistosoma mansoni is an endoparasite that lives in human blood vessels.

Social parasites take advantage of interactions between members of social organisms such as ants, termites, and bumblebees. Examples include *Phengaris arion*, a butterfly whose larvae employ mimicry to parasitize certain species of ants, *Bombus bohemicus*, a bumblebee who invades the hives of other species of bee and takes over reproduction, their young raised by host workers, and *Melipona scutellaris,* a eusocial bee where virgin queens escape killer workers and invade another colony without a queen. An extreme example of social parasitism is the ant species of *Tetramorium inquilinum* of the Alps, which spend their whole lives on the back of *Tetramorium* host ants. With tiny and deprecated bodies they have evolved for one single task: holding on to their host. If they fall off, they most likely would not have the strength to climb back on top of another ant, and eventually they will die.

In kleptoparasitism (from the Greek (kleptes), thief), parasites appropriate food gathered by the host. An example is the brood parasitism practiced by cowbirds, whydahs, cuckoos, and black-headed ducks which do not build nests of their own and leave their eggs in nests of other species. The host behaves as a "babysitter" as they raise the young as their own. If the host removes the cuckoo's eggs, some cuckoos will return and attack the nest to compel host birds to remain subject to this parasitism.

Intraspecific social parasitism may also occur. One example of this is *parasitic nursing*, where some individuals take milk from unrelated females. In wedge-capped capuchins, higher ranking females sometimes take milk from low ranking females without any reciprocation. The high ranking females benefit at the expense of the low ranking females.

Parasitism can take the form of isolated *cheating* or *exploitation* among more generalized mutualistic interactions. For example, broad classes of plants and fungi exchange carbon and nutrients in common mutualistic mycorrhizal relationships; however, some plant species known as myco-heterotrophs "cheat" by taking carbon from a fungus rather than donating it.

An adelpho-parasite (from the Greek (adelphos), brother) is a parasite in which the host species is closely related to the parasite, often being a member of the same family or genus. An example of this is the citrus blackfly parasitoid, *Encarsia perplexa*, unmated females of which may lay haploid eggs in the fully developed larvae of their own species. These result in the production of male offspring. The marine worm *Bonellia viridis* has a similar reproductive strategy, although the larvae are planktonic.

Autoinfection is the infection of a primary host with a parasite, particularly a helminth, in such a way that the complete life cycle of the parasite happens in a single organism, without the involvement of another host. Therefore, the primary host is at the same time the secondary host of the parasite. Some of the organisms where autoinfection occurs are *Strongyloides stercoralis, Enterobius vermicularis, Taenia solium*, and *Hymenolepis nana*. Strongyloidiasis for example involves premature transformation of noninfective larvae in infective larvae, which can then penetrate the intestinal mucosa (internal autoinfection) or the skin of the perineal area (external autoinfection). Infection can be maintained by repeated migratory cycles for the remainder of the person's life.

Host Defenses

Animal Defenses

Skin

The first line of defense against invading parasites is the skin. Skin is made up of layers of dead cells and acts as a physical barrier to invading organisms. These dead cells contain the protein keratin, which makes skin tough and waterproof. Most microorganisms needs a moist environment to survive. By keeping the skin dry, it prevents invading organisms from colonizing. Furthermore, human skin also secretes sebum, which is toxic to most microorganisms.

Mouth

The mouth contains saliva, which prevents foreign organisms from getting into the body orally. Furthermore, the mouth also contains lysozyme, an enzyme found in tears and the saliva. This enzyme breaks down cell walls of invading microorganisms.

Stomach

Should the organism pass the mouth, the stomach is the next line of defense. The stomach contains hydrochloric acid and gastric acids, which makes its pH level around 2. In this environment, the acidity of the stomach helps kill most microorganisms that try to invade the body through the gastric intestinal tract.

Eyes

Parasites can also invade the body through the eyes. The lashes on the eyelid prevents invading microorganisms from entering the eye in the first place. Even if the microorganism does get into the eye, tears contain the enzyme lysozyme, which will kill most invading microorganisms.

Immune System

Should the parasite enter the body, the immune system is a vertebrate's major defense against parasitic invasion. The immune system is made up of different families of molecules. These include serum proteins and pattern recognition receptors (PRRs). PRRs are intracellular and cellular receptors that activate dendritic cells, which in turn activate the adaptive immune system's lymphocytes. Lymphocytes such as the T cells and antibody producing B cells with variable receptors that recognize parasites.

Insect Defenses

Insects often adapt their nests to aid in parasite defense. For example, one of the key reasons the *Polistes canadensis* nests across multiple combs rather than building a single comb like much of the rest of its genus is as a defense mechanism against the infestation of tineid moths. The tineid moth lays its eggs within the wasps' nests and then these eggs hatch into larvae that can burrow from cell to cell and prey on wasp pupae. Adult wasps attempt to remove and kill moth eggs and larvae by chewing down the edges of cells, coating the cells with an oral secretion that gives the nest a dark brownish appearance.

Plant Defenses

In response to parasitic attack, plants undergo a series of metabolic and biochemical reaction pathways that will enact defensive responses. For example, parasitic invasion causes an increase in the jasmonic acid-insensitivel (JA) and NahG (SA) pathway. These pathways produce chemicals that induce defensive responses, such as the production of chemicals or defensive molecules to fight off the attack. Different biochemical pathways are activated by different parasites. In general, there are two types of responses that can be activated by the pathways. Plants can either initiate a specific or non-specific response. Specific responses involve gene-gene recognition of the plant and parasite. This can be mediated by the ability of the plant's cell receptors recognizing and binding molecules that are located on the cell surface of parasites. Once the plant's receptors recognizes the parasite, the plant localizes the defensive compounds to that area creating a hypersensitive response. This form of defense mechanism localizes the area of attack and keeps the parasite from spreading. Furthermore, a specific response against parasitic attack prevents the plants from wasting its energy by increasing defenses where it's not need. However, specific defensive responses only target specific parasites. If the plant lacks the ability to recognize a parasite, specific defense responses won't be activated. Nonspecific defensive responses work against all parasites. These responses are active over time and are systematic, meaning that the responses are not confined to an area of the plant, but rather spread throughout the entirety of the organism. However, nonspecific responses are energy costly, since the plant has to ensure that the genes producing the nonspecific responses are always expressed.

Evolutionary Aspects

Parasitism has arisen independently many times. Depending on the definition used, as many as half of all animals have at least one parasitic phase in their life cycles, and is frequent in plants and fungi. Almost all free-living animals are host to one or more parasitic taxa.

Restoration of a *Tyrannosaurus* with parasite infections. A 2009 study showed that holes in the skulls of several specimens might have been caused by *Trichomonas*-like parasites

Parasites evolve in response to their hosts' defences, sometimes in a manner specific to a particular host taxon and specializing to the point where they infect only a single species. Such narrow host specificity can be costly over evolutionary time, however, if the host species becomes extinct. Therefore, many parasites can infect a variety of more or less closely related host species, with different success rates.

In turn, host defenses coevolve in response to attacks by parasites. Theoretically, parasites may have an advantage in this evolutionary arms race because their generation time commonly is shorter. Hosts reproduce less quickly than parasites, and therefore have fewer chances to adapt than their parasites do over a given span of time.

Long-term coevolution sometimes leads to a relatively stable relationship tending to commensalism or mutualism, as, all else being equal, it is in the evolutionary interest of the parasite that its host thrives. A parasite may evolve to become less harmful for its host or a host may evolve to cope with the unavoidable presence of a parasite—to the point that the parasite's absence causes the host harm. For example, although animals infected with parasitic worms are often clearly harmed, and therefore parasitized, such infections may also reduce the prevalence and effects of autoimmune disorders in animal hosts, including humans. In a more extreme example, some nematode worms cannot reproduce, or even survive, without infection by Wolbachia bacteria.

Competition between parasites tends to favor faster reproducing and therefore more virulent parasites. Parasites whose life cycle involves the death of the host, to exit the present host and sometimes to enter the next, evolve to be more virulent or even alter the behavior or other properties of the host to make it more vulnerable to predators. Parasites that reproduce largely to the offspring of the previous host tend to become less virulent or mutualist, so that its hosts reproduce more effectively.

The presumption of a shared evolutionary history between parasites and hosts can sometimes elucidate how host taxa are related. For instance, there has been dispute about whether flamingos are more closely related to the storks and their relatives, or to ducks, geese and their relatives. The fact that flamingos share parasites with ducks and geese is evidence these groups may be more closely related to each other than either is to storks.

Parasitism is part of one explanation for the evolution of secondary sex characteristics seen in breeding males throughout the animal world, such as the plumage of male peacocks and manes of male lions. According to this theory, female hosts select males for breeding based on such characteristics because they indicate resistance to parasites and other disease.

Co-speciation

In rare cases, a parasite may even undergo co-speciation with its host. One particularly remarkable example of co-speciation exists between the simian foamy virus (SFV) and its primate hosts. In one study, the phylogenies of SFV polymerase and the mitochondrial cytochrome oxidase subunit II from African and Asian primates were compared. Surprisingly, the phylogenetic trees were very congruent in branching order and divergence times. Thus, the simian foamy viruses may have co-speciated with Old World primates for at least 30 million years.

Evolutionary events like host switch, host shift, the duplication or extinction of parasite species (without similar events on the host phylogeny) often erode topographical similarities between host and parasite phylogenies.

Ecology

Quantitative Ecology

A single parasite species usually has an aggregated distribution across host individuals, which means that most hosts harbor few parasites, while a few hosts carry the vast majority of parasite individuals. This poses considerable problems for students of parasite ecology: the use of parametric statistics should be avoided. Log-transformation of data before the application of parametric test, or the use of non-parametric statistics is recommended by several authors. However, this can give rise to further problems. Therefore, modern day quantitative parasitology is based on more advanced biostatistical methods.

Diversity Ecology

Hosts represent discrete habitat patches that can be occupied by parasites. A hierarchical set of terminology has come into use to describe parasite assemblages at different host scales.

Infrapopulation

> All the parasites of one species in a single individual host.

Metapopulation

> All the parasites of one species in a host population.

Infracommunity

> All the parasites of all species in a single individual host.

Component community

> All the parasites of all species in a host population.

Compound community

> All the parasites of all species in all host species in an ecosystem.

The diversity ecology of parasites differs markedly from that of free-living organisms. For free-living organisms, diversity ecology features many strong conceptual frameworks including Robert MacArthur and E. O. Wilson's theory of island biogeography, Jared Diamond's assembly rules and, more recently, null models such as Stephen Hubbell's unified neutral theory of biodiversity and biogeography. Frameworks are not so well-developed for parasites and in many ways they do not fit the free-living models. For example, island biogeography is predicated on fixed spatial relationships between habitat patches ("sinks"), usually with reference to a mainland ("source"). Parasites inhabit hosts, which represent mobile habitat patches with dynamic spatial relationships. There is no true "mainland" other than the sum of hosts (host population), so parasite component communities in host populations are metacommunities.

Nonetheless, different types of parasite assemblages have been recognized in host individuals and populations, and many of the patterns observed for free-living organisms are also pervasive among parasite assemblages. The most prominent of these is the interactive-isolationist continuum. This proposes that parasite assemblages occur along a cline from interactive communities, where niches are saturated and interspecific competition is high, to isolationist communities, where there are many vacant niches and interspecific interaction is not as important as stochastic factors in providing structure to the community. Whether this is so, or whether community patterns simply reflect the sum of underlying species distributions (no real "structure" to the community), has not yet been established.

Adaptation

Parasites infect hosts that exist within their same geographical area (sympatric) more effectively. This phenomenon supports the "Red Queen hypothesis—which states that interactions between species (such as host and parasites) lead to constant natural selection for adaptation and counter adaptation." The parasites track the locally common host phenotypes, therefore the parasites are less infective to allopatric (from different geographical region) hosts.

Experiments published in 2000 discuss the analysis of two different snail populations from two different sources—Lake Ianthe and Lake Poerua in New Zealand. The populations were exposed to two pure parasites (digenetic trematode) taken from the same lakes. In the experiment, the snails were infected by their sympatric parasites, allopatric parasites and mixed sources of parasites. The results suggest that the parasites were more highly effective in infecting their sympatric snails than their allopatric snails. Though the allopatric snails were still infected by the parasites, the infectivity was much less when compared to the sympatric snails. Hence, the parasites were found to have adapted to infecting local populations of snails.

Transmission

Parasites have a variety of methods to infect hosts. For example, the *Acanthamoeba* enters the body when the environment is not hostile, and *Strongyloides stercoralis* enters the body when a host steps on infected ground while barefoot. Many parasites enter the food of their hosts and wait to be eaten. *Plasmodium malariae* uses a mosquito host to transmit malaria, and *Loa loa* parasites use deer flies to enter hosts.

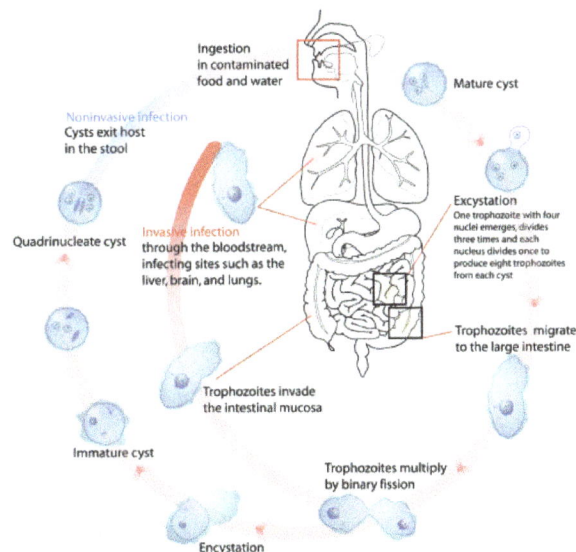

Life cycle of *Entamoeba histolytica*, an anaerobic parasitic protozoan.

Parasites inhabit living organisms and therefore face problems that free-living organisms do not. Hosts, the only habitats in which parasites can survive, actively try to avoid, repel, and destroy parasites. Parasites employ numerous strategies for getting from one host to another, a process sometimes referred to as parasite *transmission* or *colonization*.

Some endoparasites infect their host by penetrating its external surface, while others must be ingested. Once inside the host, adult endoparasites need to shed offspring into the external environment to infect other hosts. Many adult endoparasites reside in the host's gastrointestinal tract, where offspring can be shed along with host excreta. Adult stages of tapeworms, thorny-headed worms and most flukes use this method.

Among protozoan endoparasites, such as the malarial parasites and trypanosomes, infective stages in the host's blood are transported to new hosts by biting-insects, or vectors.

Larval stages of endoparasites often infect sites in the host other than the blood or gastrointestinal tract. In many such cases, larval endoparasites require their host to be consumed by the next host in the parasite's life cycle in order to survive and reproduce. Alternatively, larval endoparasites may shed free-living transmission stages that migrate through the host's tissue into the external environment, where they actively search for or await ingestion by other hosts. The foregoing strategies are used, variously, by larval stages of tapeworms, thorny-headed worms, flukes and parasitic roundworms.

Some ectoparasites, such as monogenean worms, rely on direct contact between hosts. Ectoparasitic arthropods may rely on host-host contact (e.g. many lice), shed eggs that survive off the host (e.g. fleas), or wait in the external environment for an encounter with a host (e.g. ticks). Some aquatic leeches locate hosts by sensing movement and only attach when certain temperature and chemical cues are present.

Some parasites modify host behavior in order to increase the transmission between hosts, often in relation to predator and prey (parasite increased trophic transmission). For example, in California salt marshes, the fluke *Euhaplorchis californiensis* reduces the ability of its killifish host to avoid predators. This parasite matures in egrets, which are more likely to feed on infected killifish than on uninfected fish. Another example is the protozoan *Toxoplasma gondii*, a parasite that matures in cats but can be carried by many other mammals. Uninfected rats avoid cat odors, but rats infected with *T. gondii* are drawn to this scent, which may increase transmission to feline hosts.

Roles in Ecosystems

Modifying the behavior of infected hosts, to make transmission to other hosts more likely to occur, is one way parasites can affect the structure of ecosystems. For example, in the case of *Euhaplorchis californiensis* (discussed above) it is plausible that the local predator and prey species might be different if this parasite were absent from the system.

Although parasites are often omitted in depictions of food webs, they usually occupy the top position. Parasites can function like keystone species, reducing the dominance of superior competitors and allowing competing species to co-exist.

Many parasites require multiple hosts of the different species to complete their life cycles and rely on predator-prey or other stable ecological interactions to get from one host to another. In this sense, the parasites in an ecosystem reflect the health of that system.

Value

Although parasites are generally considered to be harmful, the eradication of all parasites would not necessarily be beneficial. Parasites account for as much as or more than half of life's diversity; they perform an important ecological role (by weakening prey) that ecosystems would take some time to adapt to; and without parasites organisms may eventually tend to asexual reproduction, diminishing the diversity of sexually dimorphic traits. Parasites provide an opportunity for the transfer of genetic material between species. On rare, but significant, occasions this may facilitate evolutionary changes that would not otherwise occur, or that would otherwise take even longer.

Host (Biology)

In biology, a host is an organism that harbors a parasitic, a mutual, or a commensal symbiont, typically providing nourishment and shelter. Examples include animals playing host to *parasitic* worms (e.g. nematodes), cells harbouring a *parasitic* virus, a bean plant hosting *mutualistic* (helpful) nitrogen-fixing bacteria. More specifically in botany, a host plant supplies food resources and acts as a substrate for *commensalist* insects or other fauna.

The roof rat (*Rattus rattus*) is a reservoir host for bubonic plague: the oriental rat fleas that infest these rats are a prime source for the disease.

Guest is the generic term used for parasites, mutualists and commensals.

Definitions

A host cell is a living cell in which a virus reproduces.

A primary host or definitive host is a host in which the parasite reaches maturity and, if possible, reproduces sexually.

A reservoir host can harbour a pathogen indefinitely with no ill effects. A single reservoir host may be reinfected several times.

A secondary host or intermediate host is a host that harbors the parasite only for a short transition period, during which (usually) some developmental stage is completed. For trypanosomes, the cause of sleeping sickness, strictly, humans are the secondary host, while the tsetse fly is the primary host, given that it has been shown that reproduction occurs in the insect. Cestodes (tapeworms) and other parasitic flatworms have complex life-cycles, in which specific developmental stages are completed in a sequence of several different hosts.

As the life cycles of many parasites are not well understood, sometimes the "more important" organism is arbitrarily defined as definitive, and this designation may continue even after it is determined to be incorrect. For example, sludge worms are sometimes considered "intermediate hosts" for whirling disease, even though it is known that the parasite causing the disease reproduces sexually inside them .

In *Trichinella spiralis,* the roundworm that causes trichinosis, a host has both reproductive adults in its digestive tract and immature juveniles in its muscles, and is therefore considered both an intermediate host and a definitive host.

A paratenic host is similar to an intermediate host, only that it is not needed for the parasite's development cycle to progress. Paratenic hosts serve as "dumps" for non-mature stages of a parasite in which they can accumulate in high numbers.

A dead-end host or incidental host is an intermediate host that generally does not allow transmission to the definitive host, thereby preventing the parasite from completing its development. For example, humans are dead-end hosts for *Echinococcus canine* tapeworms. As infected humans are not usually eaten by dogs, foxes etc., the immature *Echinococcus* - although it causes serious disease in the dead-end host - is unable to infect the primary host and mature.

A host of predilection is the host preferred by a parasite.

An amplifying host is a host in which the level of pathogen can become high enough that a vector such as a mosquito that feeds on it will probably become infectious.

Host Range

The host range of a parasite is the collection of hosts that an organism can utilize as a partner. In the case of human parasites, the host range influences the epidemiology of the parasitism or disease. For instance, the production of antigenic shifts in Influenza A virus can result from pigs being infected with the virus from several different hosts (such as human and bird). This co-infection provides an opportunity for mixing of the viral genes between existing strains, thereby producing a new viral strain. An influenza vaccine produced against an existing viral strain might not be effective against this new strain, which then requires a new influenza vaccine to be prepared for the protection of the human population

References

- Wolff, Ewan D. S.; Steven W. Salisbury; John R. Horner; David J. Varricchio (2009). Hansen, Dennis Marinus, ed. "Common Avian Infection Plagued the Tyrant Dinosaurs". PLoS ONE. 4 (9): e7288. doi:10.1371/journal. pone.0007288. PMC 2748709. PMID 19789646. Retrieved 2013-07-08.

Major Areas of Parasitology

The major areas related to parasitology have been discussed in the following text. Some of these areas are human parasites, veterinary parasitogy, quantitative parasitology and structural parasitology. Human parasites include parasites that cause parasitic diseases. They can be divided into either endoparasites or ectoparasites. Endoparasites cause infections inside the body whereas ectoparasites cause infections rapidly within the skin. The section closely examines the major areas related to parasitology.

Human Parasite

Human parasites include various protozoa and worms which may infect humans, causing parasitic diseases.

Human parasites are divided into endoparasites, which cause infection inside the body, and ectoparasites, which cause infection superficially within the skin.

The cysts and eggs of endoparasites may be found in feces which aids in the detection of the parasite in the human host while also providing the means for the parasitic species to exit the current host and enter other hosts.- Although there are number of ways in which humans can contract parasitic infections, observing basic hygiene and cleanliness tips can reduce its probability. The most accurate diagnosis is by qPcr DNA antigen assay, not generally available by primary care physicians in the USA: most labs offer research only service

History

Archaeological Evidence

It was assumed that early human ancestors generally had parasites, but until recently there was no evidence to support this claim. Generally, the discovery of parasites in ancient humans relies on the study of feces and other fossilized material. The earliest known parasite in a human was eggs of the lung fluke found in fossilized feces in northern Chile and is estimated to be from around 5900BC. There are also claims of hookworm eggs from around 5000BC in Brazil and large roundworm eggs from around 2330BC in Peru. Tapeworm eggs have also been found present in Egyptian mummies dating from around 2000BC, 1250BC, and 1000BC along with a well preserved and calcified female worm inside of a mummy.

Written Evidence

The first written records of parasites date from 3000 to 400BC in Egyptian papyrus records. They identify parasites such as roundworms, Guinea worms, threadworms, and some tapeworms of unknown varieties. In ancient Greece, Hippocrates and Aristotle documented several parasites in his collection of works Corpus Hippocraticus. In this book, they documented the presence of worms and other parasites inside of fish, domesticated animals, and humans. The bladder worm is well documented in its presence in pigs along with the larval stages of a tapeworm (Taenia Solium). These tapeworms were mentioned in a play by Aristophanes as "hailstones" with Aristotle in the section about pig diseases in his book History of Animals. The cysts of the Echinococcus granulosus tapeworm were also well known in ancient cultures mainly because of their presence in slaughtered and sacrificed animals. The major parasitic disease which has been documented in early records is dracunculiasis. This disease is caused by the Guinea worm and is characterized by the female worm emerging from the leg. This symptom is so specific to the disease that it is mentioned in many texts and plays which predate 1000AD.

Greece and Rome

In Greece, Hippocrates and Aristotle created considerable medical documentation about parasites in the Corpus Hippocraticus. In this work, they documented the presence of parasitic worms in many animals ranging from fish to domesticated animals and humans. Among the most extensively documented was the Bladder Worm (Taenia solium). This condition was called "measly pork" when present in pigs and was characterized by the presence of the larval stages of the Bladder Worm in muscle tissue. This disease was also mentioned by the playwright Aristophanes when he referred to "hailstones" in one of his plays. This naming convention is also reflected by Aristotle when he refers to "bladders that are like hailstones." Another worm which was commonly written about in ancient Greek texts was the tapeworm Echinococcus granulosus. This worm was distinguished by the presence of "massive cysts" in the liver of animals. This condition was documented so well mainly because of its presence in slaughtered and sacrificed animals. It was documented by several different cultures of the time other than the Greeks including the Arabs, Romans, and Babylonians. Not many parasitic diseases were identified in ancient Greek and Roman texts mainly because the symptoms for parasitic diseases are shared with many other illnesses such as the flu, the common cold, and dysentery. However, several diseases such as Dracunculiasis (Guinea worm disease), Hookworm, Elephantiasis, Schistosomiasis, Malaria, and Amebiasis cause unique and specific symptoms and are well documented because of this. The most documented by far was Guinea worm disease mainly because the grown female worm emerges from the skin which causes considerable irritation and which cannot really be ignored. This particular disease is widely accepted to also be the "fiery serpents" written about in the Old Testament of the Bible. This disease was mentioned by Hippocrates in Greece along with Pliny the Elder, Galen, Aetius of Amida, and Paulus Aegineta of Alexandria in Rome. Strangely, this disease was never present in Greece even though it was documented.

Northern Africa, The Middle East, and Mesopotamia

The medieval Persian doctor Avicenna records the presence of several parasites in animals and in his patients including Guinea worm, threadworms, tapeworms, and the Ascaris worm. This followed a tradition of Arab medical writings spanning over 1000 years in the area near the Red Sea. However, the Arabs never made the connection between parasites and the diseases they caused. As with Greek and Roman texts, the Guinea worm is very well documented in Middle Eastern medical texts. Several Assyrian documents in the library of King Ashurbanipal refer to an affliction which has been interpreted as Guinea worm disease. In Egypt, the Ebers Papyrus contains one of the few references to hookworm disease in ancient texts. This disease does not have very specific symptoms and was vaguely mentioned. However vague the reference, it is one of the few that connect the disease to the hookworm parasite. Another documented disease is elephantiasis. Symptoms of this disease are highly visible, since it causes extreme swelling in the limbs, breasts, and genitals. A number of surviving statues indicate that Pharaoh Mentuhotep II is likely to have suffered from elephantiasis. This disease was well known to Arab physicians and Avicenna, who noted specific differences between elephantiasis and leprosy. That the disease schistosomiasis was extremely common in Ancient Egyptis suggested by mummified evidence, but it is not specifically documented in surviving texts. Other names for this disease include bilharzia, Katayama disease, Red Water fever, snail fever, and big belly. The only really defining symptom is bloody urine, but this can easily be overlooked as several other diseases exhibit the same symptom. However, the main reason it was not documented is probably because it was simply so common. In the same way, the Greeks and Romans did not acknowledge the existence of colds and coughs because of how common they were.

China

The Chinese mostly documented diseases rather than the parasites associated with them. Chinese texts contain one of the few references to Hookworm disease found in ancient records, but no connection to hookworm is made. The Emperor Huang Ti recorded the earliest mentioning (2700BC) of malaria in his text Nei Ching. He lists chills, headaches, and fevers as the main symptoms and distinguished between the different kinds of fevers.

India

In India, the Charaka Samhita and Sushruta Samhita document Malaria. These documents list the main symptoms as fever and enlarged spleens. The Bhrigu Samhita from 1000 BCE makes the earliest reference to Amebiasis. The symptoms were given as bloody and mucosal diarrhea.

Commonly Documented Parasites

Endoparasites

Protozoa

- *Plasmodium* spp. - causes Malaria

- *Entamoeba* - causes amoebiasis, amoebic dysentery

- *Giardia* - causes Giardiasis

- *Trypanosoma brucei* - causes African sleeping sickness

- *Toxoplasma gondii* - causes Toxoplasmosis

- *Acanthamoeba* - causes Acanthamoeba keratitis

- *Leishmania* - causes Leishmaniasis

- *Babesia* - causes Babesiosis

- *Balamuthia mandrillaris* - causes Granulomatous amoebic encephalitis

- *Cryptosporidium* - causes Cryptosporidiosis

- *Cyclospora* - causes Cyclosporiasis

- *Naegleria fowleri* - causes Primary amoebic meningoencephalitis

Parasitic Worms (Helminths)

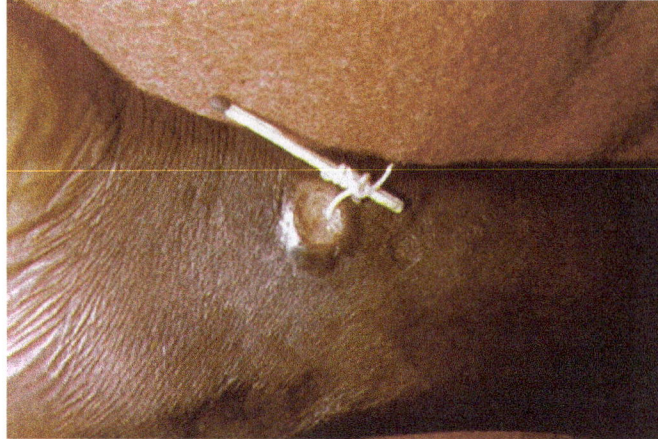

Guinea worm (Dracunculus) wrapped around a match stick

- *Ascaris lumbricoides* - cause Ascariasis

- *Pinworm* - causes Enterobiasis

- *Strongyloides stercoralis* - causes Strongyloidiasis

- *Toxocara* - causes Toxocariasis

- Guinea Worm - causes Dracunculiasis

- Hookworm

- Tapeworm

- Whipworm

Parasitic Flukes

- *Schistosoma* - causes Schistosomiasis

- *Gnathostoma* - causes Gnathostomiasis

- *Paragonimus* - causes Paragonimiasis

- *Fasciola hepatica* - causes Fascioliasis

- *Trichobilharzia regenti* - causes Swimmer's itch

Ectoparasites

- *Sarcoptes scabiei* - causes scabies

- *Pediculus humanus capitis* - causes headlice

- *Phthirus pubis* - causes pubic lice

- *Human botfly maggots* - causes Myiasis

- Chigoe flea *Tunga penetrans* - causes Tungiasis

- Ticks (*Ixodoidea*)

Veterinary Parasitology

Veterinary parasitology is the study of animal parasites, especially relationships between parasites and animal hosts. Parasites of domestic animals, (livestock and pet animals), as well as wildlife animals are considered. Veterinary parasitologists study the genesis and development of parasitoses in animal hosts, as well as the taxonomy and systematics of parasites, including the morphology, life cycles, and living needs of parasites in the environment and in animal hosts. Using a variety of research methods, they diagnose, treat, and prevent animal parasitoses. Data obtained from parasitological research in animals helps in veterinary practice and improves animal breeding. The major goal of veterinary parasitology is to protect animals and improve their health, but because a number of animal parasites are transmitted to humans, veterinary parasitology is also important for public health.

Diagnostic Methods

Various methods are used to identify parasites in animals, using feces, blood, and tissue samples from the host animal.

Coprological

Coprological examinations involve examining the feces of animals to identify and count parasite eggs. Some common methods include fecal flotation and sedimentation to separate eggs from fecal matter. Others include the McMaster method, which uses a special two-chamber slide that allows parasite eggs to be more clearly visible and easily counted. It is most commonly used to monitor parasites in horses and other grazing and livestock animals. The Baermann method is similar but requires more specialized equipment and more time and is typically used to diagnose lungworm and threadworm.

Haematological

Haematological examinations involve examining the blood of animals to determine the presence of parasites. Blood parasites tend to inhabit the erythrocytes or white blood cells and are most likely to be detected during the acute phase of infection. Veterinary parasitologists use blood smears, which involve placing a drop of blood onto a slide and spreading it over the surface in a thin film in order to examine it under a microscope. The blood is stained with a dye in order for the cells to be easily distinguished.

Histopathological

Histopathological examinations involve examining tissue samples from animals. A small slice of the organ suspected of being infected by parasites is mounted on a slide, stained, and examined under a microscope.

Though not technically considered a histopathological technique, skin scraping – which involves taking a small sample of the epidermal cells of a dog, cat, or other household pet – is commonly used to detect the presence of mites.

Immunological

Immunological examinations, such as indirect immunofluorescence, ELISA, Immunoblotting (Western blot), and Complement fixation test are methods of identifying different kinds of parasites by detecting the presence of their antigens on or within the parasite itself. These diagnostic methods are used in conjunction with coprological examinations for more specific identification of different parasite species in fecal samples.

Molecular Biological

Molecular biological methods involve studying the DNA of the parasite in order to identify it. PCR and RFLP are used to detect and amplify parasite DNA found in the feces, blood, or tissue of the host. These techniques are very sensitive, which is useful for diagnosing parasites even when they are present in very low numbers; they are also useful for identifying parasites not only in large animal hosts but smaller insect vectors.

Divisions of Veterinary Parasitology

Veterinary Protozoology

- focused on veterinary important protozoans

Examples of protozoan parasites:

- *Trypanosoma brucei*
- *Trypanosoma equiperdum*
- *Leishmania donovani*
- *Leishmania infantum*
- *Giardia duodenalis*
- *Trichomonas gallinae*
- *Tritrichomonas foetus*

- *Histomonas meleagridis*
- *Cryptosporidium parvum*
- *Balantidium coli*
- *Eimeria acervulina*
- *Eimeria tenella*
- *Isospora canis*
- *Toxoplasma gondii*
- *Neospora caninum*
- *Hammondia hammondi*
- *Besnoitia besnoiti*
- *Babesia divergens*

Veterinary Helminthology

- focused on veterinary important helminth parasites

Examples of helminth parasites:

- *Ancylostoma duodenale*
- *Ascaris suum*
- *Dicrocoelium dendriticum*
- *Dictyocaulus bovis*
- *Dipylidium caninum*
- *Echinococcus granulosus*
- *Fasciola hepatica*
- *Fascioloides magna*
- *Habronema muscae*
- *Habronema majus*
- *Haemonchus contortus*
- *Metastrongylus*
- *Muellerius capillaris*
- *Ostertagia ostertagi*
- *Paragonimus westermani*
- *Schistosoma bovis*

- *Strongyloides canis*
- *Strongylus vulgaris*
- *Syngamus trachea*
- *Taenia pisiformis*
- *Taenia saginata*
- *Taenia solium*
- *Toxocara canis*
- *Trichinella spiralis*
- *Trichobilharzia regenti*
- *Trichostrongylus axei*
- *Trichuris suis*

Veterinary Entomology (Arachnoentomology)

- focused on veterinary important Arachnids, Insects, and Crustaceans

Examples of arachnid, insect, and crustacean parasites:

- *Sarcoptes equi*
- *Psoroptes ovis*
- *Ixodes ricinus*
- *Dermacentor marginatus*
- *Caligus clemensi*
- *Caligus cuneifer*
- *Caligus elongatus*
- *Caligus rogercresseyi*
- *Cimex colombarius.*
- *Cimex lectularius*
- *Culex pipiens*
- *Culicoides imicola*
- *Demodex bovis*
- *Gasterophilus intestinalis*
- *Haematobia irritans*
- *Hypoderma bovis*

- *Knemidocoptes mutans*
- *Lepeophtheirus salmonis (sea louse)*
- *Lucilia sericata*
- *Musca domestica*
- *Nosema apis*
- *Notoedres cati*
- *Oestrus ovis*
- *Otodectes cynotis*
- *Phlebotomus*
- *Pulex irritans*
- *Rhipicephalus sanguineus*
- *Sarcophaga carnaria*
- *Tabanus atratus*
- *Triatoma*
- *Ctenocephalides canis*
- *Ctenocephalides felis*

Quantitative Parasitology

In parasitology, the quantitative study of parasitism in a host population involves the use of statistics to draw meaningful conclusions from observations of the prevalence and intensity of parasitic infection.

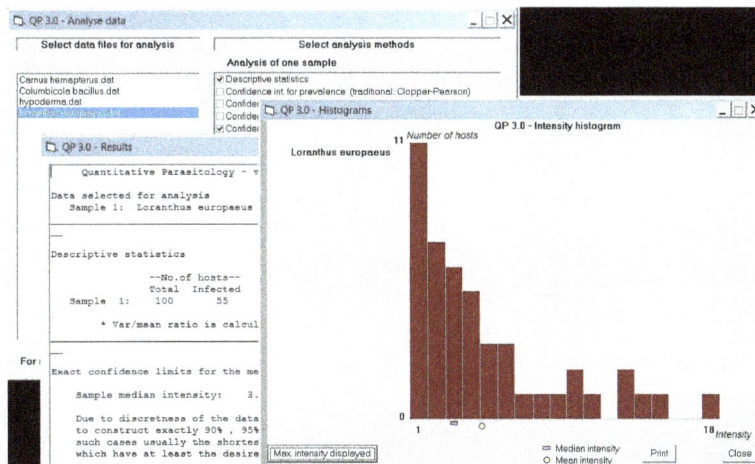

Working view of QP 3.0

Counting Parasites

Quantifying parasites in a sample of hosts or comparing measures of infection across two or more samples can be challenging.

The parasitic infection of a sample of hosts inherently exhibits a complex pattern that cannot be adequately quantified by a single statistical measure. As the use of two or more separate indices is advisable, only two or more separate statistical tests can reliably compare infections different samples of hosts.

A few of the available statistical measures have markedly different biological interpretations, while others have more-or-less overlapping interpretations or no interpretations at all. Therefore, one should apply measures that have clear and separate biological interpretations thus do not predict each other.

Parasite individuals typically exhibit an aggregated (right-skewed) distribution among host individuals; most hosts harbour few if any parasites and a few hosts harbour many of them. This quantitative feature of parasitism makes the application of many traditional statistical methods inappropriate by violating assumptions about the underlying data distribution, requiring the use of more advanced computationally-intensive methods.

Describing The Parasitic Infection of A Sample of Hosts

Always give the host sample size. In most cases, this is expressed as the number of hosts individuals examined. (Exceptionally, other units may also be used for special cases.)

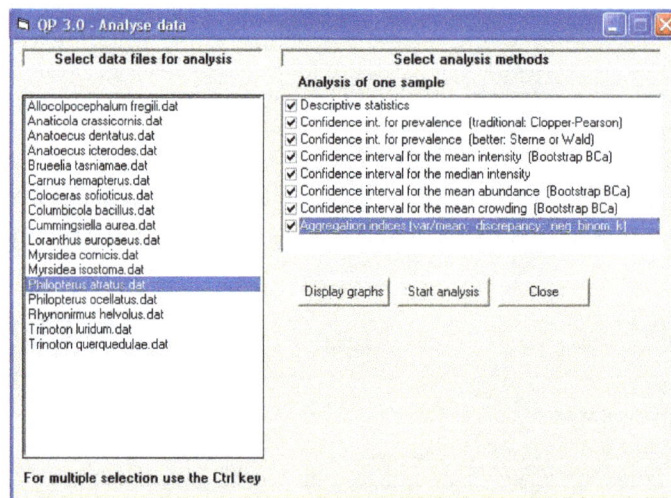

Statistical procedures to characterize the infection/infestation of a sample of hosts.

Describe prevalence. This is the proportion of infected hosts among all the hosts examined. Give the confidence interval (CI) of prevalence (either as a Clopper-Pearson interval or as adjusted Wald/Sterne's interval) to indicate the accuracy of the estimation (use of the confidence intervals belonging to the 95% probability is advisable).

Describe mean intensity. This is the mean number of parasites found in the infected hosts (the zeros of uninfected hosts are excluded). Since sample size and prevalence are known, mean intensity defines the quantity of parasites found in the sample of hosts. Given the typical aggregated (right-skewed) distribution of parasites, its actual value is highly dependent on a few extremely infected hosts. Also give CI to indicate the accuracy of the estimation. Use bias-corrected and accelerated bootstrap (BCa Bootstrap) to get this confidence interval.

Describe median intensity. This is the median number of parasites found in infected hosts (the zeros of uninfected hosts are excluded). Median intensity shows a typical level of infection among the infected hosts. Use exact CI to indicate the accuracy of the estimation.

In certain cases one may prefer to use mean abundance instead of mean intensity. This is the mean number of parasites found in all hosts (involves the zero values of uninfected hosts). Give BCa Bootstrap confidence interval to indicate the accuracy of this estimation. This measure unifies two of the former ones: prevalence and mean intensity. Do not use it, unless you have a clearly specified a reason why to prefer it.

Describing mean crowding (intensity values averaged across parasite individuals) and its confidence interval is essential only for those who study density-dependent characters of parasites. BCa Bootstrap CI can be used to indicate the accuracy of the estimation.

Finally, quantify levels of skewness of the parasites' distribution among hosts. There are 3 indices widely used for this purpose, but their interpretation is quite similar. They predict each other rather well, thus it is not necessary to use all the 3 of them.

Comparing Parasite Burdens Across Two or More Samples

Compare prevalences by Fisher's exact test. This will show whether the proportion of infected individuals differs significantly between the two (or more) samples. The time need of this test may increase dramatically when several samples are involved. Using Chi-squared test for the same purpose may be advisable in such cases.

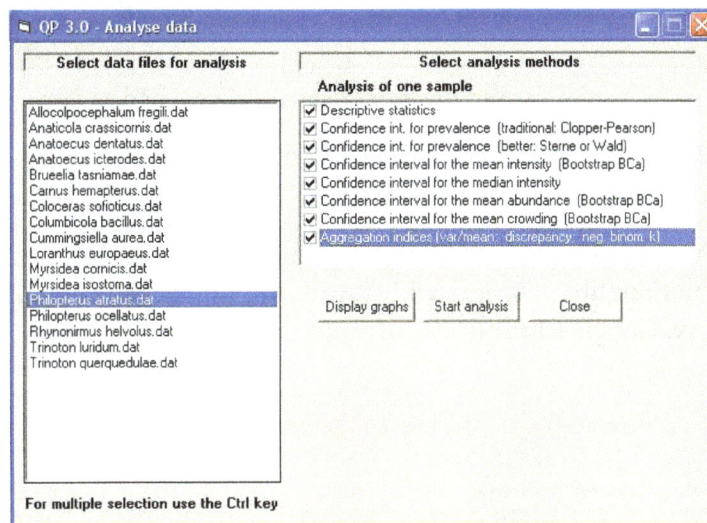

Statistical procedures to compare levels of infection/infestation across two or more samples of hosts.

Compare mean intensities by a Bootstrap t-test. This will show whether parasite quantities differ significantly between the infected proportions of the two samples.

Compare median intensities by Mood's median test. This will show whether the typical level of infection differs significantly between the infected proportions of the two samples.

One can also compare the frequency distributions of intensities by a Stochastic equality test. It compares several random pairs of individual values taken from the two samples to test whether or not there is a significant tendency to get higher values from one sample than from the other.

In certain cases, one may also decide to compare mean abundances by a Bootstrap t-test. This will show whether parasite quantities differ significantly between two samples. This comparison unifies two of the former ones: the comparison of prevalences and the comparison of mean intensities.

Finally, mean crowding can be compared across samples by a simple method: provided that the two 97.5% confidence intervals do not overlap, we conclude that the two values are different at a 95% level of significance.

Available Software

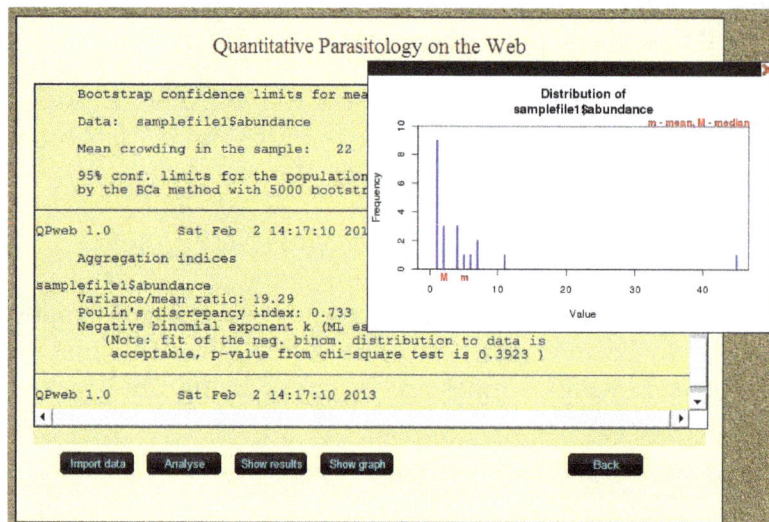

QPweb1.0: an interactive web surface to upload and analyse your data through the web

- *Quantitative Parasitology 3.0* is a zipped folder containing an exe file that runs under Microsoft Windows.

- *Quantitative Parasitology on the Web 1.0* is an interactive web surface that comes with an extended functionality. It works with Windows, Linux, Mac etc. operating systems and Explorer, Firefox, Google Chrome etc. browsers.

Structural Parasitology

Structural parasitology is the study of the structures of proteins of parasites. It applies the techniques of structural biology (such as X-ray crystallography or NMR) to determine the 3-D structures of protein molecules involved in a parasitic relationship. One goal is to distinguish the workings of functional pathways in these organisms in comparison to humans. More importantly, it is hoped that structures of parasite proteins will lead to faster discovery of drugs for diseases neglected by pharmaceutical companies.

This is a challenging field because parasite proteins are often more difficult to express using a heterologous system. The challenge is particularly great for proteins from eukaryotic parasites.

Parasites of interest include Plasmodium, Trypanosoma, Leishmania, Giardia, Entamoeba, Cryptosporidium, Helminth and Toxoplasma, many of which are agents for Neglected Diseases.

Many academic labs around the world study structural parasitology. Two groups in particular have contributed many parasite structures: the SGPP (Structural Genomnics of Pathogenic Protozoa) and the SGC (Structural Genomics Consortium).

References

- Elsheikha, HM; Khan, NA (editor) (2011). Essentials of Veterinary Parasitology. Caister Academic Press. ISBN 978-1-904455-79-0.

- Leland S. Shapiro; Patricia Mandel (22 May 2009). Pathology & Parasitology for Veterinary Technicians. Cengage Learning. pp. 237–239. ISBN 978-1-4354-3855-2. Retrieved 11 September 2011.

- A. Zajac; Gary A. Conboy; American Association of Veterinary Parasitologists (24 March 2006). Veterinary clinical parasitology. Wiley-Blackwell. p. 162. ISBN 978-0-8138-1734-7. Retrieved 11 September 2011.

- John E. Hyde (1993). Protocols in molecular parasitology. Humana Press. pp. 213–214. ISBN 978-0-89603-239-2. Retrieved 11 September 2011.

Protozoa: An Overview

Protozoa is a single-celled organism eukaryotic organism. Some of the topics listed in this chapter are plasmodium, plasmodium falciparum, plasmodium vivax, plasmodium knowlesi and trypanosome cruzi. This chapter is written in a manner that will introduce briefly all the significant aspects of protozoa.

Protozoa

In some systems of biological classification, the Protozoa are a diverse group of unicellular eukaryotic organisms. Historically, protozoa were defined as single-celled organisms with animal-like behaviors, such as motility and predation. The group was regarded as the zoological counterpart to the "protophyta", which were considered to be plant-like, as they are capable of photosynthesis. The terms *protozoa* and *protozoans* are also used informally to designate single-celled, non-photosynthetic protists, such as ciliates, amoebae and flagellates.

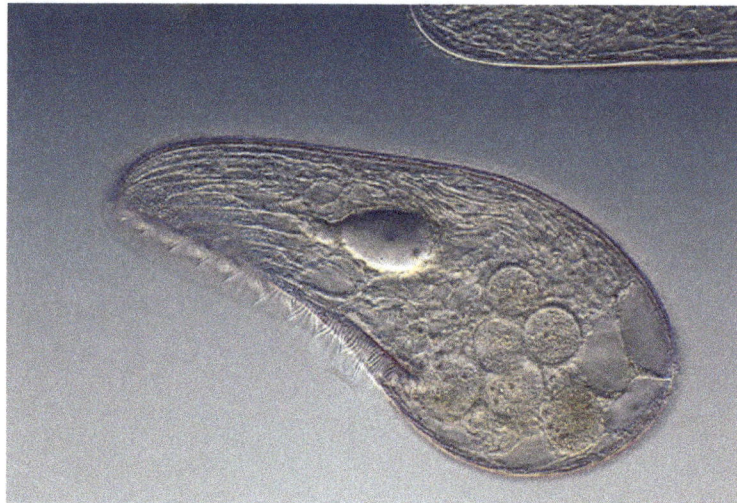

Blepharisma japonicum, a free-living ciliated protozoan.

The term Protozoa was introduced in 1818 for a taxonomic class, but in later classification schemes the group was elevated to higher ranks, including phylum, subkingdom and kingdom. In several classification systems proposed by Thomas Cavalier-Smith and his collaborators since 1981, Protozoa is ranked as a kingdom. The seven-kingdom scheme proposed by Ruggiero et al. in 2015, places eight phyla under Protozoa: Euglenozoa, Amoebozoa, Metamonada, Choanozoa, Loukozoa, Percolozoa, Microsporidia and Sulcozoa. This kingdom does not form a clade, but an evolutionary grade or paraphyletic group, from which the fungi and animals are specifically excluded.

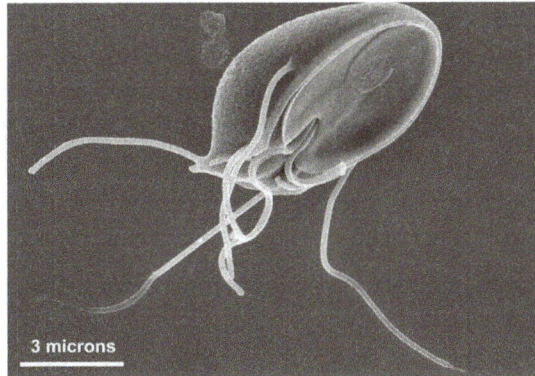

Giardia muris, a flagellate protozoan, is an intestinal parasite found in rodents, birds and reptiles.

The use of Protozoa as a formal taxon has been discouraged by some researchers, mainly because the term, which is formed from the Greek *protos* "first" + *zoia*, plural of *zoion*, "animal", misleadingly implies kinship with animals (metazoa) and promotes an arbitrary separation of "animal-like" from "plant-like" organisms. Modern ultrastructural, biochemical, and genetic techniques have shown that protozoa, as traditionally defined, belong to widely divergent lineages, and can no longer be regarded as "primitive animals." For this reason, the terms "protists," "Protista" or "Protoctista" are sometimes preferred for the high-level classification of eukaryotic microbes. In 2005, members of the Society of Protozoologists voted to change the name of that organization to the International Society of Protistologists.

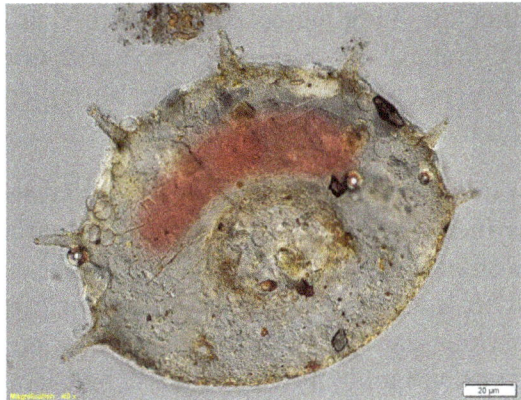

Centropyxis aculeata, a testate (shelled) amoeba

History and Terminology

The word "protozoa" (singular *protozoon* or *protozoan*) was coined in 1818 by zoologist Georg August Goldfuss, as the Greek equivalent of the German *Urthiere*, meaning "primitive, or original animals" (*ur-* 'proto-' + *Thier* 'animal'). Goldfuss erected Protozoa as a class containing what he believed to be the simplest animals. Originally, the group included not only microbes, but also some "lower" multicellular animals, such as rotifers, corals, sponges, jellyfish, bryozoa and polychaete worms.

In 1848, in light of advancements in cell theory pioneered by Theodore Schwann and Matthias Schleiden, the anatomist and zoologist C.T. von Siebold proposed that the bodies of microbial organisms such as ciliates and amoebae were made up of single cells, similar to those from which the multicellular tissues of plants and animals were constructed. Von Siebold redefined Protozoa to include only such unicellular forms, to the exclusion of all metazoa. At the same time, he raised the group to the level of a phylum containing two broad classes of microbes: Infusoria (mostly ciliates and flagellated algae), and Rhizopoda (amoeboid organisms). The definition of Protozoa as a phylum or sub-kingdom made up of "unicellular animals" was adopted by the zoologist Otto Bütschli—celebrated at his centenary as the "architect of protozoology"—and the term came into wide use.

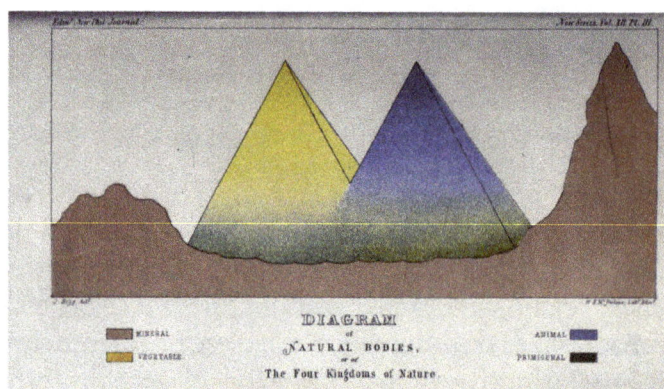

John Hogg's illustration of the Four Kingdoms of Nature, showing "Primigenal" as a greenish haze at the base of the Animals and Plants, 1860

As a phylum under Animalia, the Protozoa were firmly rooted in the old "two-kingdom" classification of life, according to which all living beings were classified as either animals or plants. As long as this scheme remained dominant, the protozoa were understood to be animals and studied in departments of Zoology, while photosynthetic microbes and microscopic fungi—the so-called Protophyta—were assigned to the Plants, and studied in departments of Botany.

Criticism of this system began in the latter half of the 19th century, with the realization that many organisms met the criteria for inclusion among both plants and animals. For example, the algae *Euglena* and *Dinobryon* have chloroplasts for photosynthesis, but can also feed on organic matter and are motile. In 1860, John Hogg argued against the use of "protozoa", on the grounds that "naturalists are divided in opinion—and probably some will ever continue so—whether many of these organisms, or living beings, are animals or plants." As an alternative, he proposed a new kingdom called Primigenum, consisting of both the protozoa and unicellular algae (protophyta), which he combined together under the name "Protoctista". In Hoggs's conception, the animal and plant kingdoms were likened to two great "pyramids" blending at their bases in the Kingdom Primigenum.

Six years later, Ernst Haeckel also proposed a third kingdom of life, which he named Protista. At first, Haeckel included a few multicellular organisms in this kingdom, but in later work he restricted the Protista to single-celled organisms, or simple colonies whose individual cells are not differentiated into different kinds of tissues.

Despite these proposals, Protozoa emerged as the preferred taxonomic placement for heterotrophic microbes such as amoebae and ciliates, and remained so for more than a century. In the course of the 20th century, however, the old "two kingdom" system began to weaken, with the growing awareness that fungi did not belong among the plants, and that most of the unicellular protozoa were no more closely related to the animals than they were to the plants. By mid-century, some biologists, such as Herbert Copeland, Robert H. Whittaker and Lynn Margulis, advocated the revival of Haeckel's Protista or Hogg's Protoctista as a kingdom-level eukaryotic group, alongside Plants, Animals and Fungi. A variety of multi-kingdom systems were proposed, and Kingdoms Protista and Protoctista became well established in biology texts and curricula.

While many taxonomists have abandoned Protozoa as a high-level group, Thomas Cavalier-Smith has retained it as a kingdom in the various classifications he has proposed. As of 2015, Cavalier-Smith's Protozoa excludes several major groups of organisms traditionally placed among the protozoa, including the ciliates, dinoflagellates and foraminifera (all members of the SAR supergroup). In its current form, his kingdom Protozoa is a paraphyletic group which includes a common ancestor and most of its descendents, but excludes two important clades that branch within it: the animals and fungi.

Characteristics

Protozoa, as traditionally defined, are mainly microscopic organisms, ranging in size from 10 to 52 micrometers. Some, however, are significantly larger. Among the largest are the deep-sea–dwelling xenophyophores, single-celled foraminifera whose shells can reach 20 cm in diameter. Free-living forms are restricted to moist environments, such as soils, mosses and aquatic habitats, although many form resting cysts which enable them to survive drying. Many protozoan species are symbionts, some are parasites, and some are predators of bacteria, algae and other protists.

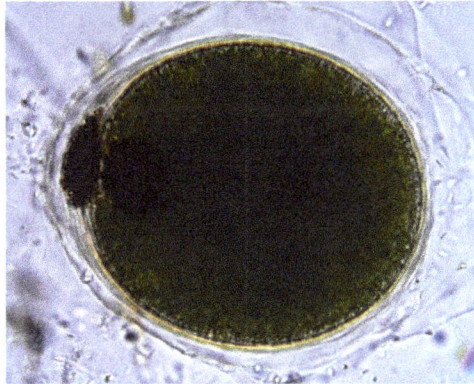

Resting cyst of ciliated protozoan *Dileptus viridis*.

Motility and Feeding

Organisms traditionally classified as protozoa are abundant in aqueous environments and soil, occupying a range of trophic levels. The group includes flagellates (which move with the help of whip-like structures called flagella), ciliates (which move by using hair-like structures called cilia) and amoebae (which move by the use of foot-like structures called pseudopodia). Some protozoa are sessile, and do not move at all.

Protozoa may take in food by osmotrophy, absorbing nutrients through their cell membranes; or they may feed by phagocytosis, either by engulfing particles of food with pseudopodia (as amoebae do), or taking in food through a mouth-like aperture called a cytostome. All protozoa digest their food in stomach-like compartments called vacuoles.

Pellicle

The pellicle is a thin layer supporting the cell membrane in various protozoa, such as ciliates, protecting them and allowing them to retain their shape, especially during locomotion, allowing the organism to be more hydrodynamic. The pellicle varies from flexible and elastic to rigid. Although somewhat stiff, the pellicle is also flexible and allows the protist to fit into tighter spaces. In ciliates and Apicomplexa, it is formed from closely packed vesicles called alveoli. In euglenids, it is formed from protein strips arranged spirally along the length of the body. Familiar examples of protists with a pellicle are the euglenoids and the ciliate *Paramecium*. In some protozoa, the pellicle hosts epibiotic bacteria that adhere to the surface by their fimbriae (attachment pili).

Life Cycle

Some protozoa have life phases alternating between proliferative stages (e.g., trophozoites) and dormant cysts. As cysts, protozoa can survive harsh conditions, such as exposure to extreme temperatures or harmful chemicals, or long periods without access to nutrients, water, or oxygen for a period of time. Being a cyst enables parasitic species to survive outside of a host, and allows their transmission from one host to another. When protozoa are in the form of trophozoites (Greek *tropho* = to nourish), they actively feed. The conversion of a trophozoite to cyst form is known as encystation, while the process of transforming back into a trophozoite is known as excystation. Protozoa reproduce asexually by binary fission or multiple fission. Many protozoan species exchange genetic material by sexual means (typically, through conjugation); however, sexuality is generally decoupled from the process of reproduction, and does not immediately result in increased population.

Classification

The classification of protozoa has been and remains a problematic area of taxonomy. Where they are available, DNA sequences are used as the basis for classification; however, for the majority of described protozoa, such material is not available. They have been and still are mostly on the basis of their morphology and for the parasitic species their hosts. Protozoa have been divided traditionally on the basis of their means of locomotion.

- Flagellates (e.g., *Giardia lamblia*)

- Amoeboids (e.g., *Entamoeba histolytica*)

- Sporozoans (e.g., *Plasmodium knowlesi*)

    o   Apicomplexa (now in Alveolata)

    o   Microsporidia (now in Fungi)

    o   Ascetosporea (now in Rhizaria)

    o   Myxosporidia (now in Cnidaria)

- Ciliates (e.g., *Balantidium coli*)

As a phylum the Protozoa were, historically, divided into four subphyla reflecting the means of locomotion:

- Subphylum Sarcomastigophora

    o Superclass Mastigophora (includes flagellates)

    o Superclass Sarcodina

    o Superclass Opalinata

- Subphylum Sporozoa (includes apicomplexans)

- Subphylum Cnidospora

    o Class Microsporidea

- Subphylum Ciliophora (includes ciliates)

These systems are no longer considered to be valid.

Ecological Role

As components of the micro- and meiofauna, protozoa are an important food source for microinvertebrates. Thus, the ecological role of protozoa in the transfer of bacterial and algal production to successive trophic levels is important. As predators, they prey upon unicellular or filamentous algae, bacteria, and microfungi. Protozoan species include both herbivores and consumers in the decomposer link of the food chain. They also control bacteria populations and biomass to some extent. On average, protozoa eat ~ 100 to 1,000 bacteria per hour. Protozoa can stimulate decomposition of organic matter, digest cellulose in the rumen of cows and termite guts, and can play a role in nutrient mobilization.

Disease

In Humans

A number of protozoan pathogens are human parasites, causing diseases such as malaria (by *Plasmodium*), amoebiasis, giardiasis, toxoplasmosis, cryptosporidiosis, trichomoniasis, Chagas disease, leishmaniasis, African trypanosomiasis (sleeping sickness), amoebic dysentery, acanthamoeba keratitis, and primary amoebic meningoencephalitis (naegleriasis).

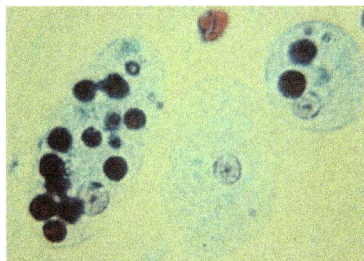

Trophozoites of the amoebic dysentery pathogen *Entamoeba histolytica* with ingested human red blood cells (dark circles)

In Other Animals

The protozoan *Ophryocystis elektroscirrha* is a parasite of butterfly larvae, passed from female to caterpillar. Severely infected individuals are weak, unable to expand their wings, or unable to eclose, and have shortened lifespans, but parasite levels vary in populations. Infection creates a culling effect, whereby infected migrating animals are less likely to complete the migration. This results in populations with lower parasite loads at the end of the migration. This is not the case in laboratory or commercial rearing, where after a few generations, all individuals can be infected.

Plasmodium

The life-cycles of *Plasmodium* species involve several different stages both in the insect and the vertebrate host. These stages include sporozoites, which are injected by the insect vector into the vertebrate host's blood. Sporozoites infect the host liver, giving rise to merozoites and (in some species) hypnozoites. These move into the blood where they infect red blood cells. In the red blood cells, the parasites can either form more merozoites to infect more red blood cells, or produce gametocytes which are taken up by insects which feed on the vertebrate host. In the insect host, gametocytes merge to sexually reproduce. After sexual reproduction, parasites grow into new sporozoites, which move to the insect's salivary glands, from which they can infect a vertebrate host bitten by the insect.

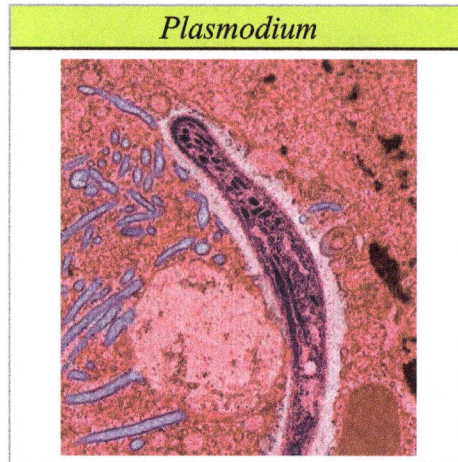

Plasmodium is a genus of parasitic protozoa, many of which cause malaria in their hosts. The parasite always has two hosts in its life cycle: a Dipteran insect host and a vertebrate host. Sexual reproduction always occurs in the insect, making it the definitive host.

The genus *Plasmodium* was first described in 1885. It now contains about 200 species, which are spread across the world where both the insect and vertebrate hosts are present. Four species regularly infect humans, while many others infect birds, reptiles, rodents, and various primates.

History

Plasmodia were first identified when Charles Louis Alphonse Laveran described parasites in the blood of malaria patients in 1880. He named the parasite *Oscillaria malariae*. The fact that several species may be involved in causing different forms of malaria was first recognized by Camillo Golgi in 1886. Soon thereafter, Giovanni Batista Grassi and Raimondo Filetti named the parasites causing two different types of human malaria *Plasmodium vivax* and *Plasmodium malariae*. In 1897, William Welch identified and named *Plasmodium falciparum*. This was followed by the recognition of the other two species of *Plasmodium* which infect humans: *Plasmodium ovale* (1922) and *Plasmodium knowlesi* (identified in long-tailed macaques in 1931; in humans in 1965). The contribution of insect hosts to the *Plasmodium* life cycle was described in 1897 by Ronald Ross and in 1899 by Giovanni Batista Grassi, Amico Bignami and Giuseppe Bastianelli.

Life Cycle

The life cycle of *Plasmodium* involves several distinct stages in the insect and vertebrate hosts. In infected mosquitoes, parasites in the salivary gland are called sporozoites. When the mosquito bites a vertebrate host, sporozoites are injected into the host with the saliva. From there, the sporozoites enter the bloodstream and are transported to the liver, where they invade and replicate within hepatocytes. At this point, some species of *Plasmodium* can form a long-lived dormant stage called a hypnozoite which can remain in the liver for many years. The parasites that emerge from infected hepatocytes are called merozoites, and these return to the blood to infect red blood cells.

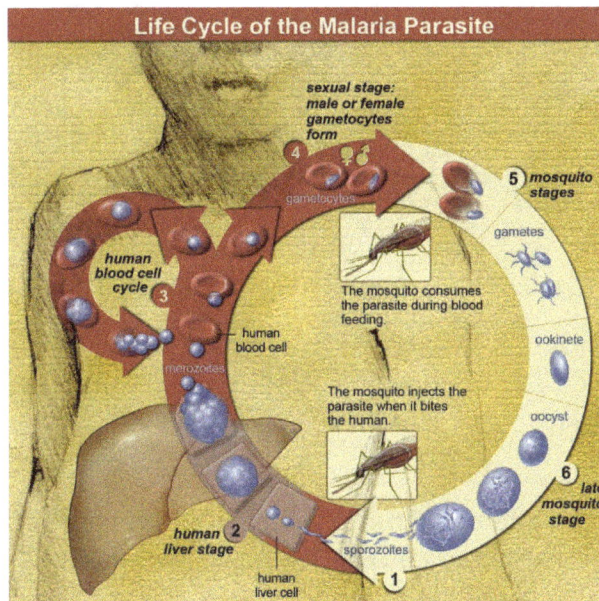

The life cycle of malaria parasites. Parasites enter the vertebrate host through a mosquito bite. Sporozoites enter the skin and travel through the bloodstream to the liver, where they multiply into merozoites, which return to the bloodstream. Merozoites infect red blood cells, where they develop through several stages to produce either more merozoites, or gametocytes. Gametocytes are taken up by a mosquito and infect the insect, continuing the life cycle.

Within the red blood cells, the merozoites grow first to a ring-shaped form and then to a larger form called a trophozoite. Trophozoites then mature to schizonts which divide several times to produce new merozoites. The infected red blood cell eventually bursts, allowing the new merozoites to travel within the bloodstream to infect new red blood cells. Most merozoites continue this replicative cycle, however some merozoites differentiate into male or female sexual forms called gametocytes. These gametocytes circulate in the blood until they are taken up when a mosquito feeds on the infected vertebrate host, taking up blood which includes the gametocytes.

In the mosquito, the gametocytes move along with the blood meal to the mosquito's midgut. Here the gametocytes develop into male and female gametes which fertilize each other, forming a zygote. Zygotes then develop into a motile form called an ookinete, which penetrates the wall of the midgut. Upon traversing the midgut wall, the ookinete embeds into the gut's exterior membrane and develops into an oocyst. Oocysts divide many times to produce large numbers of small elongated sporozoites. These sporozoites migrate to the salivary glands of the mosquito where they can be injected into the blood of the next host the mosquito bites, repeating the cycle.

Description

Plasmodium species each have 14 chromosomes in the nucleus, as well as genetic material in the mitochondrion and in the apicoplast. The chromosomes vary from 500 kilobases to 3.5 megabases in length. The apicoplast is involved in isoprenoid metabolism, Fe-S cluster synthesis, fatty acid synthesis, and phospholipid biosynthesis.

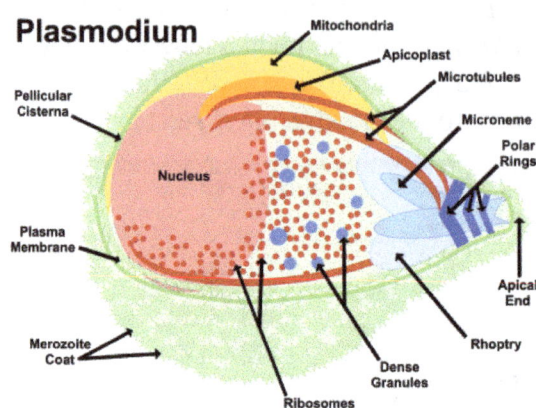

Plasmodium is a Eukaryote but with unusual features

On a molecular level, the parasite damages red blood cells using plasmepsin enzymes — aspartic acid proteases which degrade hemoglobin.

Taxonomy

Plasmodium is a member of the family Plasmodiidae, order Haemosporidia and phylum Apicomplexa which, along with dinoflagellates and ciliates, make up the taxonomic group Alveolata.

Plasmodium species were originally classified into subgenera based on their morphology, location, and host specificity. However, more recent studies of *Plasmodium* species using molecular methods have occasionally given results which conflict with the original taxonomic system.

Evolution

Evolutionary relationships of species within the genus *Plasmodium* have been controversial. *Plasmodium* species were originally divided by morphology, life-cycle characteristics, and host species. However, modern molecular approaches for determining evolutionary relationships have given results which conflict with older classification methods. Many attempts to clarify *Plasmodium* taxonomy with molecular methods have also run into technical challenges. Ribosomal RNA sequencing, which is often used in other organisms to determine evolutionary relationships, is challenging to interpret from *Plasmodium* species as *Plasmodia* maintain several different copies of ribosomal RNA which are expressed at different stages of the life cycle and which may be able to recombine with one another. Another commonly used marker for evolutionary studies has been the circumsporozoite protein (CSP) which is present in all *Plasmodium* species. However, analyses of CSP sequences are complicated by the fact that the sequence of CSP, which is present on the surface of the parasite during infection, is under heavy selective pressure from the host immune system. This could potentially obscure relevant changes.

Environmental factors play a considerable role in the evolution of *Plasmodium* and the transmission of malaria. The genetic information of *Plasmodium falciparum* has signaled a recent expansion that coincides with the agricultural revolution. It is likely that the development of extensive agriculture increased mosquito population densities by giving rise to more breeding sites, which may have triggered the evolution and expansion of *Plasmodium falciparum*.

There are over one hundred species of mosquito-transmitted *Plasmodium*. The phylogeny of these malarial parasites suggests that the *Plasmodium* of mammalian hosts forms a well-defined clade strongly associated with the specialization to the *Anopheles* mosquito vector. This was a major evolutionary transition that allowed Plasmodium to exploit humans and other mammals.

P. falciparum, the most lethal malaria parasite of humans, evolved from a "nearly identical" parasite of western gorillas.

The high mortality and morbidity caused by malaria—especially that caused by *P. falciparum*—has placed the greatest selective pressure on the human genome in recent history. Several genetic factors provide some resistance to *Plasmodium* infection, including sickle cell trait, thalassaemia traits, glucose-6-phosphate dehydrogenase deficiency, and the absence of Duffy antigens on red blood cells.

Although there are therapeutic medications to treat malaria, *Plasmodium* has accumulated increasing drug resistance over time. A recent examination has shown that even artemisinin, one of the most powerful anti-malarial drugs, has been experiencing decreased efficacy due to the development of resistance.

Subgenera

Plasmodium species have been subdivided into subgenera, which group species of similar morphology and with similar hosts.

- o Subgenus Asiamoeba (lizards)
- o Subgenus Bennettinia (birds)
- o Subgenus Carinamoeba (reptiles)
- o Subgenus Giovannolaia (birds)
- o Subgenus Haemamoeba (birds)
- o Subgenus Huffia (birds)

- o   Subgenus Lacertamoeba (reptiles)

- o   Subgenus Laverania (higher primates)

- o   Subgenus Novyella (birds)

- o   Subgenus Paraplasmodium (lizards)

- o   Subgenus Plasmodium (monkeys, higher primates)

- o   Subgenus Sauramoeba (reptiles)

- o   Subgenus Vinckeia (non-primate mammals)

Species infecting monkeys and apes (the higher primates) with the exceptions of *P. falciparum* and *P. reichenowi* (which together make up the subgenus *Laverania*) are classified in the subgenus *Plasmodium*. Parasites infecting other mammals including lower primates (lemurs and others) are classified in the subgenus *Vinckeia*. The distinction between *P. falciparum* and *P. reichenowi* and the other species infecting higher primates was based on morphological findings but have since been confirmed by DNA analysis.

The four subgenera *Giovannolaia*, *Haemamoeba*, *Huffia* and *Novyella* were created by Corradetti et al. for the known avian malarial species. A fifth — *Bennettinia* — was created in 1997 by Valkiunas. The relationships between the subgenera are a matter of current investigation.

P. juxtanucleare is the only member of the subgenus *Bennettinia*.

Unlike the mammalian and bird malarias those affecting reptiles have been more difficult to classify. In 1966 Garnham classified those with large schizonts as *Sauramoeba*, those with small schizonts as *Carinamoeba* and the single then-known species infecting snakes (*Plasmodium wenyoni*) as *Ophidiella*. In 1988, Telford used this scheme as the basis for the current system.

Hosts

All *Plasmodia* are parasitic and require both a vertebrate host and an insect host to reproduce. Known vertebrate hosts include various primates (including humans), birds, rodents, bats, porcupines and squirrels. Mosquitoes of the genera *Culex*, *Anopheles*, *Culiseta*, *Mansonia* and *Aedes* often serve as insect hosts for various *Plasmodium* species.

Humans

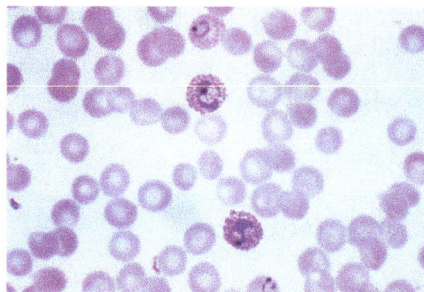

Trophozoites of the *Plasmodium vivax* parasite among human red blood cells

The species of *Plasmodium* that infect humans include:

- *Plasmodium falciparum* (the cause of malignant tertian malaria)

- *Plasmodium vivax* (the most frequent cause of benign tertian malaria)

- *Plasmodium ovale* (the other, less frequent, cause of benign tertian malaria)

- *Plasmodium malariae* (the cause of benign quartan malaria)

- *Plasmodium knowlesi* (the cause of severe quotidian malaria in South East Asia since 1965)

P. falciparum, *P. vivax*, *P. ovale*, and *P. malariae* together account for nearly all human infections with *Plasmodium* species, with *P. falciparum* accounting for the overwhelming majority of malaria deaths. An increasing number of cases of severe malaria in Southeast Asia have been attributed to *P. knowlesi*.

With the use of the polymerase chain reaction additional species have been identified in humans, although whether these species can regularly infect humans is not known. An experimental infection of a human volunteer with *Plasmodium eylesi* and *Plasmodium cynomolgi* has been reported, although the infection was not able to be transferred to another susceptible host, suggesting the parasites may not be able to productively infect humans. A possible infection with *Plasmodium tenue* has also been reported, however doubts have been raised as to the validity of this diagnosis. Other species which have been reportedly isolated from humans include *Plasmodium brasilianum*, *Plasmodium inui*, *Plasmodium rhodiani*, *Plasmodium schweitzi*, *Plasmodium semiovale*, and *Plasmodium simium*.

In addition to humans, *P. vivax* can infect chimpanzees and orangutans. In these hosts, infection tends to not cause severe disease, but may persist for some time.

P. ovale can be transmitted to chimpanzees, and is found in Africa, the Philippines and New Guinea.

Non-human Primates

Species of *Plasmodium* infect many primates across the world, such as the brown lemur, *Eulemur fulvus*, of Madagascar.

The species that infect primates other than humans include: *P. bouillize*, *P. brasilianum*, *P. bucki*, *P. cercopitheci*,*P. coatneyi*, *P. coulangesi*, *P. cynomolgi*, *P. eylesi*, *P. fieldi*, *P. foleyi*, *P. fragile*, *P. girardi*, *P. georgesi*, *P. gonderi*, *P. hylobati*, *P. inui*, *P. jefferyi*, *P. joyeuxi*, *P. knowlesi*, *P. lemuris*, *P. percygarnhami*, *P. petersi*, *P. reichenowi*, *P. rodhaini*, *P. sandoshami*, *P. semnopitheci*, *P. sil-*

vaticum, *P. simiovale*, *P. simium*, *P. uilenbergi*, *P. vivax* and *P. youngei*.

Many *Plasmodium* species infect more than one primate host species. Primates which have been found to be infected with *Plasmodium* include species of the genera *Alouatta* (also known as howler monkeys), *Ateles* (spider monkeys), *Brachyteles* (muriqui or wooly spider monkeys), *Callicebus* (titi monkeys), *Chiropotes* (bearded sakis), *Lagothrix* (woolly monkeys), *Lemur* (lemurs), *Macaca* (macaques), *Pan* (chimpanzees), *Pongo* (orangutans), and *Saimiri* (squirrel monkeys).

The insect hosts of the *Plasmodium* species which infect primates are various species of Anopheles mosquitoes. Different species of *Plasmdodium* generally infect different species of mosquito, although some mosquito species can carry several *Plasmodium* species.

Non-primate Mammals

The subgenus *Vinckeia* was created by Cyril Garnham to accommodate the mammalian parasites other than those infecting primates. Species infecting lemurs have also been included in this subgenus. *Plasmodium* species can infect a wide variety of mammals including rodents, ungulates, and bats. Several of the species which infect rodents have been shown to also infect the mosquito *Anopheles stephensi*.

Many non-primate mammals, such as mouse-deer (*Tragulus kanchil*) can carry malaria parasites.

P. aegyptensis, *P. bergei*, *P. chabaudi*, *P. inopinatum*, *P. yoelli* and *P. vinckei* infect rodents. *P. bergei*, *P. chabaudi*, *P. yoelli* and *P. vinckei* have been used to study malarial infections in the laboratory. Other members of this subgenus infect other mammalian hosts.

Birds

Many bird species, from raptors to passerines like the red-whiskered bulbul (*Pycnonotus jocosus*), can carry malaria.

Species in five *Plasmodium* subgenera infect birds — *Bennettinia, Giovannolaia, Haemamoeba, Huffia* and *Novyella*. *Giovannolaia* appears to be a polyphyletic group and may be sudivided in the future. Species that infect birds include *P. accipiteris, P. alloelongatum, P. anasum, P. ashfordi, P. bambusicolai, P. bigueti, P. biziurae, P. buteonis, P. cathemerium, P. circumflexum, P. coggeshalli, P. corradettii, P. coturnix, P. dissanaikei, P. durae, P. elongatum, P. fallax, P forresteri, P. gallinacium, P. garnhami, P. giovannolai, P. griffithsi, P. gundersi, P. guangdong, P. hegneri, P. hermani, P. hexamerium, P. huffi, P. jiangi, P. juxtanucleare, P. kempi, P. lophurae, P.lutzi, P. matutinum, P. nucleophilum, P. papernai, P. paranucleophilum, P. parvulum, P. pediocetti, P. paddae, P. pinotti, P. polare, P. relictum, P. rouxi, P. tenue, P. tejerai, P. tumbayaensis* and *P. vaughani.*

These infect a variety of bird species. In general each species of *Plasmodium* infects one to a few species of birds. Each species is also generally transmitted by a single insect species. Insect hosts include *Aedes aegypti, Mansionia crassipes,* and various *Culex* species.

Reptiles

Species in the subgenera *Asiamoeba, Carinamoeba, Lacertaemoba, Paraplasmodium* and *Sauramoeba* infect reptiles. These species of *Plasmodium* include: *P. achiotense, P. aeuminatum, P. agamae, P. arachniformis, P. attenuatum,P. aurulentum, P. australis, P. azurophilum, P. balli, P. basilisci, P. beebei, P. beltrani , P. brumpti, P. brygooi, P. chiricahuae, P. circularis, P. cnemaspi, P. cnemidophori, P. colombiense, P. cordyli, P. diminutivum, P. diploglossi, P. egerniae, P. fairchildi, P. floridense, P. gabaldoni, P. giganteum, P. gologoense, P. gracilis, P. guyannense, P. heischi, P. holaspi, P. icipeensis, P. iguanae, P.josephinae, P. kentropyxi, P. lacertiliae, P. lainsoni, P. lepidoptiformis, P. lionatum, P. loveridgei, P. lygosomae, P. mabuiae, P. mackerrasae, P. maculilabre, P. marginatum, P. mexicanum, P. michikoa, P. minasense, P. pelaezi, P. pessoai, P. pifanoi, P. pitmani, P. rhadinurum, P. sasai,P. saurocaudatum, P. scorzai, P. siamense, P. robinsoni, P. sasai, P. scorzai, P. tanzaniae, P. tomodoni, P. torrealbai, P. tribolonoti, P. tropiduri, P. uluguruense, P. uzungwiense, P. vacuolatum, P. vastator, P. volans, P. wenyoni* and *P. zonuriae.*

Over 3000 species of lizard, including the Carolina anole (*Anolis carolinensis*), carry some 90 kinds of malaria.

Plasmodium species have been reported from over 3200 species of lizard and 29 species of snake. Only three species — *P. pessoai*, *P. tomodoni* and *P. wenyoni* — infect snakes. Species infecting lizards have been reported in relatively few insect hosts, including *Lutzomyia* and *Culicoides* species, *Culex fatigans* and *Aedes aegypti*.

Insects

Mosquitoes of the genera *Culex*, *Anopheles*, *Culiseta*, *Mansonia* and *Aedes* may act as insect hosts for various *Plasmodia* species. The best studies of these have been the *Anopheles* mosquitoes which host *Plasmodia* which cause human malaria, as well as *Culex* mosquitoes which host the *Plasmodia* that cause malaria in birds. In all cases, only female mosquitoes can be infected with *Plasmodia* species, since only the females feed on the blood of vertebrate hosts.

The survivorship and relative fitness of mosquitoes are not adversely affected by *Plasmodium* infection, which indicates the importance of vector fitness in shaping the evolution of *Plasmodium*. *Plasmodium* has evolved the capability to manipulate mosquito feeding behavior. Mosquitoes harboring *Plasmodium* have a higher propensity to bite than uninfected mosquitoes. This tendency has facilitated the spread of *Plasmodium* to the various hosts.

Species Reclassified into Other Genera

Several species of *Plasmodium* have been reclassified, mostly to *Hepatocystis*. These include:

- *Plasmodium epomophori* to *Hepatocystis epomophori*
- *Plasmodium kochi* to *Hepatocystis kochi*
- *Plasmodium limnotragi* to *Hepatocystis limnotragi* (Van Denberghe 1937)
- *Plasmodium pteropi* to *Hepatocystis pteropi* (Breinl 1911)
- *Plasmodium ratufae* to *Hepatocystis ratufae* (Donavan 1920)
- *Plasmodium vassali* to *Hepatocystis vassali* (Laveran 1905)
- *Plasmodium gonatodi* to *Garnia gonatodi*

Nomenclature

As with many other genera, the genus name also yields a common noun; thus species of the genus are known as **plasmodia**.

Plasmodium Falciparum

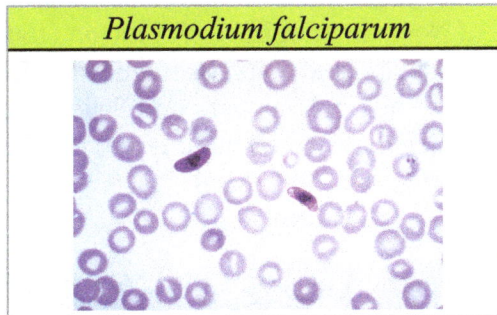

Plasmodium falciparum

Plasmodium falciparum is a protozoan parasite, one of the species of *Plasmodium* that cause malaria in humans. It is transmitted by the female *Anopheles* mosquito. This species causes the disease's most dangerous form, malignant or falciparum malaria. It has the highest complication rates and mortality. Around the world, malaria is the most significant parasitic disease of humans and claims the lives of more children worldwide than any other infectious disease.

The 2015 World Health Organization report found 214 million cases of malaria worldwide. This resulted in an estimated 438,000 deaths. Rates of infection decreased from 2000 to 2015 by 37%, but increased from 2014's 198 million cases. In sub-Saharan Africa, over 75% of cases were due to *P. falciparum*, whereas in most other malarial countries, other, less virulent plasmodial species predominate. Almost every malarial death is caused by *P. falciparum*.

Of the six malarial parasites, *P. falciparum* causes the most-often fatal and medically severe form of disease. Malaria is prevalent in tropical countries with an incidence of 300 million per year and mortality of 1 to 2 million per year. Roughly 50% of all malarial infections are caused by *P. falciparum*.

History

Malaria is caused by an infection with protozoa of the genus *Plasmodium*. The name malaria, from the Italian *mala aria*, meaning "bad air", comes from the linkage suggested by Giovanni Maria Lancisi (1717) of malaria with the poisonous vapours of swamps. This species name comes from the Latin *falx*, meaning "sickle" and *parere* meaning "to give birth".

The parasite was first seen by Laveran on November 6, 1880, at a military hospital in Constantine, Algeria, when he discovered a microgametocyte exflagellating. Patrick Manson (1894) hypothesised that mosquitoes could transmit malaria. This hypothesis was experimentally confirmed independently by Giovanni Battista Grassi and Ronald Ross in 1898. Grassi (1900) proposed an exerythrocytic stage in the lifecycle, later confirmed by Shortt, Garnham, Covell and Shute (1948), who found *Plasmodium vivax* in the human liver.

Since 1900, the area of the world exposed to malaria has been halved, yet two billion more people are presently exposed. Morbidity, as well as mortality, is substantial. Infection rates in children in endemic areas are on the order of 50%. Chronic infection has been shown to reduce school scores by up to 15%. Reduction in the incidence of malaria coincides with increased economic output.

While no effective vaccines are known for any of the six or more species that cause human malaria, drugs have been employed for centuries. In 1640, Huan del Vego first employed the tincture of the cinchona bark for treating malaria; the native Indians of Peru and Ecuador had been using it even earlier for treating fevers. Thompson (1650) introduced this "Jesuits' bark" to England. Its first recorded use there was by John Metford of Northampton in 1656. Morton (1696) presented the first detailed description of the clinical picture of malaria and of its treatment with cinchona. Gize (1816) studied the extraction of crystalline quinine from the cinchona bark and Pelletier and Caventou (1820) in France extracted pure quinine alkaloids, which they named quinine and cinchonine.

Treatment

Attempts to make synthetic antimalarials began in 1891. Atabrine, developed in 1933, was used widely throughout the Pacific in World War II, but was unpopular because of the yellowing of the skin it caused. In the late 1930s, the Germans developed chloroquine, which went into use in the North African campaigns. Mao Zedong encouraged Chinese scientists to find new antimalarials after seeing the casualties in the Vietnam War. Chinese scientist Tu Youyou discovered Artemisinin in the 1970s based on a medicine described in China in 340 CE. This drug became known to Western scientists in the late 1980s and early 1990s and is now a standard treatment. In 1976, *P. falciparum* was successfully cultured *in vitro* for the first time, which facilitated the development of new drugs. A 2008 study highlighted the emergence of artemisinin-resistant strains of *P.falciparum* in Cambodia. In February 2015, the WHO confirmed that *P. falciparum* resistant to artemisinin therapies were detected in five countries in Southeast Asia; Cambodia, the Lao People's Democratic Republic, Myanmar, Thailand and Viet Nam.

Lifecycle

Seasonal temperature suitability for transmission of P. falciparum. The Z(T) normalized index of temperature suitability for P. falciparum displayed by week across an average year.

Infection in humans begins with the bite of an infected female *Anopheles* mosquito. *Plasmodium* sporozoites released from the salivary glands of the mosquito enter the bloodstream during feeding, quickly invading liver cells (hepatocytes). The immune system clears the sporozoites from the circulation within 30 minutes.

During the next 14 days the liver-stage parasites differentiate and undergo asexual reproduction, producing up to 40,000 merozoites that burst from the hepatocyte. Other blood sporozoans, such as *P. vivax*, *P. ovale* and *P. malariae*, that infect humans and cause malaria do not have such a productive cycle for invasion. The process of bursting red blood cells does not have any symptoms, but destruction of the cells does cause anemia, since the bone marrow cannot compensate for the damage. When red blood cells rupture, hemozoin wastes cause cytokine release, chills and then fever.

P. falciparum trophozoites develop sticky knobs in red blood cells, which then adhere to endothelial cells in blood vessels, thus evading clearance in the spleen. The adhering red blood cells may cause cerebral malaria by preventing oxygenation of the brain. Symptoms of cerebral malaria include impaired consciousness, convulsions, neurological disorder and coma. Additional complications from *P. falciparum*-induced malaria include advanced immunosuppression.

Individual merozoites invade red blood cells (erythrocytes) and reproduce, producing 12-16 merozoites within a schizont. The length of this erythrocytic stage depends on the parasite species: an irregular interval for *P. falciparum*, 48 hours for *P. vivax* and *P. ovale* and 72 hours for *P. malariae*.

The clinical manifestations of malaria, fever and chills, are associated with the synchronous rupture of the infected erythrocytes. The released merozoites invade additional erythrocytes. Not all of the merozoites divide into schizonts; some differentiate into sexual forms, male and female gametocytes. These gametocytes are taken up by a female *Anopheles* mosquito during a blood meal. Within the mosquito midgut, the male gametocyte undergoes a rapid nuclear division, producing eight flagellated microgametes that fertilize the female macrogamete. The resulting ookinete traverses the mosquito gut wall and encysts on the exterior of the gut wall as an oocyst. Soon, the oocyst ruptures, releasing hundreds of sporozoites into the mosquito body cavity, where they eventually migrate to the mosquito salivary glands.

Because fusion of gametes, zygote formation and meiosis must occur in the mosquito gut for the parasite to complete its life cycle, *P. falciparum* is an obligate sexual organism. It is often self-fertilizing. Its population structure appears to predominantly reflect inbreeding.

Pathogenesis

P. falciparum works via sequestration, a distinctive property not shared by any other *Plasmodium*. Within the 48-hour asexual blood stage cycle, the mature forms change the surface properties of infected red blood cells, causing them to stick to blood vessel walls (cytoadherence). This leads to obstruction of the microcirculation and results in dysfunction of multiple organs, such as the brain in cerebral malaria.

Complicated malaria occurs more commonly in children under age 5 and sometimes in pregnant women. Women become susceptible to severe complicated malaria if infected by *P. falciparum* during their first pregnancy even if they live in hyperendemic areas. Susceptibility to severe malaria is reduced in subsequent pregnancies due to increased antibody levels against variant surface antigens that appear on infected erythrocytes.

Microscopic Appearance

The preferred method to diagnose malaria and identify the species of *Plasmodium* is by microscopic examination of a blood film. Each species has distinctive physical characteristics. In *P. falciparum*, only early (ring-form) trophozoites and gametocytes are seen in the peripheral blood. It is unusual to see mature trophozoites or schizonts in peripheral blood smears, as these are usually sequestered in the tissues. The parasitised erythrocytes are not enlarged and it is common to see cells hosting more than one parasite (multiply parasitised erythrocytes). On occasion, faint, comma-shaped, red dots called "Maurer's dots" are seen on the red cell surface. The comma-shaped dots can also appear as pear-shaped blotches.

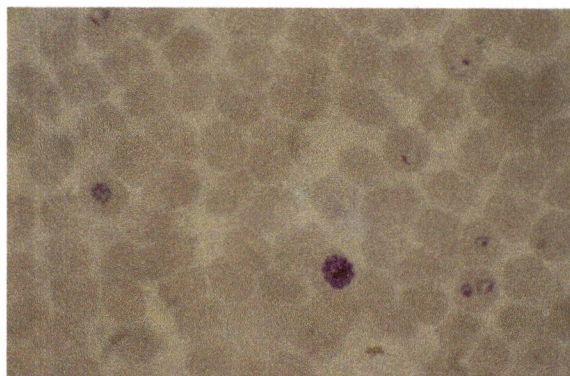

Blood smear from a *P. falciparum* culture (K1 strain - asexual forms) - several red blood cells have ring stages inside them. Close to the center is a schizont and on the left a trophozoite.

Genome

In 1995 the Malaria Genome Project was set up to sequence the genome of *P. falciparum*. The genome of its mitochondrion was reported in 1995, that of the nonphotosynthetic plastid known as the apicoplast in 1996, and the sequence of the first nuclear chromosome (chromosome 2) in 1998. The sequence of chromosome 3 was reported in 1999 and the entire genome was reported on 3 October 2002. Annotated genome data are hosted in several databases including the UCSC Malaria Genome Browser, PlasmoDB and GeneDB. The roughly 24-megabase genome is extremely AT-rich (about 80%) and is organised into 14 chromosomes. Just over 5,300 genes were described.

Influence on The Human Genome

The presence of the parasite in human populations caused selection in the human genome, as humans developed resistance to the disease. Beet, a doctor working in Southern Rhodesia (now Zimbabwe) in 1948, first suggested that sickle-cell disease could offer some protection from malaria. This suggestion was reiterated by J. B. S. Haldane in 1949, who suggested that thalassaemia could provide similar protection. This hypothesis has since been confirmed and has been extended to hemoglobin C and hemoglobin E, abnormalities in ankyrin and spectrin (ovalocytosis, elliptocytosis), in glucose-6-phosphate dehydrogenase deficiency and pyruvate kinase deficiency, loss of the Gerbich antigen (glycophorin C) and the Duffy antigen on the erythrocytes, thalassemias and variations in the major histocompatibility complex classes 1 and 2 and CD32 and CD36.

Sickle-cell Anemia

Individuals with sickle-cell anemia or sickle-cell trait have reduced parasitemia when compared to wild-type individuals for the hemoglobin protein in red blood cells. These genetic deviations of hemoglobin from normal states provide protection against the parasite.

Individuals with sickle-cell trait and sickle-cell anemia are privileged because the red blood cell sticky knobs are altered. The merozoites of each parasite species that causes malaria invade the red blood cell in three stages: contact, attachment and endocytosis. Individuals suffering from sickle-cell anemia have deformed red blood cells that interfere with the attachment phase and *P. falciparum* and the other forms of malaria have trouble with endocytosis.

These individuals have reduced attachment when compared to red blood cells with the normally functioning hemoglobin because of differing protein interactions. In normal circumstances, merozoites enter red blood cells through two PfEMP-1 protein-dependent interactions. These interactions promote the malaria inflammatory response associated with symptoms of chills and fever. When these proteins are impaired, as in sickle-cell cases, parasites cannot undergo cytoadherance interactions and cannot infect the cells; therefore, sickle-cell-anemic individuals and individuals carrying the sickle-cell trait have lower parasite loads and shorter time for symptoms than individuals expressing normal red blood cells.

Individuals with sickle-cell anemia may experience reduced symptoms of malaria because *P. falciparum* trophozoites cannot bind to hemoglobin to form sticky knobs. Without knob-binding complexes red blood cells do not stick to blood vessel walls and protecting infected individuals from e.g., cerebral malaria.

Individuals with sickle-cell deformities are less susceptible to the effects of *Plasmodium* parasite infections. Therefore, individuals expressing the genes and individuals carrying genes are selected to remain within the population. The incidence of sickle-cell anemia matches that of endemic regions for malarial infections.

Origins and Evolution

The closest relative of *P. falciparum* is *P. reichenowi*, a parasite of chimpanzees. *P. falciparum* and *P. reichenowi* are not closely related to the other *Plasmodium* species that parasitize humans, or indeed mammals in general. These two species were once thought to originate from a parasite of birds. More recent analyses instead suggest that the ability to parasitize mammals evolved only once within *Plasmodium*.

Evidence based on analysis of more than 1,100 mitochondrial, apicoplastic and nuclear DNA sequences suggested that *P. falciparum* may in fact have speciated from a lineage present in gorillas.

According to this theory, *P. falciparum* and *P. reichenowi* may both represent host switches from an ancestral line that infected primarily gorillas; *P. falciparum* went on to infect primarily humans, while *P. reichenowi* specialized in chimpanzees. The ongoing debate over the evolutionary origin of *P. falciparum* will likely be the focus of continuing genetic study.

A third potential species that appears to related to these two is *P. gaboni*. This putative species was (as of 2009) known only from two DNA sequences and is not yet accepted as distinct.

Molecular clock analyses suggest *P. falciparum* is as old as the human line; the two species diverged at the same time as humans and chimpanzees. However, low levels of polymorphism within the *P. falciparum* genome suggest a much more recent origin. It may be that this discrepancy exists because *P. falciparum* is old, but its population recently underwent a great expansion. Some evidence indicates that *P. reichenowi* was the ancestor of *P. falciparum*. The timing of this speciation is unclear at present, but it may have occurred about 10,000 years ago.

More recently, *P. falciparum* has evolved in response to human interventions. Most strains of malaria can be treated with chloroquine, but *P. falciparum* is resistant. A combination of quinine and tetracycline has been used, but some strains are resistant to this treatment, as well. Various strains of *P. falciparum* are resistant to other treatments. Resistance depends on location. Many cases of malaria that come from the Caribbean and west of the Panama Canal, as well as the Middle East and Egypt, can be treated with chloroquine. Nearly all cases contracted in Africa, India and Southeast Asia are resistant to this medication and cases in Thailand and Cambodia have shown resistance to nearly all treatments. Often, strains develop resistance in areas where treatment protocols are less tightly regulated.

Like most apicomplexans, malaria parasites harbor a plastid, an apicoplast, similar to plant chloroplasts, which they probably acquired by engulfing (or being invaded by) a eukaryotic alga and retaining the algal plastid as a distinctive organelle encased within four membranes. The apicoplast is an essential organelle, thought to be involved in the synthesis of lipids and several other compounds and provides an attractive drug target.

Gametocyte production likely has an adaptive basis: it increases when conditions for asexual reproduction of the parasite worsen (e.g. upon exposure to immunological stress and/or antimalarial chemotherapy).

Treatment

Uncomplicated Malaria

According to WHO guidelines 2010, artemisinin-based combination therapies (ACTs) are the recommended first line antimalarial treatments for uncomplicated malaria caused by *P. falciparum*. The following ACTs are recommended by the WHO:

- artemether plus lumefantrine
- artesunate plus amodiaquine
- artesunate plus mefloquine
- artesunate plus sulfadoxine-pyrimethamine
- dihydroartemisinin plus piperaquine

The choice of ACT is based on the level of resistance to the constituents in the combination. Artemisinin and its derivatives are not appropriate for monotherapy. As second-line antimalarial treatment, when initial treatment does not work, an alternative ACT known to be effective in the region is recommended, such as:

- Artesunate plus tetracycline or doxycycline or clindamycin.
- Quinine plus tetracycline or doxycycline or clindamycin

Any of these combinations is to be given for 7 days.

In Africa, the overall treatment failure was less for dihydroartemisinin-piperaquine when compared to artemether-lumefantrine and both drugs had polymerase chain reaction (PCR)-adjusted failure rates of less than 5%. However, in Asian countries, dihydroartemisinin-piperaquine was found to be better tolerated, but as effective as artesunate plus mefloquine.

For pregnant women, the recommended first-line treatment during the first trimester is quinine plus clindamycin for 7 days. Artesunate plus clindamycin for 7 days is indicated if this treatment fails. Still, an ACT is indicated only if this is the only treatment immediately available, or if treatment with 7-day quinine plus clindamycin fails or given uncertain compliance with a 7-day treatment. In second and third trimesters, the recommended treatment is an ACT known to be effective in the country/region or artesunate plus clindamycin for 7 days, or quinine plus clindamycin for 7 days. Lactating women receive standard antimalarial treatment (including ACTs) except for dapsone, primaquine and tetracyclines.

In infants and young children, the recommended first-line treatment is ACTs, with attention to accurate dosing and ensuring the administered dose is retained.

For travellers returning to nonendemic countries, any of the following is recommended:

- atovaquone-proguanil

- artemether-lumefantrine

- quinine plus doxycycline or clindamycin

Severe Malaria

In severe falciparum malaria, rapid clinical assessment and confirmation of the diagnosis is recommended, followed by administration of full doses of parenteral antimalarial treatment without delay with whichever effective antimalarial is first available.

For adults, intravenous (IV) or intramuscular (IM) artesunate is recommended. Quinine is an acceptable alternative if parenteral artesunate is not available.

For children, especially in the malaria-endemic areas of Africa, any the following antimalarial medicines is recommended:

- artesunate IV or IM,

- quinine (IV infusion or divided IM injection),

- artemether IM - should be used only if none of the alternatives is available, as its absorption may be erratic.

Parenteral antimalarials should be administered for a minimum of 24 hours, irrespective of the patient's ability to tolerate oral medication earlier. Thereafter, complete treatment is recommended including complete course of any of the following:

- an ACT

- artesunate plus clindamycin or doxycycline

- quinine plus clindamycin or doxycycline

If complete treatment is not possible, patients should be given prereferral treatment and referred immediately to an appropriate facility for further treatment. The following are options for prereferral treatment:

- rectal artesunate

- quinine IM

- artesunate IM

- artemether IM

Vaccination

Although an antimalarial vaccine is urgently needed, infected individuals never develop a sterilizing (complete) immunity, making the prospects for such a vaccine dim. The parasites live inside cells, where they are largely hidden from the immune response. Infection has a profound effect on the immune system including immune suppression. Dendritic cells suffer a maturation defect following interaction with infected erythrocytes and become unable to induce protective liver-stage immunity. Infected erythrocytes directly adhere to and activate peripheral blood B cells from nonimmune donors. The *var* gene products, a group of highly expressed surface antigens, bind the Fab and Fc fragments of human immunoglobulins in a fashion similar to protein A to *Staphylococcus aureus*, which may offer some protection to the parasite from the human immune system. Despite the poor prospects for a fully protective vaccine, it may be possible to develop a vaccine that would reduce the severity of malaria for children living in endemic areas.

Plasmodium Vivax

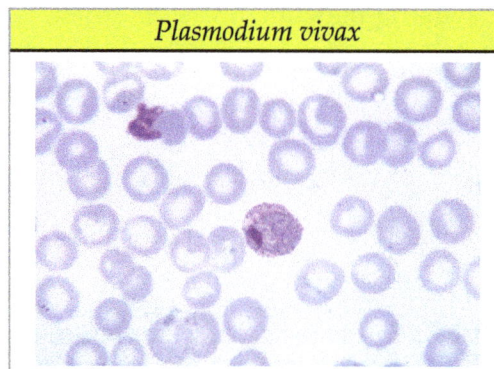

Plasmodium vivax

Plasmodium vivax is a protozoal parasite and a human pathogen. The most frequent and widely distributed cause of recurring (Benign tertian) malaria, *P. vivax* is one of the five species of malaria parasites that commonly infect humans. It is less virulent than *Plasmodium falciparum*, the deadliest of the five, but vivax malaria can lead to severe disease and death due to splenomegaly (a pathologically enlarged spleen). *P. vivax* is carried by the female *Anopheles* mosquito, since it is only the female of the species that bite.

Health

The World Health Organization (WHO) is drawing up a plan to address vivax malaria, due out in

2015.

Epidemiology

P. vivax was found mainly in the United States, Latin America, and in some parts of Africa. More recently it became a plague of low- and middle-income countries, except those in sub-Saharan Africa, where the P. vivax map has a conspicuous hole. Overall it accounts for 65% of malaria cases in Asia and South America. It is logical that plasmodium vivax is found there where humans and mosquito population are high. It is uncommon in cooler areas.

As overall malaria rates fall in a region, its proportion of cases increases. It has been estimated that 2.5 billion people are at risk of infection with this organism.

Although the Americas contribute 22% of the global area at risk, high endemic areas are generally sparsely populated and the region contributes only 6% to the total population at risk. In Africa, the widespread lack of the Duffy antigen in the population has ensured that stable transmission is constrained to Madagascar and parts of the Horn of Africa. It contributes 3.5% of global population at risk. Central Asia is responsible for 82% of global population at risk with high endemic areas coinciding with dense populations particularly in India and Myanmar. South East Asia has areas of high endemicity in Indonesia and Papua New Guinea and overall contributes 9% of global population at risk.

P. vivax is carried by at least 71 mosquito species. Many vivax vectors live happily in temperate climates—as far north as Finland. Some prefer to bite outdoors or during the daytime, hampering the effectiveness of indoor insecticide and bed nets. Several key vector species have yet to be grown in the lab for closer study, and insecticide resistance is unquantified.

Clinical Presentation

Pathogenesis results from rupture of infected red blood cells, leading to fever. Infected red blood cells may also stick to each other and to walls of capillaries. Vessels plug up and deprive tissues of oxygen. Infection may also cause the spleen to enlarge.

Unlike *P. falciparum*, *P. vivax* can populate the bloodstream with sexual-stage parasites—the form picked up by mosquitoes on their way to the next victim—even before a patient shows symptoms. Consequently, prompt treatment of symptomatic patients doesn't necessarily help stop an outbreak, as it does with falciparum malaria, in which fevers occur as sexual stages develop. Even when symptoms appear, because they are usually not immediately fatal, the parasite continues to multiply.

The parasite can go dormant in the liver for days to years, causing no symptoms and remaining undetectable in blood tests. They form what are called hypnozoites (the name derives from "sleeping organisms"), a small form that nestles inside an individual liver cell. The hypnozoites allow the parasite to survive in more temperate zones, where mosquitoes bite only part of the year.

A single infectious bite can trigger six or more relapses a year, leaving sufferers more vulnerable to other diseases. Other infectious diseases, including falciparum malaria, appear to trigger relapses.

Diagnosis

P. vivax and *P. ovale* that has been sitting in EDTA for more than 30 minutes before the blood film is made will look very similar in appearance to *P. malariae*, which is an important reason to warn the laboratory immediately when the blood sample is drawn so they can process the sample as soon as it arrives. Blood films are preferably made within 30 minutes of the blood draw and must certainly be made within an hour of the blood being drawn. Diagnosis can be done with the strip fast test of antibodies.

Treatment

Chloroquine remains the treatment of choice for *vivax* malaria, except in Indonesia's Irian Jaya (Western New Guinea) region and the geographically contiguous Papua New Guinea, where chloroquine resistance is common (up to 20% resistance). Chloroquine resistance is an increasing problem in other parts of the world, such as Korea and India.

When chloroquine resistance is common or when chloroquine is contraindicated, then artesunate is the drug of choice, except in the U.S., where it is not approved for use. Where an artemisinin-based combination therapy has been adopted as the first-line treatment for *P. falciparum* malaria, it may also be used for P. vivax malaria in combination with primaquine for radical cure. An exception is artesunate plus sulfadoxine-pyrimethamine (AS+SP), which is not effective against P. vivax in many places. Mefloquine is a good alternative and in some countries is more readily available. Atovaquone-proguanil is an effective alternative in patients unable to tolerate chloroquine. Quinine may be used to treat *vivax* malaria but is associated with inferior outcomes.

32–100% of patients will relapse following successful treatment of *P. vivax* infection if a radical cure (eradication of liver stages) is not given.

Eradication of the liver stages is achieved by giving primaquine. Patients with glucose-6-phosphate dehydrogenase risk haemolysis. G6PD is an enzyme important for blood chemistry. No field-ready test is available. Recently, this point has taken particular importance for the increased incidence of vivax malaria among travelers. At least a 14-day course of primaquine is required for the radical treatment of *P. vivax*.

Tafenoquine

In 2013 a Phase IIb trial was completed that studied a single-dose alternative drug named tafenoquine. It is an 8-aminoquinoline, of the same family as primaquine, developed by researchers at the Walter Reed Army Institute of Research in the 1970s and tested in safety trials. It languished, however, until the push for malaria elimination sparked new interest in primaquine alternatives.

Among patients who received a 600-mg dose, 91% were relapse-free after 6 months. Among patients who received primaquine, 24% relapsed within 6 months. "The data are absolutely spectacular," Wells says. Ideally, he says, researchers will be able to combine the safety data from the Army's earlier trials with the new study in a submission to the U.S. Food and Drug Administration for approval. Like primaquine, tafenoquine causes hemolysis in people who are G6PD deficient.

In 2013 researchers produced cultured human "microlivers" that supported liver stages of both *P. falciparum* and *P. vivax* and may have also created hypnozoites.

Eradication

Mass-treating populations with a primaquine can kill the hypnozoites, exempting those with G6PD deficiency. However, the standard regimen requires a daily pill for 14 days across an asymptomatic population.

Korea

P. vivax is the only indigenous malaria parasite on the Korean peninsula. In the years following the Korean War (1950–53), malaria-eradication campaigns successfully reduced the number of new cases of the disease in North Korea and South Korea. In 1979, World Health Organization declared the Korean peninsula vivax malaria-free, but the disease unexpectedly re-emerged in the late 1990s and still persists today. Several factors contributed to the re-emergence of the disease, including reduced emphasis on malaria control after 1979, floods and famine in North Korea, emergence of drug resistance and possibly global warming. Most cases are identified along the Korean Demilitarized Zone. As such, vivax malaria offers the two Koreas an opportunity to work together on an important health problem that affects both countries.

Biology

Life Cycle

Like all malaria parasites, *P. vivax* has a complex life cycle. It infects a definitive insect host, where sexual reproduction occurs, and an intermediate vertebrate host, where asexual amplification occurs. In *P. vivax*, the definitive hosts are Anopheles mosquitoes (also known as the vector), while humans are the intermediate asexual hosts. During its life cycle, *P. vivax* assumes many different physical forms.

Asexual Forms:

- Sporozoite: Transfers infection from mosquito to human

- Immature trophozoites (Ring or signet-ring shaped), about 1/3 of the diameter of a RBC.

- Mature trophozoites: Very irregular and delicate (described as *amoeboid*); many pseudopodial processes seen. Presence of fine grains of brown pigment (malarial pigment) or hematin probably derived from the haemoglobin of the infected red blood cell.

- Schizonts (also called meronts): As large as a normal red cell; thus the parasitized corpuscle becomes distended and larger than normal. There are about sixteen merozoites.

Sexual Forms:

- Gametocytes: Round. *P. vivax* gametocytes are commonly found in human peripheral blood at about the end of the first week of parasitemia.

- Gametes: Formed from gametocytes in mosquitoes.

- Zygote: Formed from combination of gametes

- Oocyst: Contains zygote, develops into sporozoites

Human Infection

P. vivax human infection occurs when an infected mosquito feeds on a human. During feeding, the mosquito injects saliva to prevent blood clotting (along with sporozoites), thousands of sporozoites are inoculated into human blood; within a half-hour the sporozoites reach the liver. There they enter hepatic cells, transform into the trophozoite form and feed on hepatic cells, and reproduce asexually. This process gives rise to thousands of merozoites (plasmodium daughter cells) in the circulatory system and the liver.

The incubation period of human infection usually ranges from ten to seventeen days and sometimes up to a year. Persistent liver stages allow relapse up to five years after elimination of red blood cell stages and clinical cure.

Liver Stage

The *P. vivax* sporozoite enters a hepatocyte and begins its exoerythrocytic schizogony stage. This is characterized by multiple rounds of nuclear division without cellular segmentation. After a certain number of nuclear divisions, the parasite cell will segment and merozoites are formed.

There are situations where some of the sporozoites do not immediately start to grow and divide after entering the hepatocyte, but remain in a dormant, hypnozoite stage for weeks or months. The duration of latency is variable from one hypnozoite to another and the factors that will eventually trigger growth are not known; this explains how a single infection can be responsible for a series of waves of parasitaemia or "relapses". Different strains of *P. vivax* have their own characteristic relapse pattern and timing. The earlier stage is exo-erythrocytic generation.

Erythrocytic Cycle

P. vivax preferentially penetrates young red blood cells (reticulocytes). In order to achieve this, merozoites have two proteins at their apical pole (PvRBP-1 and PvRBP-2). The parasite uses the Duffy blood group antigens (Fy6) to penetrate red blood cells. This antigen does not occur in the majority of humans in West Africa [phenotype Fy (a-b-)]. As a result, *P. vivax* occurs less frequently in West Africa.

The parasitised red blood cell is up to twice as large as a normal red cell and Schüffner's dots (also known as Schüffner's stippling or Schüffner's granules) are seen on the infected cell's surface. Schüffner's dots have a spotted appearance, varying in color from light pink, to red, to red-yellow, as coloured with Romanovsky stains. The parasite within it is often wildly irregular in shape (described as "amoeboid"). Schizonts of *P. vivax* have up to twenty merozoites within them. It is rare to see cells with more than one parasite within them. Merozoites will only attach to immature blood cell (reticulocytes) and therefore it is unusual to see more than 3% of all circulating erythrocytes parasitised.

Mosquito Stage

Parasite life cycle in mosquitoes includes all stages of sexual reproduction:

1. Infection and Gametogenesis

   o Microgametes

   o Macrogametes

2. Fertilization

3. Ookinite

4. Oocyst

5. Sporogony

Mosquito Infection and Gamete Formation

When a female Anopheles mosquito bites an infected person, gametocytes and other stages of the parasite are transferred to the mosquito stomach. Gametocytes ultimately develop into gametes, a process known as gametogony.

Microgametocytes become very active, and their nuclei undergo fission (i.e amitosis) to each give 6-8 daughter nuclei, which becomes arranged at the periphery. The cytoplasm develops long thin flagella like projections, then a nucleus enter into each one of these extensions. These cytoplasmic extensions later break off as mature male gametes (microgametes). This process of formation of flagella-like microgametes or male gametes is known as exflagellation. Macrogametocytes show very little change. They develop a cone of reception at one side and becomes mature as macrogametocytes (female gametes).

Fertilization

Male gametes move actively in the stomach of mosquitoes in search of female gametes. Male gametes then enter into female gametes through the cone of reception. The complete fusion of 2 gametes results in the formation of zygote. Here, fusion of 2 dissimilar gametes occurs, known as anisogamy.

The zygote remains inactive for sometime but it soon elongates, becomes vermiform (worm-like) and motile. It is now known as ookinete. The pointed ends of ookinete penetrate the stomach wall and come to lie below its outer epithelial layer. Here the zygote becomes spherical and develops a cyst wall around itself. The cyst wall is derived partly from the stomach tissues and partly produced by the zygote itself. At this stage, the zygote is known as an oocyst. The oocyst absorbs nourishment and grows in size. Oocysts protrude from the surface of stomach, giving it a blistered appearance. In a highly infected mosquito, as many as 1000 oocysts may be seen.

Sporogony

The oocyst nucleus divides repeatedly to form large number of daughter nuclei. At the same time, the cytoplasm develops large vacuoles and forms numerous cytoplasmic masses. These cytoplasmic masses then elongate and a daughter nuclei migrates into each mass. The resulting sickle-shaped bodies are known as sporozoites. This phase of asexual multiplication is known as sporogony and is completed in about 10–21 days. The oocyst then bursts and sporozoites are released into the body cavity of mosquito. Sporozoites eventually reach the salivary glands of mosquito via its hemolymph. The mosquito now becomes infectious. Salivary glands of a single infected mosquito may contain as many as 200,000 sporozoites. When the mosquito bites a healthy person, thousands of sporozoites are infected into the blood along with the saliva and the cycle starts again.

Taxonomy

P. vivax can be divided into two clades one that appears to have origins in the Old World and a second that originated in the New World. The distinction can be made on the basis of the structure of the A and S forms of the rRNA. A rearrangement of these genes appears to have occurred in the New World strains. It appears that a gene conversion occurred in an Old World strain and this strain gave rise to the New World strains. The timing of this event has yet to be established.

At present both types of *P. vivax* circulate in the Americas. The monkey parasite - *Plasmodium simium* - is related to the Old World strains rather than to the New World strains.

A specific name - *Plasmodium collinsi* - has been proposed for the New World strains but this suggestion has not been accepted to date.

Miscellaneous

It has been suggested that P. vivax has horizontally acquired genetic material from humans.

Plasmodium vivax is not known to have a particular gram stain (negative vs. positive) and may appear as either.

Therapeutic Use

P. vivax was used between 1917 and the 1940s for malariotherapy, that is, to create very high fevers to combat certain diseases such as tertiary syphilis. In 1917, the inventor of this technique, Julius Wagner-Jauregg, received the Nobel Prize in Physiology or Medicine for his discoveries. However, the technique was dangerous, killing about 15% of patients, so it is no longer in use.

Plasmodium Malariae

Plasmodium malariae is a parasitic protozoa that causes malaria in humans. It is one of several species of *Plasmodium* parasites that infect humans including *Plasmodium falciparum* and *Plasmodium vivax* which are responsible for most malarial infection. While found worldwide, it is a so-called "benign malaria" and is not nearly as dangerous as that produced by *P. falciparum* or *P. vivax*. It causes fevers that recur at approximately three-day intervals (a *quartan fever*), longer than the two-day (tertian) intervals of the other malarial parasites, hence its alternate names quartan fever and quartan malaria.

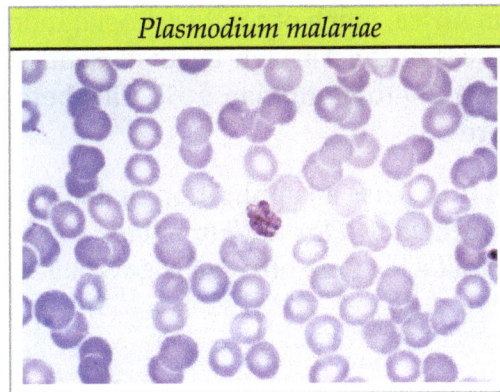

Plasmodium malariae

History

Malaria has been recognized since the Greek and Roman civilizations over 2,000 years ago, with different patterns of fever described by the early Greeks. In 1880, Alphonse Laveran discovered that the causative agent of malaria is a parasite. Detailed work of Golgi in 1886 demonstrated that in some patients there was a relationship between the 72-hour life cycle of the parasite and the chill and fever patterns in the patient. The same observation was found for parasites with 48-hour cycles. Golgi concluded that there must be more than one species of malaria parasite responsible for these different patterns of infection.

Epidemiology

Each year, approximately 500 million people will be infected with malaria worldwide Of those infected, roughly two million will die from the disease. Malaria is caused by six *Plasmodium* species: *Plasmodium falciparum*, *Plasmodium vivax*, *Plasmodium ovale curtisi*, *Plasmodium ovale wallikeri*, *Plasmodium malariae* and *Plasmodium knowlesi*. At any one time, an estimated 300 million people are said to be infected with at least one of these *Plasmodium* species and so there is a great need for the development of effective treatments for decreasing the yearly mortality and morbidity rates.

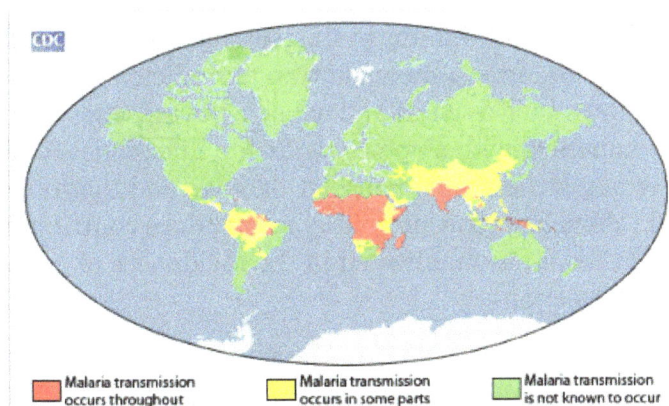

| Malaria transmission occurs throughout | Malaria transmission occurs in some parts | Malaria transmission is not known to occur |

Geographical areas of malaria transmission

P. malariae is the one of the least studied of the six species that infect humans, in part because of its low prevalence and milder clinical manifestations compared to the other species. It is widespread throughout sub-Saharan Africa, much of southeast Asia, Indonesia, on many of the islands of the western Pacific and in areas of the Amazon Basin of South America. In endemic regions, prevalence ranges from less than 4% to more than 20%, but there is evidence that *P. malariae* infections are vastly underreported.

The Center for Disease Control (CDC) has an application that allows people to view specific parts of the world and how they are affected by *Plasmodium vivax*, and other types of the *Plasmodium* parasite. It can be found at the following link: http://cdc.gov/malaria/map/index.html.

Role in Disease

P. malariae can infect several species of mosquito and can cause malaria in humans. *P. malariae* can be maintained at very low infection rates among a sparse and mobile population because unlike the other *Plasmodium* parasites, it can remain in a human host for an extended period of time and still remain infectious to mosquitoes.

Vector

zInformation about the prepatent period, or the period of time between the infection of the parasite and demonstration of that parasite within the body, of *P. malariae* associated malaria is limited, but the data suggests that there is great variation, often the length of time depending on the strain of *P. malariae* parasite. Usually, the prepatent period ranges from 16 to 59 days.

Infection in Humans

Plasmodium malariae causes a chronic infection that in some cases can last a lifetime. The *P. malariae* parasite has several differences between it and the other *Plasmodium* parasites, one being that maximum parasite counts are usually low compared to those in patients infected with *P. falciparum* or *P. vivax*. The reason for this can be accounted for by the lower number of merozoites produced per erythrocytic cycle, the longer 72-hour developmental cycle (compared to the 48-hour cycle of *P. vivax* and *P. falciparum*), the preference for development in older erythrocytes and the resulting earlier development of immunity by the human host. Another defining feature of *P. malariae* is that the fever manifestations of the parasite are more moderate relative to those of *P. falciparum* and *P. vivax* and fevers show quartan periodicity. Along with bouts of fever and more general clinical symptoms such as chills and nausea, the presence of edema and the nephrotic syndrome has been documented with some *P. malariae* infections. It has been suggested that immune complexes may cause structural glomerular damage and that renal disease may also occur. Although *P. malariae* alone has a low morbidity rate, it does contribute to the total morbidity caused by all *Plasmodium* species, as manifested in the incidences of anemia, low birth rate and reduced resistance to other infections.

Due to a similarity in the appearances of the pathogens, *P. knowlesi* infections are often misdiagnosed as *P. malariae* infections. Molecular analysis is usually required for an accurate diagnosis.

Diagnostics

The preferable method for diagnosis of *P. malariae* is through the examination of peripheral blood films stained with Giemsa stain. PCR techniques are also commonly used for diagnoses confirmation as well as to separate mixed *Plasmodium* infections. Even with these techniques, however, it may still be impossible to differentiate infections, as is the case in areas of South America where humans and monkeys coexist and *P. malariae* and *P. brasilianum* are not easily distinguishable.

Biology

As a protist, the plasmodium is a eukaryote of the phylum Apicomplexa. Unusual characteristics of this organism in comparison to general eukaryotes include the rhoptry, micronemes, and polar rings near the apical end. The plasmodium is known best for the infection it causes, malaria.

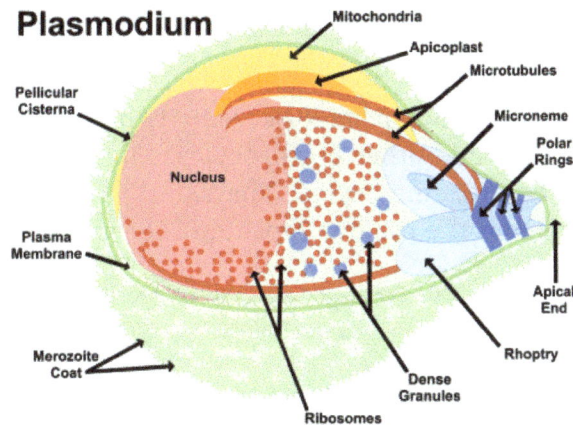

Life Cycle

P. malariae is the only human malaria parasite that causes fevers that recur at approximately three-day intervals (therefore occurring evey fourth day, a *quartan fever*), longer than the two-day (*tertian*) intervals of the other malarial parasites.

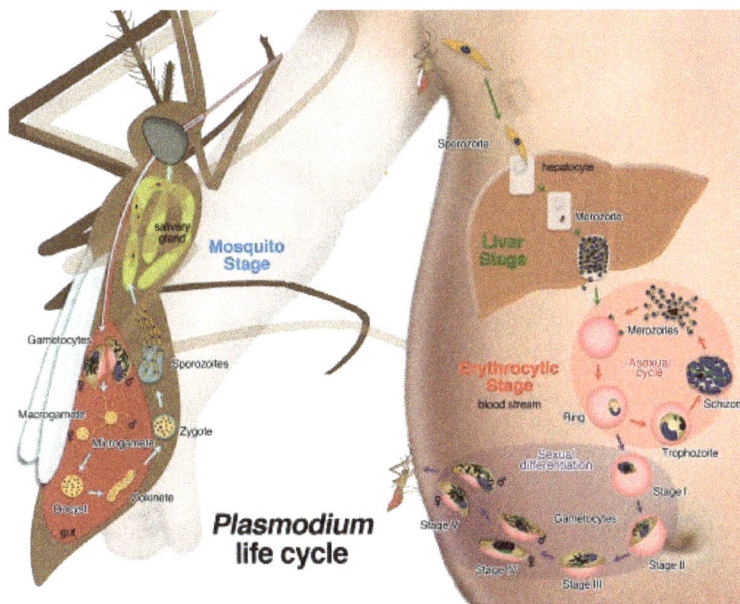

Plasmodium malariae

Human Infection

Liver Stage

In this stage, many thousands of merozoites are produced in each schizont. As the merozoites are released, they invade erythrocytes and initiate the erythrocytic cycle, where the parasite digests hemoglobin to obtain amino acids for protein synthesis.

Erythrocytic Cycle

The total length of the intraerythrocytic development is roughly 72 hours for *P. malariae*.

At the schizont stage, after schizogonic division, there are roughly 6–8 parasite cells in the erythrocyte.

Following the erythrocytic cycle, which lasts for seventy two hours on average, six to fourteen merozoites are released to reinvade other erythrocytes. Finally, some of the merozoites develop into either micro- or macrogametocytes. The two types of gametocytes are taken into the mosquito during feeding and the cycle is repeated. There are no animal reservoirs for *P. malariae*.

Mosquito Stage

Similar to the other human-infecting *Plasmodium* parasites, *Plasmodium malariae* has distinct developmental cycles in the Anopheles mosquito and in the human host. The mosquito serves as the definitive host and the human host is the intermediate. When the Anopheles mosquito takes a blood meal from an infected individual, gametocytes are ingested from the infected person. A process known as exflagellation of the microgametocyte soon ensues and up to eight mobile microgametes are formed.

Sexual Stage

Following fertilization of the macrogamete, a mobile ookinete is formed, which penetrates the peritropic membrane surrounding the blood meal and travels to the outer wall of the mid-gut of the mosquito. The oocyst then develops under the basal membrane and after a period of two to three weeks a variable amount of sporozoites are produced within each oocyst. The number of sporozoites that are produced varies with temperature and can range from anywhere between many hundreds to a few thousand. Eventually, the oocyst ruptures and the sporozoites are released into the hemocoel of the mosquito. The sporozoites are then carried by the circulation of the hemolymph to the salivary glands, where they become concentrated in the acinal cells. A small number of sporozoites are introduced into the salivary duct and injected into the venules of the bitten human. This initiates the cycle in the human liver.

Morphology

The ring stages that are formed by the invasion of merozoites released by rupturing liver stage schizonts are the first stages that appear in the blood. The ring stages grow slowly but soon fill one-fourth to one-third of the parasitized cell. Pigment increases rapidly and the half-grown parasite may have from 30 to 50 jet-black granules. The parasite changes various shapes as it grows and stretches across the host cell to form the band form.

Laboratory Considerations

P. vivax and *P. ovale* sitting in EDTA for more than 30 minutes before the blood film is made will look very similar in appearance to *P. malariae*, which is an important reason to warn the laboratory immediately when the blood sample is drawn so they can process the sample as soon as it arrives.

Microscopically, the parasitised red blood cell (erythrocyte) is never enlarged and may even appear smaller than that of normal red blood cells. The cytoplasm is not decolorized and no dots are visible on the cell surface. The food vacuole is small and the parasite is compact. Cells seldom host more than one parasite. Band forms, where the parasite forms a thick band across the width of the infected cell, are characteristic of this species (and some would say is diagnostic). Large grains of malarial pigment are often seen in these parasites: more so than any other *Plasmodium* species, 8 merozoites.

Management and Therapy

Failure to detect some *P. malariae* infections has led to modifications of the species-specific primers and to efforts towards the development of real-time PCR assays. The development of such an assay has included the use of generic primers that target a highly conserved region of the 18S rRNA genes of the four human-infecting species of *Plasmodium*. This assay was found to be highly specific and sensitive. Although serologic tests are not specific enough for diagnostic purposes, they can be used as basic epidemiologic tools. The immunofluorescent-antibody (IFA) technique can be used to measure the presence of antibodies to *P. malariae*.. A pattern has emerged in which an infection of short duration causes a rapidly declining immune response, but upon re-infection or recrudescence, the IFA level rises significantly and remains present for many months or years.

The increasing need to correctly identify *P. malariae* infection is underscored by its possible anti-malarial resistance. In a study by Müller-Stöver *et al.*, the researchers presented three patients who were found to be infected with the parasite after taking anti-malarial medications. Given the slower pre-erythrocytic development and longer incubation period compared to the other malaria causing *Plasmodium* species, the researchers hypothesized that the anti-malarials may not be effective enough against the pre-erythrocytic stages of *P. malariae*. They thought that further development of *P. malariae* can occur when plasma concentrations of the anti-malarials gradually decrease after the anti-malarial medications are taken. According to Dr. William E. Collins from the Center of Disease Control (CDC), chloroquine is most commonly used for treatment and no evidence of resistance to this drug has been found. In that event, it is possible that the results from Müller-Stöver *et al.* provided isolated incidences.

Public Health, Prevention Strategies and Vaccines

The food vacuole is the specialized compartment that degrades hemoglobin during the asexual erythrocytic stage of the parasite. It is implied that effective drug treatments can be developed by targeting the proteolytic enzymes of the food vacuole. In a paper published in 1997, Westling *et al.* focused their attention on the aspartic endopeptidase class of enzymes, simply called plasmepsins. They sought to characterize the specificity for the enzymes cloned from *P. vivax* and *P. malariae*. Using substrate specificity studies and inhibitor analysis, it was found that the plasmepsins for

P. malariae and *P. vivax* showed less specificity than that for *P. falciparum*. Unfortunately, this means that the development of a selective inhibitor for *P. malariae* may prove more challenging than the development of one for *P. falciparum*. Another study by Bruce *et al.* presented evidence that there may be regular genetic exchange within *P. malariae* populations. Six polymorphic genetic markers from *P. malariae* were isolated and analyzed in 70 samples of naturally acquired *P. malariae* infections from different parts of the world. The data showed a high level of multi-genotypic carriage in humans.

Both of these experiments illustrate that development of vaccine options will prove challenging, if not impossible. Dr. William Collins doubts that anyone is currently looking for possible vaccines for *P. malariae* and given the complexity of the parasite it can be inferred why. He states that very few studies are conducted with this parasite, perhaps as a result of its perceived low morbidity and prevalence. Collins sights the great restrictions of studies with chimpanzees and monkeys as a sizeable barrier. Since the *Plasmodium brasilianum* parasite that infects South American monkeys is thought to be an adapted form of *P. malariae,* more research with *P. brasilianum* may hold valuable insight into *P. malariae*.

The continuing work with the plasmepsin associated with *P. malariae,* plasmepsin 4, by Professor Ben Dunn and his research team from the University of Florida may provide hope for long term malaria control in the near future.

In Popular Culture

The pathogen Plasmodium malariae is mentioned in an early episode of Doctor at Large (TV series), along with chloroquine as one of its treatments, when Dr. Maxwell appears to have relapsed malaria, having been in Africa some years before, but turns out to be pretending.

Plasmodium Knowlesi

Plasmodium knowlesi is a primate malaria parasite commonly found in Southeast Asia. It causes malaria in long-tailed macaques (*Macaca fascicularis*), but it may also infect humans, either naturally or artificially.

Plasmodium knowlesi is the sixth major human malaria parasite (following the division of *Plasmodium ovale* into 2 species). It may cause severe malaria as indicated by its asexual erythrocytic cycle of about 24 hours, with an associated fever that typically occurs at the same frequency (i.e. the fever is quotidian). This is an emerging infection that was reported for the first time in humans in 1965. It accounts for up to 70% of malaria cases in certain areas in South East Asia where it is mostly found. This parasite is transmitted by the bite of an *Anopheles* mosquito. *Plasmodium knowlesi* has health, social and economic consequences for the regions affected by it.

History of Discovery

The first person to see *P. knowlesi* was probably the Italian Giuseppe Franchiti in 1927 when he was examining the blood of *Macaca fascicularis* and he noted that it differed from *Plasmo-*

dium cynomolgi and *Plasmodium inui*. It was later seen by Campbell in 1931 in a long-tailed macaque imported from Singapore to the Calcutta School of Tropical Medicine and Hygiene in India. Campbell was interested in kala azar and was working under Napier. Napier inoculated the strain into three monkeys, one of which was a rhesus macaque (*Macaca mulatta*), which developed a fulminating infection. Knowing that the Protozoological Department were looking for a monkey malaria strain, they handed the original infected monkey to Biraj Mohan Das Gupta, who was the assistant of Robert Knowles. Dr Das Gupta maintained the species by serial passage in monkeys until Dr Knowles returned from leave. In 1932, Knowles and Das Gupta described the species in detail for the first time and showed that it could be transmitted to man by blood passage, but failed to name it. It was named by Sinton and Mulligan in 1932 after Dr Knowles. From early in the 1930s to 1955, *P. knowlesi* was used as a pyretic agent for the treatment of patients with neurosyphillis.

In 1957, it was suggested by Garnham et al. that *P. knowlesi* could be the fifth species capable of causing endemic malaria in humans.

In 1965, the first case of a naturally occurring infection of knowlesi malaria in humans was reported in an American man who had returned after working in the jungle in peninsular Malaysia. Although the infecting parasite was initially identified as *P. falciparum*, one day later it was then identified as *P. malariae* and it was only confirmed to be *P. knowlesi* after infected blood was used to inoculate rhesus monkeys. A second report emerged in 1971 about the natural infection of a man in Malaysia with *Plasmodium knowlesi* followed by the description of a large focus of human infections in the Kapit Division of Sarawak, Malaysian Borneo. This was made possible due to the development of molecular detection assays which could differentiate between *Plasmodium knowlesi* and the morphologically similar *Plasmodium malariae*. Since 2004, there has been an increasing number of reports of the incidence of *P. knowlesi* among humans in various countries in South East Asia, including Malaysia, Thailand, Singapore, the Philippines, Vietnam, Myanmar and Indonesia.

Work with archival samples has shown that infection with this parasite has occurred in Malaysia at least since the 1990s and it is now known to cause 70% of the malaria cases in certain areas of Sarawak.

Evolution

Based on a Bayesian coalescent approach the most probable time of evolution of *P. knowelsi* is 257,000 years ago (95% range 98,000–478,000). Yakob and coauthors calculated the likelihood of natural host switching from the long-tailed macaque monkey to humans using an evolutionary invasion analysis and demonstrated how this switch was contingent on relative host densities and individual-level mosquito feeding preferences.

Life Cycle

Plasmodium knowlesi parasite replicates and completes its blood stage cycle in 24-hour cycles resulting in fairly high loads of parasite densities in a very short period of time. This makes it a potentially very severe disease if it remains untreated. Life cycle: merozoite → trophozoites → schizont → merozoite. These stages of *Plasmodium knowlesi* are microscopically indistinguishable from *Plasmodium malariae* and the early trophozoites are identical to those of *Plasmodium falciparum* .

Mosquito stages: A mosquito ingests gametocytes, which have been formed in the mammalian host. These are either microgametocytes (which are male gametocytes) or macrogametocytes (which are female gametetocytes). These gametocytes mature into microgametes and macrogametes respectively, and then fertilize to form zygotes within the midgut of the mosquito. The zygotes mature into ookinetes, then into oocysts. Finally, the oocysts mature to release sporozoites which move to salivary gland of the mosquito.

Summary: gametocyte → (microgamete or macrogamete) → zygote → ookinete → oocyst → sporozoites.

In man: exoerythrocytic stage (in the liver): The sporozoites are injected into humans when the mosquito bites and they travel to the liver through blood stream and undergo asexual reproduction to become merozoites through schizonts in the liver cell. Hypnozoites in the liver has not yet been found.

Summary: sporozoites → schizonts → merozoites.

In man: erythrocytic stage (in the blood): Merozoites are unleashed into the blood stream to infect erythrocytes constituting one asexual cycle of infection of the erythrocytes. Within the red blood cells some merozoites develop into trophozoites, which in turn mature into schizonts that rupture to release merozoites, while others develop into microgametocytes or macrogametocytes. These gametocytes remain in the blood to be ingested by mosquitoes.

Summary: Merozoite → trophozoite → schizont → merozoites.

Epidemiology

P. knowlesi infection is normally considered a parasite of long-tailed (*Macaca fascicularis*) and pig-tailed (*Macaca nemestrina*) macaques but humans who work at the forest fringe or enter the rainforest to work are at risk of infection. With the increasing popularity of deforestation and development efforts in South East Asia, many macaques are now coming in close and direct contact with humans. Hence more and more people who live in the semi-urban areas are being found to be infected with knowlesi malaria. 2,584 cases of this type of malaria were reported in Malaysia in 2014.

This parasite is mostly found in South East Asian countries particularly in Borneo, Cambodia, Malaysia, Myanmar, Philippines, Singapore, Thailand and neighboring countries and it appears to occur in regions that are reportedly free of the other four types of human malaria. Infective mosquitoes are restricted to the forest areas. Non-infective mosquitoes are typically found in the urban areas but transmission may occur due to the abundance of mosquitoes in this region. particularly Malaysia, but there are also reports on the Thai-Burmese border. One fifth of the cases of malaria diagnosed in Sarawak, Malaysian Borneo are due to *P. knowlesi*.

Plasmodium knowlesi is absent in Africa. This may be because there are neither long-tailed nor pig-tailed macaques (the reservoir hosts of *P. knowlesi*) in Africa, and many West Africans lack the Duffy antigen - a protein on the surface of the red blood cell that the parasite uses to invade.

P. knowlesi is the most common cause of malaria in childhood in the Kudat district of Sabah, Malaysia.

Vectors

Theoretically there are four modes of transmission: from an infected mosquito to another monkey, from an infected monkey to a human, from an infected human to another human and from an infected human back to a monkey. In practice human malaria appears to be almost entirely due to monkey to human transmission.

The known vectors belong to the genus *Anopheles*, subgenus *Cellia*, series *Neomyzomyia* and group *Leucosphyrus*. Mosquitoes of this group are typically found in forest areas in South East Asia but with a greater clearing of forest areas for farmland, humans are increasingly becoming exposed to these vectors.

Within the monkey population in Peninsular Malaysia, *Anopheles hackeri* is believed to be the main vector of *P. knowlesi*: although *A. hackeri* is capable of transmitting malaria to humans, it is not normally attracted to humans and seems unlikely to be an important vector for transmission to humans.

Anopheles latens is attracted to both macaques and humans and has been shown to be the main vector transmitting *P. knowlesi* to humans in the Kapit Division of Sarawak, Malaysian Borneo.

Anopheles cracens has also been reported as a vector of *P. knowlesi*. Both species of mosquitoes have been shown to contain as many as 1,000 sporozoites suggesting that they may be efficient vectors.

A study of potential vectors in Malayasia suggests that *Anopheles cracens* may be an important vector of *P. knowlesi*.

Clinical

Two possible modes of transmission to humans have been proposed: either from an infected monkey to a human or from an infected human to another human.

Symptoms typically begin approximately 11 days after an infected mosquito has bitten a person and the parasites can be seen in the blood between 10 – 12 days after infection. The parasite may multiply rapidly resulting in very high parasite densities that may be fatal.

Although the current infection rate with *Plasmodium knowlesi* is relatively low, one risk it presents is misdiagnosis with other forms of malarial parasites such as *P. malariae* especially when microscopy is used. *P. knowlesi* can only be accurately distinguished from *P. malariae* using PCR assay and/or molecular characterization.

Symptoms of *P. knowlesi* in humans include headache, fever, chills and cold sweats. Singh *et al.* (2004) showed clinical symptoms in 94 patients with single species *P. knowlesi* infection at Kapit Hospital, Sarawak, Malaysian Borneo. Symptoms included fever, chills, and rigor in 100% of patients, headache in 32%, cough in 18%, vomiting in 16%, nausea in 6%, and diarrhea in 4%. Asexual cycle of the parasite in humans and its natural host macaque is about 24 hours. Hence the disease may be called quotidian malaria, in concert with designation of tertian malaria and quartan malaria. In addition to a lab diagnosis using PCR assay, knowlesi malaria may also present itself with elevated levels of C-reactive protein and thrombocytopenia.

This parasite causes non-relapsing malaria due to lack of hypnozoites in its exoerythrocytic stage.

While infection with this organism is normally not serious, life-threatening complications or even death may occur in a minority of cases. The most common complications are respiratory distress, abnormal liver function including jaundice and renal failure. Mortality in one series of cases was about 2%.

Diagnosis

P. knowlesi infection is diagnosed by examining thick and thin blood films in the same way as other malarias. The appearance of *P. knowlesi* is similar to that of *P. malariae* and is unlikely to be correctly diagnosed except by using molecular detection assays in a malaria reference laboratory.

The morphology of *Plasmodium knowlesi* is similar to that of *Plasmodium malariae*. *P. malariae* is characterized by a compact parasite (all stages) and does not alter the host erythrocyte's shape or size or cause enlargement. Elongated trophozoites stretching across the erythrocyte, called band forms, are sometimes observed. Schizonts will typically have 8-10 merozoites that are often arranged in a rosette pattern with a clump of pigment in the center.

Rapid diagnostic tests kits may or may not recognize *P. knowlesi* because of their specificity.

Currently PCR assay and molecular characterization are the most reliable methods for detecting and diagnosing *P. knowlesi* infection. PCR identifies the parasite DNA but this technique is not rapid and cannot be used for routine identification. PCR is also expensive and requires very specialized equipment.

Treatment

Because *P. knowlesi* takes only 24 hours to complete its erythrocytic cycle, it can rapidly result in very high levels of parasitemia with fatal consequences. Anyone with a severe and rapidly deteriorating condition should be treated aggressively and urgently as if were infected with falciparum malaria. *P. knowlesi* responds well to treatment with chloroquine and primaquine. In a clinical study of treatment where response was observed after oral chloroquine was given for three days, and at 24 hours oral primaquine was administered for two consecutive days., it was found that this regime gave a rapid response with a median time to parasite clearance of three hours. This was more rapid that is found in *Plasmodium vivax* malaria where the median time to clearance is between six and seven hours.

Public Health, Prevention Strategies and Vaccines

1. Mosquito bed nets

2. Medication – Mefloquine, Chloroquine

3. Vector control

4. Residual spraying using insecticides

Pathology

A single post mortem case has been described to date The patient was a male who became unwell 10 days after exposure. After four days he presented acutely unwell to a hospital. He was found to have a raised eosinophil count, to be thrombocytopaenic, hyponatraemic with an elevated blood urea, potassium, lactate dehydrogenase and amino transferase values. Dengue fever was suspected but ruled out on investigation. Malarial parasites were seen on the blood film and later identified as *Plasmodium knowlesi* by PCR. At post mortem the liver and spleen were enlarged. The brain and endocardium showed multiple petechial haemorrhages. The lungs had features consistent with acute respiratory distress syndrome. Histological examination showed sequestration of pigmented parasitized red blood cells in the vessels of the cerebrum, cerebellum, heart and kidney without evidence of chronic inflammatory reaction in the brain or any other organ examined. The spleen and liver had abundant pigment containing macrophages and parasitized red blood cells. The kidney had evidence of acute tubular necrosis. Endothelial cells in heart sections were prominent. Brain sections were negative for intracellular adhesion molecule-1.

The overall post mortem picture was very similar to that found in cases of *Plasmodium falciparum*. There were important differences including the absence of coma despite petechial haemorrhages and parasite sequestration in the brain.

Notes

There are at least two subspecies of *P. knowlesi* known - *P. knowlesi edesoni* and *P. knowlesi knowlesi*. The subspecies *P. knowlesi edesoni* was described by Garnham in 1963. It was named after the parasitologist J F B Edeson. It is not known if there are any clinical differences between these two subspecies.

Rhinosporidium Seeberi

Rhinosporidium seeberi is a eukaryotic pathogen responsible for rhinosporidiosis, a disease which affects humans, horses, dogs, and to a lesser extent cattle, cats, foxes, and birds. It is most commonly found in tropical areas, especially India and Sri Lanka.

The pathogen was first identified in 1892, and was comprehensively described in 1900 by Seeber.

There are many aspects of the disease and of the pathogen *Rhinosporidium seeberi* which remain problematic and enigmatic. These include, the pathogen's natural habitat, some aspects of its 'life cycle', its immunology, some aspects of the epidemiology of the disease in humans and in animals, the reasons for the delay at *in vitro* culture and establishment of disease in experimental animals and hence paucity of information on its sensitivity to drugs, and the immunology of the pathogen. Thankamani isolated an organism believed to be *Rhinosporidium seeberi* and gave the name "UMH.48." It was originally isolated from the biopsies and nasal swabs of Rhinosporidiosis patients. The various developmental stages of UMH.48 showed a strong resemblance with the structures seen in hisopathological sections of Rhinosporidiosis in tissue samples. The spores of UMH.48 were found to be viable even after a decade of preservation in the refrigerator without any subculture, resembling the features of *Synchytridium endobioticum*, a lower aquatic fungus that causes black wart disease in potatoes. However carefully performed molecular studies would show the definitive identity of the organism.

Arseculeratne, Sarath N; Atapattu, Dhammika N. (2011). Rhinosporidiosis in Humans and Animals & Rhinosporidium seeberi. Faculty of Medicine, University of Peradeniya. ISBN 9555891575. discusses recent research developments and clinical associations of this enigmatic disease.

Incompletely resolved aspects of the disease, such as its epidemiology and pathogenesis, are expected to be understood in detail soon.

Phylogeny

For most of the 20th century, the classification of *R. seeberi* was unclear (being considered either a fungus or a protist), but it was shown to be part of a group called the Mesomycetozoea (or "DRIP clade"), which includes a number of well-known fish pathogens such as *Dermocystidium* and the Rosette Agent. The Mesomycetozoea is neither part of the fungi nor of animals, but diverged from them close to the time when they diverged from each other. On the contrary, recent research by 18S rRNA gene sequencing of UMH.48 and fungal extracts of biopsy from new cases of nasal rhinosporidiosis showed 100% identity. Thus UMH.48 has been tentatively identified as a lower aquatic fungus close to *Colletotrichum truncatum*, *Glomerulla sps* and *Synchytrium minutum*. Based on the absence of a perfect sexual phase and asexual spores, very rare microscopic morphology, life cycle and remarkable resemblance with members of lower aquatic fungi, UMH.48 is identified as a Fungus (unknown) and not an Ascomycete or Protozoan or Mesomycetozoan

Rhinosporidium is generally classified as having a single species, although there is some evidence that different host species may be infected by different strains.

Epidemiology

Infection in humans with this organism has been reported from ~70 countries, with the majority of cases (95%) reported from India and Sri Lanka: per capita, Sri Lanka has the highest incidence in the world. The disease is also found in other parts of the world.

An all-India survey conducted in 1957 found that this disease was absent from the states of Jummu and Kashmir, Himachal Pradesh, Punjab, Haryana and the North Eastern states of India. In Tamil Nadu, 4 endemic areas were identified in the survey — (Madurai, Ramnad, Rajapalayam and Sivaganga). The common denominator found in these areas was the practice of bathing in common ponds.

Transmission and Dissemination

1. Demellow's theory of infection
2. Karunarathnae's autoinoculation theory
3. Haematogenous spread — to distant sites
4. Lymphatic spread — causing lymphadenitis (rare)

Demellow postulated that while bathing in common ponds, the nasal mucosa came into contact with infectious material. Karunarathnae proposed that the satellite lesions in skin and conjuncti-val mucosa arose as a result of autoinoculation.

It is presumed that because of its relationship to fish pathogens, *Rhinosporidium* evolved from aquatic pathogens similar to the other Mesomycetozoea and evolved to infect mammal and bird hosts. It is not known if this happened once or more than once.

Natural Habitat

Karunarathnae also proposed that *Rhinosporidium* existed in a dimorphic state — a saprotroph in soil and water and a yeast form inside living tissues. Recent studies done using Fluorescent in-situ-hybridization techniques provide evidence that its natural habitat is reservoir water, and, perhaps, soil contaminated with this water.

Pathology

One report indicates that patients with rhinosporidiosis possess anti-*R. seeberi* IgG to an inner wall antigen expressed only during the mature sporangial stage. This finding suggests that the mapping of antigenic proteins may lead to important antigens with the potential as vaccine candidates.

Humoral and Cell-mediated Immune responses in human patients and in experimental mice have been defined; several mechanisms of immune evasion by *R. seeberi* have been identified.

A novel method for the determination of the viability of rhinosporidial endospores by MTT-reduc-tion led to the study of the sensitivity of endospores to biocides and anti-microbial drugs (paper in preparation for submission).

Clinical Features

This organism infects the mucosa of the nasal cavity producing a mass like lesion. This mass ap-pears to be polypoidal in nature with a granular surface speckled with whitish spores. The rhi-nosporidial mass has been classically described as a strawberry like mulberry mass. This mass may extend from the nasal cavity into the nasopharynx and present itself in the oral cavity. These lesions commonly cause bleeding from the nasal cavity.

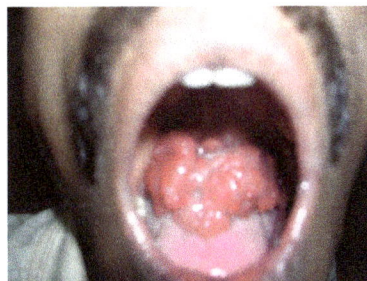

Image showing a large rhinosporidial mass in the oropharynx of a patient.

Rhinosporidium seeberi can also affect the lacrimal gland and also rarely the skin and genitalia.

Common sites affected:

1. Nose — 78%

2. Nasopharynx — 68%

3. Tonsil — 3%

4. Eye — 1%

5. Skin — very rare

Treatment

Treatment is generally by surgical removal of the infected tissues.

Povidone-iodine and anti-fungal drugs like amphotericin-B, Dapsone and silver nitrate have been suggested as possible antiseptics.

Toxoplasma Gondii

Toxoplasma gondii is an obligate intracellular, parasitic pro-tozoan that causes the disease toxoplasmosis. Found worldwide, *T. gondii* is capable of infecting virtually all warm-blooded animals, but felids such as domestic cats are the only known definitive hosts in which the parasite can undergo sexual reproduction.

In humans, *T. gondii* is one of the most common parasites in developed countries; serological studies estimate that 30–50% of the global population has been exposed to and may be chronically infected with *T. gondii*, although infection rates differ significantly from country to country. For example, previous estimates have shown the highest prevalence of persons infected to be in France, at 84%. Although mild, flu-like symptoms occasionally occur during the first few weeks following exposure, infection with *T. gondii* produces no readily observable symptoms in healthy human adults. This asymptomatic state of infection is referred to as a latent infection and has recently been associated with numerous subtle adverse or pathological behavioral alterations in humans. In infants, HIV/AIDS patients, and others with weakened immunity, infection can cause a serious and occasionally fatal illness, toxoplasmosis.

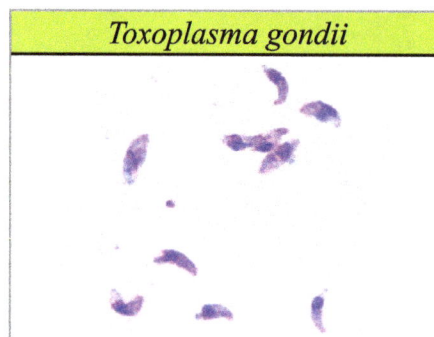
Toxoplasma gondii

T. gondii has been shown to alter the behavior of infected rodents in ways thought to increase the rodents' chances of being preyed upon by cats. Support for this "Manipulation Hypothesis" stems from studies showing *T. gondii* infected rats have a decreased aversion to cat urine. Because cats are the only hosts within which *T. gondii* can sexually reproduce to complete and begin its lifecycle, such behavioral manipulations are thought to be evolutionary adaptations that increase the parasite's reproductive success. The rats wouldn't shy away from areas where cats live and would also be less able to escape should a cat try to prey on them. The primary mechanisms of *T. gondii*–induced behavioral changes in rodents is now known to occur through epigenetic remodeling in neurons which govern the associated behaviors; for example, it modifies epigenetic methylation to cause hypomethylation of arginine vasopressin-related genes in the medial amygdala to greatly decrease predator aversion. Widespread histone-lysine acetylation in cortical astrocytes appears to be another epigenetic mechanism employed by *T. gondii*. Differences in aversion to cat urine are observed between non-infected and infected humans and sex differences within these groups were apparent as well.

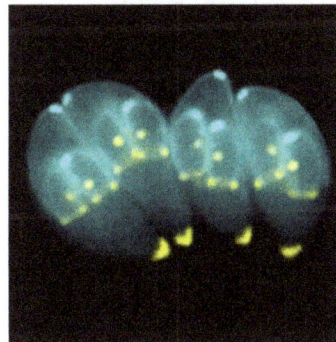

Dividing *T. gondii* parasites

A number of studies have suggested that subtle behavioral or personality changes may occur in infected humans, and infection with the parasite has recently been associated with a number of neurological disorders, particularly schizophrenia. A 2015 study also found cognitive deficits in adults to be associated with joint infection by both *T. gondii* and *Helicobacter pylori* in a regression model with controls for race-ethnicity and educational attainment. However, although a causal relationship between latent toxoplasmosis with these neurological phenomena has not yet been established, preliminary evidence suggests that *T. gondii* infection can induce some of the same alterations in the human brain as those observed in mice.

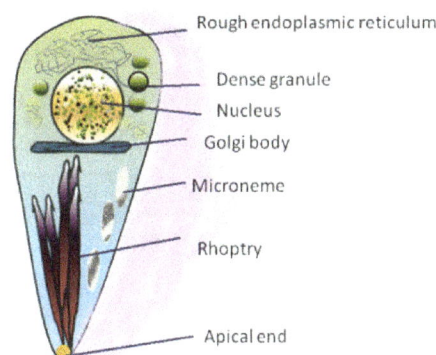

Diagram of *T. gondii* structure

Lifecycle

The lifecycle of *T. gondii* can be broadly summarized into two components: 1) a sexual component that occurs only within cats (felids, wild or domestic), and 2) an asexual component that can occur within virtually all warm-blooded animals, including humans, cats, and birds. Because *T. gondii* can sexually reproduce only within cats, they are defined as the definitive host of *T. gondii*. All other hosts – hosts in which only asexual reproduction can occur – are defined as intermediate hosts.

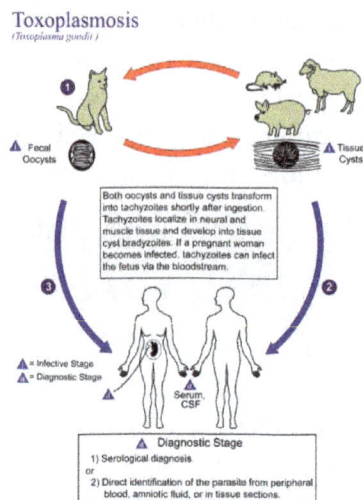

Lifecycle of *Toxoplasma gondii*

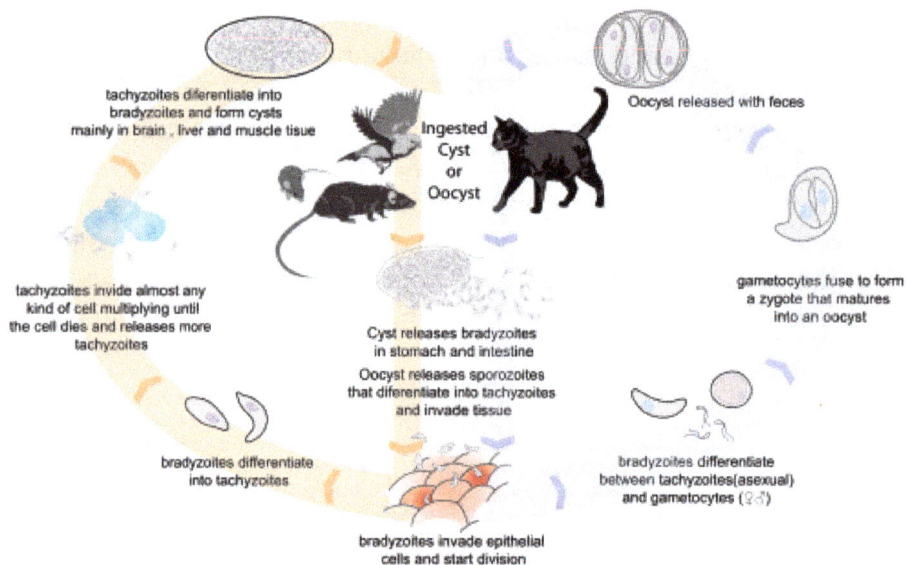

Lifecycle of *T. gondii*

Sexual Reproduction in The Feline Definitive Host

When a member of the cat family is infected with *T. gondii* (e.g. by consuming an infected mouse laden with the parasite's tissue cysts), the parasite survives passage through the stomach, eventually infecting epithelial cells of the cat's small intestine. Inside these intestinal cells, the parasites undergo sexual development and reproduction, producing millions of thick-walled, zygote-containing cysts known as oocysts.

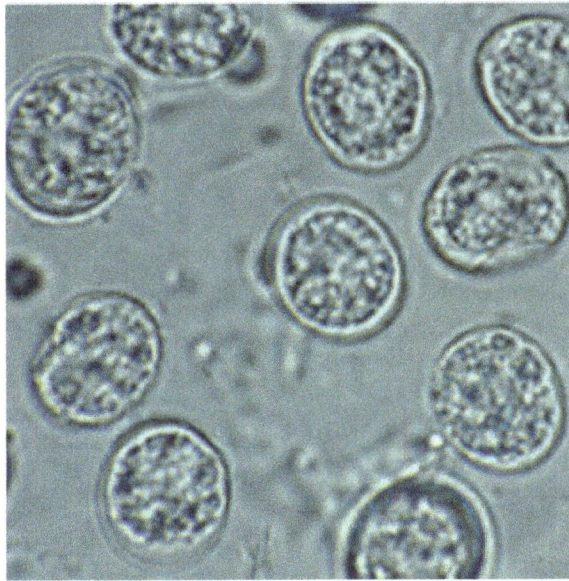

T. gondii oocysts in a fecal flotation

Feline Shedding of Oocysts

Infected epithelial cells eventually rupture and release oocysts into the intestinal lumen, whereupon they are shed in the cat's feces. Oocysts can then spread to soil, water, food, or anything potentially contaminated with the feces. Highly resilient, oocysts can survive and remain infective for many months in cold and dry climates.

Ingestion of oocysts by humans or other warm-blooded animals is one of the common routes of infection. Humans can be exposed to oocysts by, for example, consuming unwashed vegetables or contaminated water, or by handling the feces (litter) of an infected cat. Although cats can also be infected by ingesting oocysts, they are much less sensitive to oocyst infection than are intermediate hosts.

Initial Infection of The Intermediate Host

T. gondii is considered to have three stages of infection; the tachyzoite stage of rapid division, the bradyzoite stage of slow division within tissue cysts, and the oocyst environmental stage. When an oocyst or tissue cyst is ingested by a human or other warm-blooded animal, the resilient cyst wall is dissolved by proteolytic enzymes in the stomach and small intestine, freeing sporozoites from within the oocyst. The parasites first invade cells in and surrounding the intestinal epithelium, and inside these cells, the parasites differentiate into tachyzoites, the motile and quickly multiplying cellular stage of *T. gondii*. Tissue cysts in tissues such as brain and muscle tissue, form approximately 7–10 days after initial infection.

Asexual Reproduction in The Intermediate Host

Inside host cells, the tachyzoites replicate inside specialized vacuoles (called the parasitophorous vacuoles) created during parasitic entry into the cell. Tachyzoites multiply inside this vacuole until the host cell dies and ruptures, releasing and spreading the tachyzoites via the bloodstream to all organs and tissues of the body, including the brain.

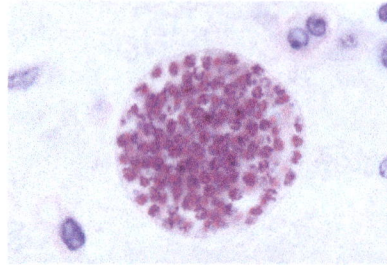

T. gondii tissue cyst in a mouse brain, individual bradyzoites can be seen within

Formation of Tissue Cysts

Following the initial period of infection characterized by tachyzoite proliferation throughout the body, pressure from the host's immune system causes *T. gondii* tachyzoites to convert into bradyzoites, the semidormant, slowly dividing cellular stage of the parasite. Inside host cells, clusters of these bradyzoites are known as tissue cysts. The cyst wall is formed by the parasitophorous vacuole membrane. Although bradyzoite-containing tissue cysts can form in virtually any organ, tissue cysts predominantly form and persist in the brain, the eyes, and striated muscle (including the heart). However, specific tissue tropisms can vary between species; in pigs, the majority of tissue cysts are found in muscle tissue, whereas in mice, the majority of cysts are found in the brain.

Cysts usually range in size between five and 50 μm in diameter, (with 50 μm being about two-thirds the width of the average human hair).

Consumption of tissue cysts in meat is one of the primary means of *T. gondii* infection, both for humans and for meat-eating, warm-blooded animals. Humans consume tissue cysts when eating raw or undercooked meat (particularly pork and lamb). Tissue cyst consumption is also the primary means by which cats are infected.

Chronic Infection

Tissue cysts can be maintained in host tissue for the lifetime of the animal. However, the perpetual presence of cysts appears to be due to a periodic process of cyst rupturing and re-encysting, rather than a perpetual lifespan of individual cysts or bradyzoites. At any given time in a chronically infected host, a very small percentage of cysts are rupturing, although the exact cause of this tissue cysts rupture is, as of 2010, not yet known.

Theoretically, *T. gondii* can be passed between intermediate hosts indefinitely via a cycle of consumption of tissue cysts in meat. However, the parasite's lifecycle begins and completes only when the parasite is passed to a feline host, the only host within which the parasite can again undergo sexual development and reproduction.

Population Structure in The Wild

Khan et al. reviewed evidence that despite the occurrence of a sexual phase in its life cycle, *T. gondii* has an unusual population structure dominated by three clonal lineages (Types I, II and III) that occur in North America and Europe. They estimated that a common ancestor founded these clonal lineages about 10,000 years ago. In a further and larger study (with 196 isolates from diverse sources including *T. gondii* found in the bald eagle, gray wolves, Arctic foxes and sea otters), Dubey et al. also found that *T. gondii* strains infecting North American wildlife have limited genetic diversity with the occurrence of only a few major clonal types. They found that 85% of strains in North America were of one of three widespread genotypes (Types II, III and Type 12). Thus *T. gondii* has retained the capability for sex in North America over many generations, producing largely clonal populations, and matings have generated little genetic diversity.

Immune Response

Initially, a *T. gondii* infection stimulates production of IL-2 and IFN-γ by the innate immune system. Continuous IFN-c production is necessary for control of both acute and chronic *T. gondii* infection. These two cytokines elicit a CD4+ and CD8+ T-cell mediated immune response. IL-12 is also produced during *T. gondii* infection to activate natural killer (NK) cells. Tryptophan is an essential amino acid for *T. gondii,* which it scavenges from host cells. IFN-γ induces the activation of indole-amine-2,3-dioxygenase (IDO) and tryptophan-2,3-dioxygenase (TDO), two enzymes that are responsible for the degradation of tryptophan. Immune pressure eventually leads the parasite to form cysts that normally are deposited in the muscles and in the brain of the hosts.

Immune Response and Behaviour Alterations

The IFN-γ-mediated activation of IDO and TDO is an evolutionary mechanism that serves to starve the parasite, but it can result in depletion of tryptophan in the brain of the host. IDO and TDO degrade tryptophan to N-formylkynurenine and administration of L-kynurenine is capable of inducing depressive-like behaviour in mice. *T. gondii* infection has been demonstrated to increase the levels of kynurenic acid (KYNA) in the brains of infected mice and KYNA has also been demonstrated to be increased in the brain of schizophrenic persons. Low levels of tryptophan and serotonin in the brain were already associated to depression.

Cellular Stages

During different periods of its life cycle, individual parasites convert into various cellular stages, with each stage characterized by a distinct cellular morphology, biochemistry, and behavior. These stages include the tachyzoites, merozoites, bradyzoites (found in tissue cysts), and sporozoites (found in oocysts).

Tachyzoites

Motile, and quickly multiplying, tachyzoites are responsible for expanding the population of the parasite in the host. When a host consumes a tissue cyst (containing bradyzoites) or an oocyst (containing sporozoites), the bradyzoites or sporozoites stage-convert into tachyzoites upon infecting the intestinal epithelium of the host. During the initial, acute period of infection, tachyzoites spread throughout the body via the blood stream. During the later, latent (chronic) stages of infection, tachyzoites stage-convert to bradyzoites to form tissue cysts.

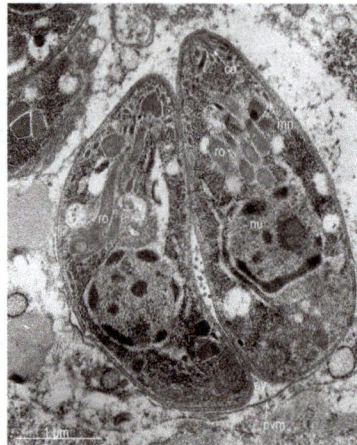
Two tachyzoites, transmission electron microscopy

Merozoites

Like tachyzoites, merozoites divide quickly, and are responsible for expanding the population of the parasite inside the cat intestine prior to sexual reproduction. When a feline definitive host consumes a tissue cyst (containing bradyzoites), bradyzoites convert into merozoites inside intestinal epithelial cells. Following a brief period of rapid population growth in the intestinal epithelium, merozoites convert into the noninfectious sexual stages of the parasite to undergo sexual reproduction, eventually resulting in the formation of zygote-containing oocysts.

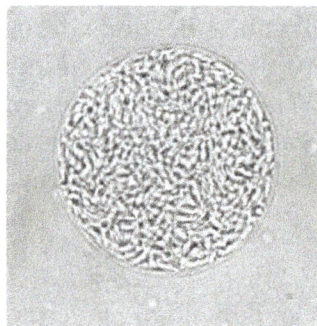
An unstained *T. gondii* tissue cyst, bradyzoites can be seen within

Bradyzoites

Bradyzoites are the slowly dividing stage of the parasite that make up tissue cysts. When an uninfected host consumes a tissue cyst, bradyzoites released from the cyst infect intestinal epithelial cells before converting to the proliferative tachyzoite stage. Following the initial period of proliferation throughout the host body, tachyzoites then convert back to bradyzoites, which reproduce inside host cells to form tissue cysts in the new host.

Sporozoites

Sporozoites are the stage of the parasite residing within oocysts. When a human or other warm-blooded host consumes an oocyst, sporozoites are released from it, infecting epithelial cells before converting to the proliferative tachyzoite stage.

Risk Factors for Human Infection

The following have been identified as being risk factors for *T. gondii* infection:

Infection in humans and other warm-blooded animals can occur:

1. by consuming raw or undercooked meat containing *T. gondii* tissue cysts. The most common threat to citizens in the United States is from eating raw or undercooked lamb or pork. It is possible, though unlikely, to ingest the parasite through other products:

2. by ingesting water, soil, vegetables, or anything contaminated with oocysts shed in the feces of an infected animal. Cat fecal matter is particularly dangerous: Just one cyst consumed by a cat can result in thousands of oocysts. This is why physicians recommend pregnant or ill persons do not clean the cat's litter box at home. These oocysts are resilient to harsh environmental conditions and can survive over a year in contaminated soil.

3. from a blood transfusion or organ transplant

4. from transplacental transmission from mother to fetus, particularly when *T. gondii* is contracted during pregnancy

5. from drinking unpasteurized goat milk

6. by contact with soil

7. from eating unwashed raw vegetables or fruits

Cleaning cat litter boxes is a potential route of infection; however, numerous studies have shown living in a household with a cat is not a significant risk factor for *T. gondii* infection, though living with several kittens has some significance.

In warm-blooded animals, such as brown rats, sheep, and dogs, *T. gondii* has also been shown to be sexually transmitted, and it is hypothesized that it may be sexually transmitted in humans, although not yet proven. Although *T. gondii* can infect, be transmitted by, and asexually reproduce within humans and virtually all other warm-blooded animals, the parasite can sexually reproduce only within the intestines of members of the cat family (felids). Felids are therefore defined as the definitive hosts of *T. gondii*, with all other hosts defined as intermediate hosts like human or other mammals.

Sewage has been identified as a carriage medium for the organism.

Preventing Infection

The following precautions are recommended to prevent or greatly reduce the chances of becoming infected with *T. gondii*. This information has been adapted from the websites of United States Centers for Disease Control and Prevention and the Mayo Clinic.

From Food

Basic food handling safety practices can prevent or reduce the chances of becoming infected with *T. gondii*, such as washing unwashed fruits and vegetable and avoiding raw or undercooked meat, poultry, and seafood. Other unsafe practices such as drinking unpasteurized milk or untreated water can increase odds of infection. Because *T. gondii* is typically transmitted through cysts that reside in the tissues of infected animals, meat that is not properly prepared can present an increased risk of infection. Freezing meat for several days at subzero temperatures (0 °F or −18 °C) before cooking eliminates tissue cysts, which can rarely survive these temperatures. During cooking, whole cuts of red meat should be cooked to an internal temperature of 145 °F (63 °C). Medium rare meat is generally cooked between 130 and 140 °F (55 and 60 °C), so cooking whole cuts of meat to medium is recommended. After cooking, a rest period of 3 min should be allowed before consumption. However, ground meat should be cooked to an internal temperature of at least 160 °F (71 °C) with no rest period. All poultry should be cooked to an internal temperature of at least 165 °F (74 °C). After cooking, a rest period of 3 min should be allowed before consumption.

From Environment

Oocysts in cat feces take at least a day to sporulate and become infectious after they are shed, so disposing of cat litter daily greatly reduces the chances of infectious oocysts being present in litter. As infectious oocysts from cat feces can spread and survive in the environment for months, humans should wear gloves when gardening or working with soil, and should wash their hands promptly after disposing of cat litter. The same precautions apply to outdoor sandboxes, which should be covered when not in use.

Furthermore, pregnant or immunocompromised people are at higher risk of becoming infected or transmitting the parasite to their fetus. Because of this, they should not change or handle cat litter boxes. Ideally, cats should be kept indoors and only fed food that has low to no risk of carrying oocysts, such as commercial cat food or well-cooked table food.

Vaccination

As of 2015, no approved human vaccine exists against *Toxoplasma gondii*. Research on human vaccines is ongoing.

For sheep there is an approved live vaccine available called Toxovax (MSD Animal Health) which provides lifetime protection.

History

In 1908, while working at the Pasteur Institute in Tunis, Charles Nicolle and Louis Manceaux discovered a protozoan organism in the tissues of a hamster-like rodent known as the gundi, *Ctenodactylus gundi*. Although Nicolle and Manceaux initially believed the organism to be a member of the genus *Leishmania* that they described as *"Leishmania gondii"*, they soon realized they had discovered a new organism entirely. They named it *Toxoplasma gondii*, a reference to its morphology (*Toxo*, from Greek (toxon); arc, bow, and (plasma); i.e., anything shaped or molded) and the host in which it was discovered, the gundi (gondii). The same year Nicolle and Mancaeux discovered *T. gondii*, Alfonso Splendore identified the same organism in a rabbit in Brazil. However, he did not give it a name.

The first conclusive identification of *T. gondii* in humans was in an infant girl delivered full term by Caesarean section on May 23, 1938, at Babies' Hospital in New York City. The girl began having seizures at three days of age, and doctors identified lesions in the maculae of both of her eyes. When she died at one month of age, an autopsy was performed. Lesions discovered in her brain and eye tissue were found to have both free and intracellular *T. gondii*'. Infected tissue from the girl was homogenized and inoculated intracerebrally into rabbits and mice; the animals subsequently developed encephalitis. Later, congenital transmission was found to occur in numerous other species, particularly in sheep and rodents.

The possibility of *T. gondii* transmission via consumption of undercooked meat was first proposed by D. Weinman and A.H Chandler in 1954. In 1960, the cyst wall of tissue cysts was shown to dissolve in the proteolytic enzymes found in the stomach, releasing infectious bradyzoites into the stomach (and subsequently into the intestine). The hypothesis of transmission via consumption of undercooked meat was tested in an orphanage in Paris in 1965; yearly acquisition rates of *T. gondii* rose from 10% to 50% after adding two portions of barely cooked beef or horse meat to the orphans' daily diets, and to 100% after adding barely cooked lamb chops.

In 1959, a study in Bombay found the prevalence of *T. gondii* in strict vegetarians to be similar to that found in nonvegetarians. This raised the possibility of a third major route of infection, beyond congenital and carnivorous transmission. In 1970, the existence of oocysts was discovered in cat feces, and the fecal-oral route of infection via oocysts was demonstrated.

Throughout the 1970s and 1980s, a vast number of species were tested for the ability to shed oocysts upon infection. Whereas at least 17 different species of felids have been confirmed to shed oocysts, no nonfelid has been shown to be permissive for *T. gondii* sexual reproduction and subsequent oocyst shedding.

Behavioral Differences of Infected Hosts

Once healthy humans are infected with *Toxoplasma gondii*, it may lead to altered behavioral differences like psychomotor performance and neurological disorders like depression and/or suicide and even schizophrenia. There are many instances where behavioural changes were reported in rodents with *T. gondii*. The changes seen were a reduction in their innate dislike of cats, which made it easier for cats to prey on the rodents. In an experiment conducted by Berdoy and colleagues the infected rats showed preference for the cat odour area versus the area with the rabbit scent, therefore making it easier for the parasite to take its final step in its definitive feline host. This is an example of the extended phenotype concept, that is, the idea that the behaviour of the infected animal changes in order to maximize survival of the genes that increase predation of the intermediate rodent host.

Differences in behaviour observed in infected hosts compared to non-infected individuals have been shown to be sex dependent. Looking at humans, studies using the Cattell's 16 Personality Factor questionnaire, found that infected men scored lower on Factor G (superego strength/rule consciousness) and higher on Factor L (vigilance) while the opposite pattern was observed for infected women. This means that men were more likely to disregard rule and were more expedient, suspicious and jealous. On the other hand, women were more warm hearted, outgoing, conscientious and moralistic. However, human studies have not been able to show causation as they have all been observational studies. Mice infected with *T. gondii* have motor performance worse than non-infected mice. Thus, a computerized simple reaction test was given to both infected and non-infected adults. It was found that the infected adults performed much more poorly and lost their concentration more quickly than the control group. But, the effect of the infection only explains less than 10% of the variability in performance. (i.e. could be other confounding factors) Correlation has also been observed between seroprevalence of *T. gondii* in humans and increased risk of traffic accidents. Infected subjects have a 2.65 times higher risk of getting into a traffic accident. A similar study done in Turkey showed that there is a higher incidence of *Toxoplasma gondii* antibodies among drivers who have been involved in traffic accidents. Furthermore, this parasite has been associated with many neurological disorders such as schizophrenia. In a meta-analysis, 23 studies met inclusion criteria. The results demonstrate that the seroprevalence of antibodies to *T.gondii* in people with schizophrenia is significantly higher than in control populations (OR=2.73, P<0.000001). In more recent studies, it was found that suicide attempters has significantly higher IgG antibody levels to *T. gondii* as compared with patients without a suicide attempt. Infection was also shown to be associated with suicide in women over the age of 60. (P<0.005) As mentioned before, these results of increased proportions of people seropositive for the parasite in cases of these neurological disorders does not indicate a causal relationship between the infection and disorder. It is important to mention that in 2016 a population-representative birth cohort study which was done, to test a hypothesis that toxoplasmosis is related to impairment in brain and behaviour measured by a range of phenotypes including neuropsychiatric disorders, poor impulse control, personality and neurocognitive deficits. Unfortunately, the results do not support any previously mentioned study. None of the P-values showed significance for any outcome measure. Therefore, they concluded that *T. gondii* antibodies does not result in increase susceptibility to any of the phenotypes. This team did not observe any significant association between *T. gondii* serpositivity and schizophrenia. The team reports that the null findings may be a false negative due to low statistical power because of small sample sizes.

Trypanosoma Brucei

Trypanosoma brucei is a species of parasitic protozoan belonging to the genus *Trypanosoma*. It causes African trypanosomiasis, known also as sleeping sickness in humans and nagana in other animals. *T. brucei* has traditionally been grouped into three subspecies: *T. b. brucei*, *T. b. gambiense* and *T. b. rhodesiense*. The latter two are typically parasites of humans, while the first is that of other animals. Only rarely can the *T.b.brucei* infect a human.

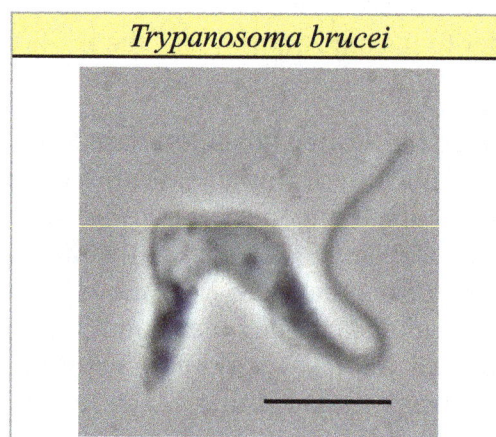

Trypanosoma brucei

T. brucei is transmitted between mammal hosts by an insect vector belonging to the species of tsetse fly. Transmission occurs by biting during the insect's blood meal. The parasites undergo complex morphological changes as they move between insect and mammal over the course of their life cycle. The mammalian bloodstream forms are notable for their variant surface glycoprotein (VSG) coats, which undergo remarkable antigenic variation, enabling persistent evasion of host adaptive immunity and chronic infection. *T. brucei* is one of only a few pathogens that can cross the blood brain barrier. There is an urgent need for the development of new drug therapies, as current treatments can prove fatal to the patient.

False colour SEM micrograph of procyclic form *Trypanosoma brucei* as found in the tsetse fly midgut. The cell body is shown in orange and the flagellum is in red. 84 pixels/μm.

Whilst not historically regarded as *T. brucei* subspecies due to their different means of transmission, clinical presentation, and loss of kinetoplast DNA, genetic analyses reveal that *T. equiperdum* and *T. evansi* are evolved from parasites very similar to *T. b. brucei*, and are thought to be members of the *brucei* clade.

The parasite was discovered in 1894 by Sir David Bruce, after whom the scientific name was given in 1899.

Species

T. brucei comprises a species complex that includes:

- *T. brucei gambiense* — Causes slow onset chronic trypanosomiasis in humans. Most common in central and western Africa, where humans are thought to be the primary reservoir.

- *T. brucei rhodesiense* — Causes fast onset acute trypanosomiasis in humans. Most common in southern and eastern Africa, where game animals and livestock are thought to be the primary reservoir.

- *T. brucei brucei* — Causes animal African trypanosomiasis, along with several other species of trypanosoma. *T. b. brucei* is not infective to humans due to its susceptibility to lysis by trypanosome lytic factor-1 (TLF-1). However, as it is closely related to, and shares fundamental features with the human infective subspecies.

Structure

T. brucei is a typical unicellular eukaryotic cell, and measures 8 to 50 μm in length. It has an elongated body having a streamlined and tapered shape. Its cell membrane (called pellicle) encloses the cell organelles, including the nucleus, mitochondria, endoplasmic reticulum, Golgi apparatus, and ribosomes. In addition, there is an unusual organelle called the kinetoplast, which is made up of numerous circular DNA (mitochondrial DNA) and functions as a single large mitochondrion. The kinetoplast lies near the basal body with which it is indistinguishable under microscope. From the basal body arises a single flagellum that run towards the anterior end. Along the body surface, the flagellum is attached to the cell membrane forming an undulating membrane. Only the tip of the flagellum is free at the anterior end. The cell surface of the bloodstream form features a dense coat of variant surface glycoproteins (VSGs) which is replaced by an equally dense coat of procyclins when the parasite differentiates into the procylic in the tsetse fly midgut.

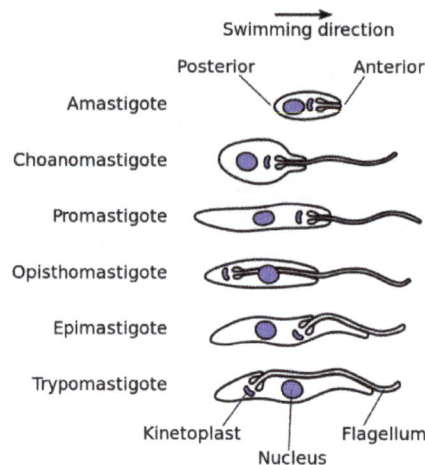

The six main morphologies of trypanosomatids. The different life cycle stages of *Trypanosoma brucei* fall into the trypomastigote and epimastigote morphological categories.

Trypanosomatids show several different classes of cellular organisation of which two are adopted by *Trypanosoma brucei* at different stages of the life cycle:

- Epimastigote, which is found in tsetse fly. Its kinetoplast and basal body lie anterior to the nucleus, with a long flagellum attached along the cell body. The flagellum starts from the centre of the body.

- Trypomastigote, which is found in mammalian hosts. The kinetoplast and basal body are posterior of nucleus. The flagellum arises from the posterior end of the body.

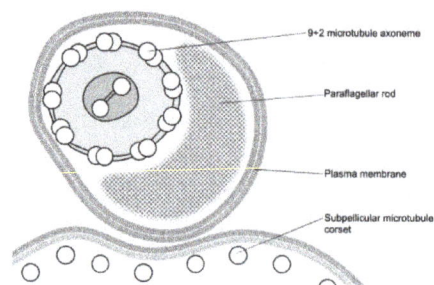

Trypanosoma Brucei flagellar structure.

These names are derived from the Greek *mastig-* meaning whip, referring to the trypanosome's whip-like flagellum. The trypanosome flagellum has two main structures. It is made up of a typical flagellar axoneme which lies parallel to the paraflagellar rod, a lattice structure of proteins unique to the kinetoplastida, euglenoids and dinoflagellates.

The microtubules of the flagellar axoneme lie in the normal 9+2 arrangement, orientated with the + at the anterior end and the - in the basal body. The a cytoskeletal structure extends from the basal body to the kinetoplast. The flagellum is bound to the cytoskeleton of the main cell body by four specialised microtubules, which run parallel and in the same direction to the flagellar tubulin.

The flagellar function is twofold — locomotion via oscillations along the attached flagellum and cell body, and attachment to the fly gut during the procyclic phase.

Life Cycle

T. brucei completes its life cycle between tsetsefly (of the genus *Glossina*) and mammalian hosts, including humans, cattle, horses, and wild animals.

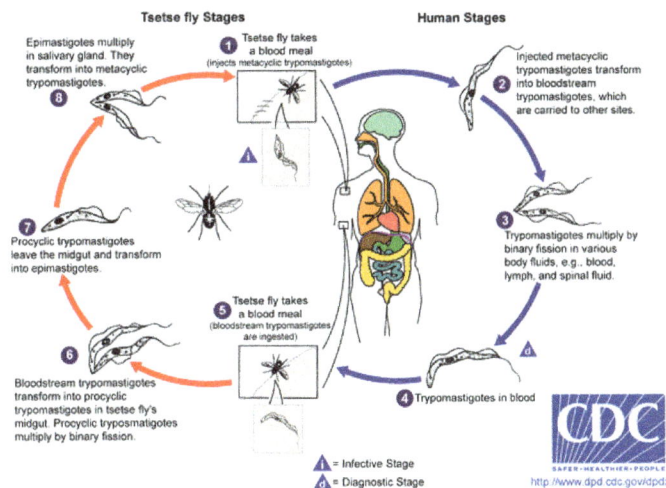

Life cycle of the *Trypanosoma brucei*

In Mammalian Host

Infection occurs when a vector tsetse fly bites a mammalian host. The fly injects the metacyclic trypomastigotes into the skin tissue. The trypomastigotes enter the lymphatic system and into the bloodstream. The initial trypomastigotes are short and stumpy. Once inside the bloodstream, they grow into long and slender forms. Then, they multiply by binary fission. The daughter cells then become short and stumpy again. The long slender forms are able to penetrate the blood vessel endothelium and invade extravascular tissues, including the central nervous system (CNS).

Sometimes, wild animals can be infected by the tsetsefly and they act as reservoirs. In these animals, they do not produce the disease, but the live parasite can be transmitted back to the normal hosts.

In Tsetse Fly

The short and stumpy trypomastigotes are taken up by tsetse fly during blood meal. The trypomastigotes enter the midgut of the fly where they become procyclic trypomastigotes. These rapidly divide to become epimastigotes. The epimastigotes migrate from the gut via the proventriculus to the salivary glands where they get attached to the salivary gland epithelium. In the salivary glands, some parasites detach and undergo transformation into short and stumpy trypomastigotes. These become the infective metacyclic trypomastigotes. They are injected into the mammalian host along with the saliva on biting. Complete development in the fly takes about 20 days.

Reproduction

Binary Fission

The reproduction of *T. brucei* is unusual compared to most eukaryotes. The nuclear membrane remains intact and the chromosomes do not condense during mitosis. The basal body, unlike the centrosome of most eukaryotic cells, does not play a role in the organisation of the spindle and instead is involved in division of the kinetoplast. The events of reproduction are:

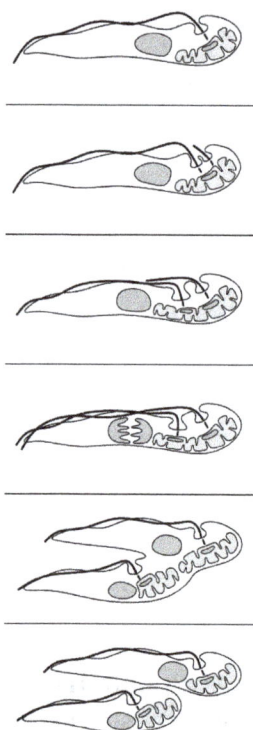

Trypanosome cell cycle (procyclic form).

1. The basal body duplicates and both remain associated with the kinetoplast. Each basal body forms a separate flagellum.

2. Kinetoplast DNA undergoes synthesis then the kinetoplast divides coupled with separation of the two basal bodies.

3. Nuclear DNA undergoes synthesis while a new flagellum extends from the younger, more posterior, basal body.

4. The nucleus undergoes mitosis.

5. Cytokinesis progresses from the anterior to posterior.

6. Division completes with abscission.

Meiosis

In the 1980s, DNA analyses of the developmental staged of *T. brucei* started to indicate the trypomastigote in tsetse fly undergo meiosis, i.e. sexual reproduction stage. But it is not always necessary for a complete life cycle. The existence of meiosis-specific proteins was reported in 2011. The haploid gametes (daughter cells produced after meiosis) were discovered in 2014. The haploid trypomastigote-like gametes can interact with each other via their flagella and undergo cell fusion (the process called syngamy). Thus, in addition to binary fission, *T. brucei* can multiply by sexual reproduction. Trypanosomes belong to the supergroup Excavata and are one of the earliest diverging lineages among eukaryotes. The discovery of sexual reporduction in *T. brucei* supports the hypothesis that meiosis and sexual reproduction are ancestral and ubiquitous features of eukaryotes.

Infection and Pathogenicity

The insect vector for *T. brucei* is the tsetse fly (genus *Glossina*). In later stages of a *T. brucei* infection of a mammalian host the parasite may migrate from the bloodstream to also infect the lymph and cerebrospinal fluids. It is under this tissue invasion that the parasites produce the sleeping sickness.

In addition to the major form of transmission via the tsetse fly *T. brucei* may be transferred between mammals via bodily fluid exchange, such as by blood transfusion or sexual contact, although this is thought to be rare.

Distribution

T. brucei is found where the tsetse fly vectors are prevalent. It is present in tropical and subtropical areas of Africa north of the equator, covering East, Central and West Africa. Hence, the equatorial region of Africa is called the "sleeping sickness" belt. However, the specific type of the trypanosome differs according to geography. *T. b. rhodesiense* is found primarily in East Africa (Botswana, Ethiopia, Kenya, Malawi, Tanzania, Uganda, Zaire, and Zimbabwe), while *T. b. gambiense* is found in Central and West Africa.

Evolution

Trypanosoma brucei gambiense evolved from a single progenitor ~10,000 years ago. It is evolving asexually and its genome shows the Meselson effect.

Genetics

There are two subpopulations of *T. b. gambiense* that possesses two distinct groups that differ in genotype and phenotype. Group 2 is more akin to *T. b. brucei* than group 1 *T. b. gambiense*.

All *T. b. gambiense* are resistant to killing by a serum component — trypanosome lytic factor (TLF) of which there are two types: TLF-1 and TLF-2. Group 1 *T. b. gambiense* parasites avoid uptake of the TLF particles while those of group 2 are able to either neutralize or compensate for the effects of TLF.

In contrast *T. b. rhodesiense* is dependent upon the expression of a serum resistance associated (SRA) gene. This gene is not found in *T. b. gambiense*.

Genome

The genome of *T. brucei* is made up of:

- 11 pairs of large chromosomes of 1 to 6 megabase pairs.

- 3–5 intermediate chromosomes of 200 to 500 kilobase pairs.

- Around 100 minichromosomes of around 50 to 100 kilobase pairs. These may be present in multiple copies per haploid genome.

Most genes are held on the large chromosomes, with the minichromosomes carrying only *VSG* genes. The genome has been sequenced and is available online .

The mitochondrial genome is found condensed into the kinetoplast, an unusual feature unique to the kinetoplastea class. The kinetoplast and the basal body of the flagellum are strongly associated via a cytoskeletal structure.

In 1993, a new base, beta-d-glucopyranosyloxymethyluracil (base J), was identified in the nuclear DNA of Trypanosoma brucei

VSG Coat

The surface of the trypanosome is covered by a dense coat of ~5×10^6 molecules of variant surface glycoprotein (VSG). This coat enables an infecting *T. brucei* population to persistently evade the host's immune system, allowing chronic infection. VSG is highly immunogenic, and an immune response raised against a specific VSG coat rapidly kills trypanosomes expressing this variant. Antibody-mediated trypanosome killing can also be observed in vitro by a complement-mediated lysis assay. However, with each cell division there is a possibility that one or both of the progeny will switch expression to change the VSG that is being expressed. The frequency of VSG switching has been measured to be approximately 0.1% per division. As *T. brucei* populations can peak at a size of 10^{11} within a host this rapid rate of switching ensures that the parasite population is typically highly diverse. Because host immunity against a specific VSG does not develop immediately, some parasites will have switched to an antigenically-distinct VSG variant, and can go on to multiply and continue the infection. The clinical effect of this cycle is successive 'waves' of parasitaemia (trypanosomes in the blood).

Expression of *VSG* genes occurs through a number of mechanisms yet to be fully understood. The expressed VSG can be switched either by activating a different expression site (and thus changing to express the *VSG* in that site), or by changing the *VSG* gene in the active site to a different variant. The genome contains many hundreds if not thousands of *VSG* genes, both on minichromosomes and in repeated sections ('arrays') in the interior of the chromosomes. These are transcriptionally silent, typically with omitted sections or premature stop codons, but are important in the evolution of new VSG genes. It is estimated up to 10% of the *T. brucei* genome may be made up of VSG genes or pseudogenes. It is thought that any of these genes can be moved into the active site by recombination for expression.

Killing by Human Serum and Resistance to Human Serum Killing

Trypanosoma brucei brucei (as well as related species *T. equiperdum* and *T. evansi*) is not human infective because it is susceptible to innate immune system 'trypanolytic' factors present in the serum of some primates, including humans. These trypanolytic factors have been identified as two serum complexes designated trypanolytic factors (TLF-1 and -2) both of which contain haptoglobin related protein (HPR) and apolipoprotein LI (ApoL1). TLF-1 is a member of the high density lipoprotein family of particles while TLF-2 is a related high molecular weight serum protein binding complex. The protein components of TLF-1 are haptoglobin related protein (HPR), apolipoprotein L-1 (apoL-1) and apolipoprotein A-1 (apoA-1). These three proteins are colocalized within spherical particles containing phospholipids and cholesterol. The protein components of TLF-2 include IgM and apolipoprotein A-I.

Trypanolytic factors are found only in a few species including humans, gorillas, mandrills, baboons and sooty mangabeys. This appears to be because haptoglobin related protein and apolipoprotein L-1 are unique to primates. This suggests these gene appeared in the primate genome 25 million years ago-35 million years ago.

Human infective subspecies *T. b. gambiense* and *T. b. rhodesiense* have evolved mechanisms of resisting the trypanolytic factors, described below.

ApoL1

ApoL1 is a member of a six gene family, ApoL1-6, that have arisen by tandem duplication. These proteins are normally involved in host apoptosis or autophagic death and possess a Bcl-2 homology domain 3. ApoL1 has been identified as the toxic component involved in trypanolysis. ApoLs have been subject to recent selective evolution possibly related to resistance to pathogens

The gene encoding ApoL1 is found on the long arm of chromosome 22 (22q12.3). Variants of this gene, termed G1 and G2, provide protection against *T. b. rhodesiense*. These benefits are not without their downside as a specific ApoL1 glomeropathy has been identified. This glomeropathy may help to explain the greater prevalence of hypertension in African populations.

The gene encodes a protein of 383 residues, including a typical signal peptide of 12 amino acids. The plasma protein is a single chain polypeptide with an apparent molecular mass of 42 kiloDaltons. ApoL1 has a membrane pore forming domain functionally similar to that of bacterial colicins. This domain is flanked by the membrane addressing domain and both these domains are required for parasite killing.

Within the kidney, ApoL1 is found in the podocytes in the glomeruli, the proximal tubular epithelium and the arteriolar endothelium. It has a high affinity for phosphatidic acid and cardiolipin and can be induced by interferon gamma and tumor necrosis factor alpha.

Hpr

Hpr is 91% identical to haptoglobin (Hp), an abundant acute phase serum protein, which possesses a high affinity for haemoglobin (Hb). When Hb is released from erythrocytes undergoing intravascular hemolysis Hp forms a complex with the Hb and these are removed from the circulation by the CD163 scavenger receptor. In contrast to Hp–Hb, the Hpr–Hb complex does not bind CD163 and the Hpr serum concentration appears to be unaffected by haemolysis.

Killing Mechanism

The association HPR with haemoglobin allows TLF-1 binding and uptake via the trypanosome haptoglobin-hemoglobin receptor (TbHpHbR). TLF-2 enters trypanosomes independently of TbHpHbR. TLF-1 uptake is enhanced in the low levels of haptoglobin which competes with haptoglobin related protein to bind free haemoglobin in the serum. However the complete absence of haptoglobin is associated with a decreased killing rate by serum.

The trypanosome haptoglobin-hemoglobin receptor is an elongated three a-helical bundle with a small membrane distal head. This protein extends above the variant surface glycoprotein layer that surrounds the parasite.

The first step in the killing mechanism is the binding of TLF to high affinity receptors—the haptoglobin-hemoglobin receptors—that are located in the flagellar pocket of the parasite. The bound TLF is endocytosed via coated vesicles and then trafficked to the parasite lysosomes. ApoL1 is the main lethal factor in the TLFs and kills trypanosomes after insertion into endosomal / lysosomal membranes. After ingestion by the parasite, the TLF-1 particle is trafficked to the lysosome wherein Apo1 is activated by a pH mediated conformational change. After fusion with the lysosome the pH drops from ~7 to ~5. This induces a conformational change in the ApoL1 membrane addressing domain which in turn causes a salt bridge linked hinge to open. This releases ApoL1 from the HDL particle to insert in the lysosomal membrane. The ApoL1 protein then creates anionic pores in the membrane which leads to depolarization of the membrane, a continuous influx of chloride and subsequent osmotic swelling of the lysosome. This influx in its turn leads to rupture of the lysosome and the subsequent death of the parasite.

Resistance Mechanisms: T. B. Gambiense

Trypanosoma brucei gambiense causes 97% of human cases of sleeping sickness. Resistance to ApoL1 is principally mediated by the hydrophobic ß-sheet of the *T. b. gambiense* specific glycoprotein. Other factors involved in resistance appear to be a change in the cysteine protease activity and TbHpHbR inactivation due to a leucine to serine substitution (L210S) at codon 210. This is due to a thymidine to cytosine mutation at the second codon position.

These mutations may have evolved due to the coexistence of malaria where this parasite is found. Haptoglobin levels are low in malaria because of the haemolysis that occurs with the release of the merozoites into the blood. The rupture of the erythrocytes results in the release of free haem into the blood where it is bound by haptoglobin. The haem is then removed along with the bound haptoglobin from the blood by the reticuloendothelial system.

Resistance Mechanisms: T. b. Rhodesiense

Trypanosoma brucei rhodesiense relies on a different mechanism of resistance: the serum resistance associated protein (SRA). The SRA gene is a truncated version of the major and variable surface antigen of the parasite, the variant surface glycoprotein. It has a low sequence homology with the VSGc (<25%). SRA is an expression site associated gene in *T. b. rhodesiense* and is located upstream of the VSGs in the active telomeric expression site. The protein is largely localized to small cytoplasmic vesicles between the flagellar pocket and the nucleus. In *T. b. rhodesiense* the TLF is directed to SRA containing endosomes while some dispute remain on its presence in the lysosome. SRA binds to ApoL1 using a coiled–coiled interaction at the ApoL1 SRA interacting domain while within the trypanosome lysosome. This interaction prevents the release of the ApoL1 protein and the subsequent lysis of the lysosome and death of the parasite.

Baboons are known to be resistant to *Trypanosoma brucei rhodesiense*. The baboon version of the ApoL1 gene differs from the human gene in a number of respects including two critical lysines near the C terminus that are necessary and sufficient to prevent baboon ApoL1 binding to SRA. Experimental mutations allowing ApoL1 to be protected from neutralization by SRA have been shown capable of conferring trypanolytic activity on *T. b. rhodesiense*. These mutations resemble those found in baboons, but also resemble natural mutations conferring protection of humans against *T. b. rhodesiense* which are linked to kidney disease .

Trypanosoma Cruzi

Trypanosoma cruzi is a species of parasitic euglenoid protozoan. Amongst the protozoa, the trypanosomes characteristically bore tissue in another organism and feed on blood, and lymph. This behaviour causes disease or the likelihood of disease that varies with the organism: for example, trypanosomiasis in humans (Chagas disease in South America and sleeping sickness in Africa), dourine and surra in horses, and a brucellosis-like disease in cattle. Parasites need a host body and the haematophagous insect triatomine (descriptions "assassin bug", "cone-nose bug", and "kissing bug") is the major vector in accord with a mechanism of infection. The triatomine likes the nests of vertebrate animals for shelter, where it bites and sucks blood for food. Individual triatomines infected with protozoa from other contact with animals transmit trypanosomes when the triatomine deposits its faeces on the host's skin surface and then bites. Penetration of the infected faeces is further facilitated by the scratching of the bite area by the human or animal host.

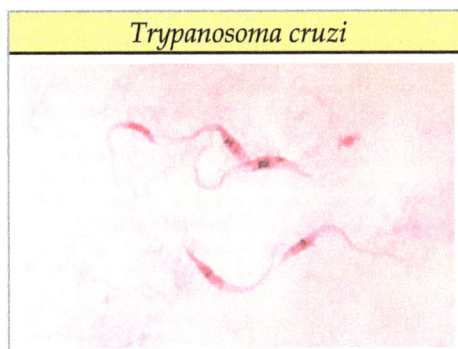

Trypanosoma cruzi

Life Cycle

The *Trypanosoma cruzi* life cycle starts in an animal reservoir, usually mammals, wild or domestic, including humans. A triatomine bug serves as the vector. While taking a blood meal, it ingests *T. cruzi*. In the triatomine bug (*Triatoma infestans*) the parasite goes into the epimastigote stage, making it possible to reproduce. After reproducing through binary fission, the epimastigotes move onto the rectal cell wall, where they become infectious. Infectious *T. cruzi* are called metacyclic trypomastigotes. When the triatomine bug subsequently takes a blood meal from a human, it defecates. The trypomastigotes are in the feces and are capable of swimming into the host's cells using flagella, a characteristic swimming tail dominant in the Euglenoid class of protists.

The trypomastigotes enter the human host through the bite wound or by crossing mucous membranes. The host cells contain macromolecules such as laminin, thrombospondin, heparin sulphate, and fibronectin that cover their surface. These macromolecules are essential for adhesion between parasite and host and for the process of host invasion by the parasite. The trypomastigotes must cross a network of proteins that line the exterior of the host cells in order to make contact and invade the host cells. The molecules and proteins on the cytoskeleton of the cell also bind to the surface of the parasite and initiate host invasion.

Pathophysiology

Trypanosomiasis in humans progresses with the development of the trypanosome into a trypomastigote in the blood and into an amastigote in tissues. The acute form of trypanosomiasis is usually unnoticed, although it may manifest itself as a localized swelling at the site of entry. The chronic form may develop 30 to 40 years after infection and affect internal organs (e.g., the heart, the oesophagus, the colon, and the peripheral nervous system). Affected people may die from heart failure.

Acute cases are treated with nifurtimox and benznidazole, but no effective therapy for chronic cases is currently known.

Cardiac Manifestations

Researchers of Chagas' disease have demonstrated several processes that occur with all cardiomyopathies. The first event is an inflammatory response. Following inflammation, cellular damage occurs. Finally, in the body's attempt to recover from the cellular damage, fibrosis begins in the cardiac tissue.

Another cardiomyopathy found in nearly all cases of chronic Chagas' disease is thromoembolic syndrome. Thromboembolism describes thrombosis, the formation of a clot, and its main complication is embolism, the carrying of a clot to a distal section of a vessel and causing blockage there. This occurrence contributes to the death of a patient by four means: arrhythmias, stasis secondary to cardiac dilation, mural endocarditis, and cardiac fibrosis. These thrombi also affect other organs such as the brain, spleen and kidney.

Myocardial Biochemical Response

Subcellular findings in murine studies with induced *T. cruzi* infection revealed that the chronic state is associated with the persistent elevation of phosphorylated (activated) extracellular-signal-regulated kinase (ERK), AP-1, and NF-κB. Also, the mitotic regulator for G1 progression, cyclin D1 was found to be activated. Although there was no increase in any isoform of ERK, there was an increased concentration of phosphorylated ERK in mice infected with *T. cruzi*. It was found that within seven days the concentration of AP-1 was significantly higher in *T. cruzi*–infected mice when compared to the control. Elevated levels of NF-κB have also been found in myocardial tissue, with the highest concentrations being found in the vasculature. It was indicated through Western blot that cyclin D1 was upregulated from day 1 to day 60 post-infection. It was also indicated through immunohistochemical analysis that the areas that produced the most cyclin D1 were the vasculature and interstitial regions of the heart.

Rhythm Abnormalities

Conduction abnormalities are also associated with *T. cruzi*. At the base of these conduction abnormalities is a depopulation of parasympathetic neuronal endings on the heart. Without proper parasympathetic innervations, one could expect to find not only chronotropic but also ionotropic abnormalities. It is true that all inflammatory and non-inflammatory heart disease may display forms of parasympathetic denervation; this denervation presents in a descriptive fashion in Chagas' disease. It has also been indicated that the loss of parasympathetic innervations can lead to sudden death due to a severe cardiac failure that occurs during the acute stage of infection.

Another conduction abnormality presented with chronic Chagas' disease is a change in ventricular repolarization, which is represented on an electrocardiogram as the T-wave. This change in repolarization inhibits the heart from relaxing and properly entering diastole. Changes in the ventricular repolarization in Chagas' disease are likely due to myocardial ischemia. This ischemia can also lead to fibrillation. This sign is usually observed in chronic Chagas' disease and is considered a minor electromyocardiopathy.

Epicardial Lesions

Villous plaque is characterized by exophytic epicardial thickening, meaning that the growth occurs at the border of the epicardium and not the center of mass. Unlike milk spots and chagasic rosary, inflammatory cells and vasculature are present in villous plaque. Since villous plaque contains inflammatory cells it is reasonable to suspect that these lesions are more recently formed than milk spots or chagasic rosary.

Epidemiology

T. cruzi transmission has been documented in the Southwestern U.S., and warming trends may allow vector species to move north. U.S. domestic and wild animals are reservoirs for T. cruzi. Triatomine species in the southern U.S. have taken human blood meals, but because triatomines do not favor typical U.S. housing risk to the U.S. population is very low.

Chagas' disease's geographical occurrence happens worldwide but high-risk individuals include those who don't have access to proper housing. Its reservoir is in wild animals but its vector is a kissing bug. This is a contagious disease and can be transmitted through a number of ways: congenital transmission, blood transfusion, organ transplantation, consumption of uncooked food that has been contaminated with feces from infected bugs, and accidental laboratory exposure.

The incubation period is five to fourteen days, after a host becomes in contact with feces. Chagas disease undergoes two phases which are the acute and chronic phase. The acute phase can last from two weeks to two months but can go unnoticed because symptoms are minor and short-lived. Symptoms of the acute phase include swelling, fever, fatigue, and diarrhea. The chronic phase causes digestive problems, constipation, heart failure, and pain in the abdomen. Diagnostic methods include microscopic examination, serology, or the isolation of the parasite by inoculating blood into a guinea pig, mouse, or rat. No vaccines are available but there are ways to be protected from this disease. Taking preventative measures such as applying bug repellent on the skin, wearing protective clothing, and staying in higher quality hotels—when traveling. Investing in quality housing would be ideal to decrease risk of contracting this disease. Consider installing plaster walls or new flooring to decrease the crevasses that bugs can hide in.

References

- I. Edward Alcamo; Jennifer M. Warner (28 August 2009). Schaum's Outline of Microbiology. McGraw Hill Professional. pp. 144–. ISBN 978-0-07-162326-1. Retrieved 14 November 2010.

- Margulis, Lynn (1974). Dobzhansky, Theodosius; Hecht, Max K.; Steere, William C., eds. Five-Kingdom Classification and the Origin and Evolution of Cells. Springer US. pp. 45–78. ISBN 978-1-4615-6946-6.

- Guidelines for the treatment of malaria, second edition Authors: WHO. Number of pages: 194. Publication date: 2010. Languages: English. ISBN 978-92-4-154792-5

- Dardé, ML; Ajzenberg, D; Smith, J (2011). "3 – Population structure and epidemiology of Toxoplasma gondii". In Weiss, LM; Kim, K. Toxoplasma Gondii: The Model Apicomplexan. Perspectives and Methods. London: Academic Press/Elsevier. pp. 49–80. doi:10.1016/B978-012369542-0/50005-2. ISBN 978-0-12-369542-0.

- Clarence R. Robbins (24 February 2012). Chemical and Physical Behavior of Human Hair. Springer. p. 585. ISBN 978-3-642-25610-3. Retrieved 12 March 2013.

- Chatterjee, K.D. (2009). Parasitology (Protozoology and Helminthology) in relation to clinical medicine (13 ed.). New Delhi: CBC Publishers. pp. 56–57. ISBN 978-8-12-39-1810-5.

- Barry JD, McCulloch R (2001). "Antigenic variation in trypanosomes: enhanced phenotypic variation in a eukaryotic parasite". Adv Parasitol. 49: 1–70. doi:10.1016/S0065-308X(01)49037-3. ISBN 978-0-12-031749-3.

- Kohl, Linda; Bastin, Philippe (2005). "The Flagellum of Trypanosomes". In Jeon, Kwang W. A Survey of Cell Biology. International Review of Cytology. 244. pp. 227–85. doi:10.1016/S0074-7696(05)44006-1. ISBN 978-0-08-045779-6. PMID 16157182.

- "Details - Lehrbuch der vergleichenden Anatomie der Wirbellosen Thiere / von C. Th. v. Siebold. - Biodiversity Heritage Library". www.biodiversitylibrary.org. Retrieved 2015-05-20.

- Cavalier-Smith T (August 1998). "A revised six-kingdom system of life". Biological Reviews. 73: 203–266. doi:10.1111/j.1469-185X.1998.tb00030.x. PMID 9809012. Retrieved 2015-06-09.

Tapeworm and Its Types

Tapeworms are parasitic and live in the digestive systems of adults. Tapeworms are hermaphroditic; they have both male and female reproductive systems in their bodies. Some of the types of tapeworms are diphyllobothrium, echinococcus granulosus, hymenolepis nana and taenia solium. In order to understand tapeworms, it is necessary to understand all the types of tapeworms.

Cestoda

Cestoda (Cestoidea) is a class of parasitic flatworms, of the phylum Platyhelminthes. Biologists informally refer to them as cestodes. The best-known species are commonly called tapeworms. All cestodes are parasitic and their life histories vary, but typically they live in the digestive tracts of vertebrates as adults, and often in the bodies of other species of animals as juveniles. Over a thousand species have been described, and all vertebrate species may be parasitised by at least one species of tapeworm.

Humans are subject to infection by several species of tapeworms if they eat undercooked meat such as pork (*Taenia solium*), beef (*T. saginata*), and fish (*Diphyllobothrium* spp.), or if they live in, or eat food prepared in, conditions of poor hygiene (*Hymenolepis* or *Echinococcus* species).

T. saginata, the beef tapeworm, can grow up to 20 m (65 ft); the largest species, the whale tapeworm *Polygonoporus giganticus*, can grow to over 30 m (100 ft). Species using small vertebrates as hosts, though, tend to be small. For example, vole and lemming tapeworms are only 13–240 mm (0.51–9.45 in) in length, and those parasitizing shrews only 0.8–60 mm (0.031–2.362 in).

Tapeworm parasites of vertebrates have a long history: recognizable clusters of cestode eggs, one with a developing larva, have been discovered in fossil feces (coprolites) of a shark dating to the mid- to late Permian, some 270 million years ago.

Anatomy

Scolex

The worm's scolex ("head") attaches to the intestine of the definitive host. In some species, the scolex is dominated by bothria, or "sucking grooves" that function like suction cups. Other species have hooks and suckers that aid in attachment. Cyclophyllid cestodes can be identified by the presence of four suckers on their scolices.

While the scolex is often the most distinctive part of an adult tapeworm, it is often unnoticed in a clinical setting, as it is inside the host. Therefore, identifying eggs and proglottids in feces is the simplest way to diagnose an infection.

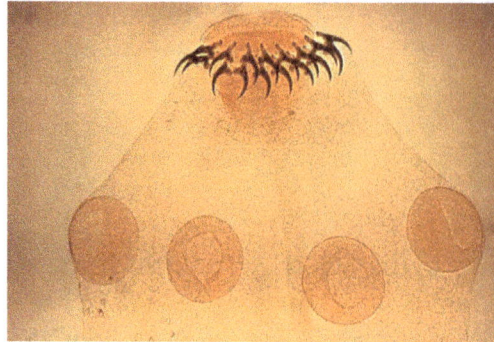

Scolex of *Taenia solium* with hooks and suckers.

Body Systems

The main nerve centre of a cestode is a cerebral ganglion in its scolex. Motor and sensory innervation depends on the number of nerves in and complexity of the scolex. Smaller nerves emanate from the ganglion to supply the general body muscular and sensory ending. The cirrus and vagina are innervated, and sensory endings around the genital pore are more plentiful than other areas. Sensory function includes both tactoreception (touch) and chemoreception (smell or taste). Some nerves are only temporary.

Proglottids

The body is composed of successive segments called proglottids. The sum of the proglottids is called a strobila, which is thin, and resembles a strip of tape. From this is derived the common name "tapeworm". Proglottids are continually produced by the neck region of the scolex, as long as the scolex is attached and alive. Like some other flatworms, cestodes use flame cells (protonephridia), located in the proglottids, for excretion. Mature proglottids are released from the tapeworm's end segment and leave the host in feces or migrate as independent motile proglottids. The proglottids farthest away from the scolex are the mature ones containing eggs. Mature proglottids are essentially bags of eggs, each of which is infective to the proper intermediate host.

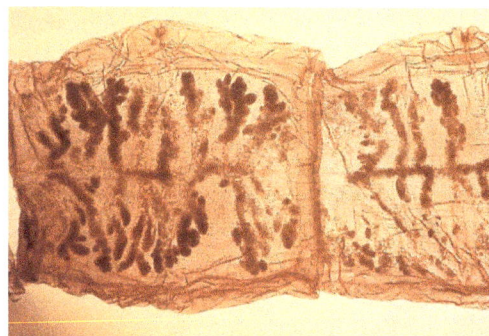

Two proglottids of *Taenia solium*

The layout of proglottids comes in two forms: craspedote, meaning any given proglottid is overlapped by the previous proglottid, and acraspedote, indicating the proglottids are not overlapping.

Cestodes are unable to synthesise lipids and are entirely dependent on their host, although lipids are not used as an energy reserve, but for reproduction. Once anchored to the host's intestinal wall, the tapeworm absorbs nutrients through its skin as the food being digested by the host flows over and around it. Soon, it begins to grow a tail composed of a series of segments, with each segment containing an independent digestive system and reproductive tract. Older segments are pushed toward the tip of the tail as new segments are produced by the neckpiece. By the time a segment has reached the end of the worm's tail, only the reproductive tract is left. The segment then separates, carrying the tapeworm eggs out of the definitive host as what is basically a sack of eggs.

Lifecycle of The Tapeworms

True tapeworms are exclusively hermaphrodites; they have both male and female reproductive systems in their bodies. The reproductive system includes one or more testes, cirri, vas deferens, and seminal vesicles as male organs, and a single lobed or unlobed ovary with the connecting oviduct and uterus as female organs. The common external opening for both male and female reproductive systems is known as the genital pore, which is situated at the surface opening of the cup-shaped atrium. Though they are sexually hermaphroditic, self-fertilization is a rare phenomenon. To permit hybridization, cross-fertilization between two individuals is often practiced for reproduction. During copulation, the cirri of one individual connect with those of the other through the genital pore, and then spermatozoa are exchanged.

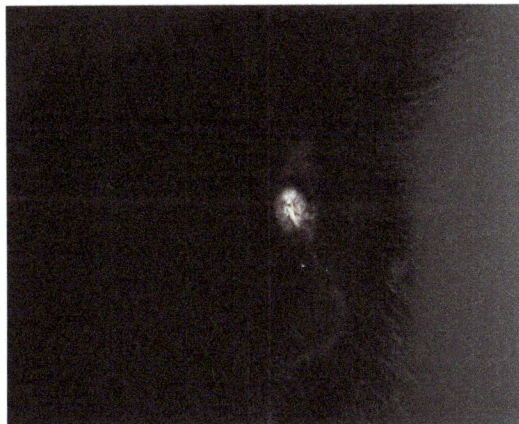

Image of a tapeworm proglottid leaving its definitive host

The lifecycle of tapeworms is simple in the sense that no asexual phases occur as in other flatworms, but complicated in that at least one intermediate host is required as well as the definitive host. This lifecycle pattern has been a crucial criterion for assessing evolution among Platyhelminthes. Many tapeworms have a two-phase lifecycle with two types of hosts. The adult *Taenia saginata* lives in the gut of a primate such as a human, but more alarming is *Taenia solium*, which can form cysts in the human brain. Proglottids leave the body through the anus and fall onto the ground, where they may be eaten with grass by an animal such as a cow. If the tapeworm is compatible with the eating animal, this animal becomes an intermediate host. The juvenile form of the worm enters through the mouth, but then migrates and establishes as a cyst in the intermediate host's body tissues such as muscles, rather than the gut. This can cause more damage to the intermediate host than it does to its definitive host. The parasite completes its lifecycle when the intermediate host passes on the parasite to the definitive host. This is usually done by the definitive host eating a suitably infected intermediate host, e.g., a human eating raw or undercooked meat.

Infection and Treatment

Symptoms vary widely, as do treatment options, and these issues are discussed in detail in the individual articles on each worm. Praziquantel is an effective treatment for tapeworm infection, and is preferred over the older niclosamide. Cestodes can also be treated with certain kinds of antibiotics. While accidental tapeworm infections in developed countries are quite rare, some US dieters have risked intentional infection for the purpose of weight loss.

Certain medicines are used to remove it, such as praziquantel or albendazole. Physicians also give enema treatment to the patient to completely remove intestinal flatworms.

Taxonomy

The taxonomy of the Cestoda has been clarified with molecular data. The Gyrocotylidea are a sister group to all other Cestoda (Nephroposticophora): the Amphilinidea form the sister group to the Eucestoda. The Caryophyllidea are the sister group to Spathebothriidea and remaining Eucestoda. The Haplobothriidea are the sister group to Diphyllobothriidae. The Diphyllidea and Trypanorhyncha may be sister groups, but this is not definite.

At the more derived groups, the taxonomy appears to be:

- Bothriocephalidea

o Litobothriidea

- Lecanicephalidea

- Rhinebothriidea

- Tetraphyllidea

- Acanthobothrium, Proteocephalidea

- Cyclophyllidea, Mesocestoididae, Nippotaeniidea, Tetrabothriidea

The Tetraphyllidea appear to be paraphyletic. The relations between Nippotaeniidea, Mesocestoididae, Tetrabothriidea, and Cyclophyllidea require further clarification.

The taxonomy of the Eucestoda has been also clarified. The current taxonomy is

- Monogenea

o Amphilinidea

- Caryophyllidea

- Spathebothriidea

- Trypanorhyncha

- Pseudophyllidea

- Tetraphyllidea

- Diphyllidea, Proteocephalidea

- Nippotaeniidea

- Cyclophyllidea, Tetrabothriidea

The Tetraphyllidea, Pseudophyllidea (because of the Diphyllobothriidae) and Cyclophyllidea (because of the Mesocestoididae) are paraphyletic.

The Taeniidae may be the most basal of the 12 orders of the Cyclophyllidea.

The Tetraphyllidea, Lecanicephalidea, Proteocephalidea, Nippotaeniidea, Tetrabothriidea, and Cyclophyllidea are considered to be the 'higher' tapeworms.

The 277 known species in the marine order Trypanorhyncha are placed in five superfamilies - Tentacularioidea, Gymnorhynchoidea, Otobothrioidea, Eutetrarhynchidae, and Lacistorhynchidae.

Society and Culture

In Popular Culture

Deliberately implanted tapeworms that take over their human hosts to form either voracious, zombie-like "sleepwalkers" or human-tapeworm chimerae are the basis of Mira Grant's science-fiction horror trilogy, Parasitology.

Diphyllobothrium

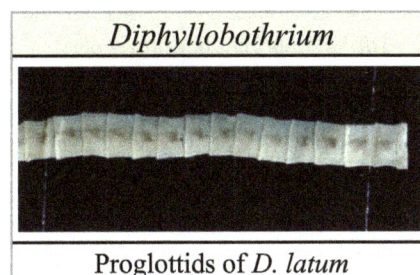

Diphyllobothrium

Proglottids of *D. latum*

Diphyllobothrium is a genus of tapeworm which can cause diphyllobothriasis in humans through consumption of raw or undercooked fish. The principal species causing diphyllobothriosis is *Diphyllobothrium latum*, known as the broad or fish tapeworm, or broad fish tapeworm. *D. latum* is a pseudophyllid cestode that infects fish and mammals. *D. latum* is native to Scandinavia, western Russia, and the Baltics, though it is now also present in North America, especially the Pacific Northwest. In Far East Russia, *D. klebanovskii*, having Pacific salmon as its second intermediate host, was identified. Other members of the genus *Diphyllobothrium* include *Diphyllobothrium dendriticum* (the salmon tapeworm), which has a much larger range (the whole northern hemisphere), *D. pacificum*, *D. cordatum*, *D. ursi*, *D. lanceolatum*, *D. dalliae*, and *D. yonagoensis*, all of which infect humans only infrequently. In Japan, the most common species in human infection is *D. nihonkaiense*, which was only identified as a separate species from *D. latum* in 1986. More recently, a molecular study found *D. nihonkaiense* and *D. klebanovskii* to be a single species.

History

The fish tapeworm has a long documented history of infecting people who regularly consume fish and especially those whose customs include the consumption of raw or undercooked fish. In the 1970s, most of the known cases of diphyllobothriasis came from Europe (5 million cases), and Asia (4 million cases) with fewer cases coming from North America and South America, and no reliable data on cases from Africa or Australia. Interestingly, despite the relatively small number of cases seen today in South America, some of the earliest archeological evidence of diphyllobothriasis comes from sites in South America. Evidence of *Diphyllobothrium spp.* has been found in 4,000- to 10,000-year-old human remains on the western coast of South America. There is no clear point in time when *Diphyllobothrium latum* and related species were "discovered" in humans, but it is clear that diphyllobothriasis has been endemic in human populations for a very long time. Due to the changing dietary habits in many parts of the world, autochthonous, or locally acquired, cases of diphyllobothriasis have recently been documented in previously non-endemic areas, such as Brazil. In this way, diphyllobothriasis represents an emerging infectious disease in certain parts of the world where cultural practices involving eating raw or undercooked fish are being introduced.

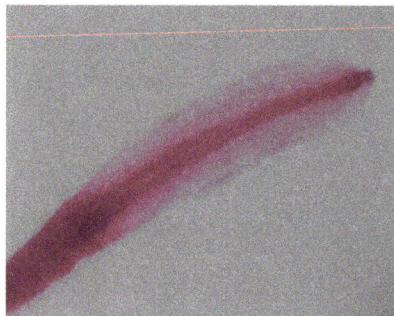

Diphyllobothrium latum scolex

Morphology

The adult worm is composed of three fairly distinct morphological segments: the scolex (head), the neck, and the lower body. Each side of the scolex has a slit-like groove, which is a bothrium for attachment to the intestine. The scolex attaches to the neck, or proliferative region. From the neck grow many proglottid segments which contain the reproductive organs of the worm. *D. latum* is the longest tapeworm in humans, averaging ten meters long. Adults can shed up to a million eggs a day.

In adults, proglottids are wider than they are long (hence the name *broad tapeworm*). As in all pseudophyllid cestodes, the genital pores open midventrally.

Life Cycle

Adult tapeworms may infect humans, canids, felines, bears, pinnipeds, and mustelids, though the accuracy of the records for some of the nonhuman species is disputed. Immature eggs are passed in feces of the mammal host (the definitive host, where the worms reproduce). After ingestion by a suitable freshwater crustacean such as a copepod (the first intermediate host), the coracidia develop into procercoid larvae. Following ingestion of the copepod by a suitable second intermediate host, typically a minnow or other small freshwater fish, the procercoid larvae are released from the crustacean and migrate into the fish's flesh where they develop into a plerocercoid larvae (sparganum). The plerocercoid larvae are the infective stage for the definitive host (including humans).

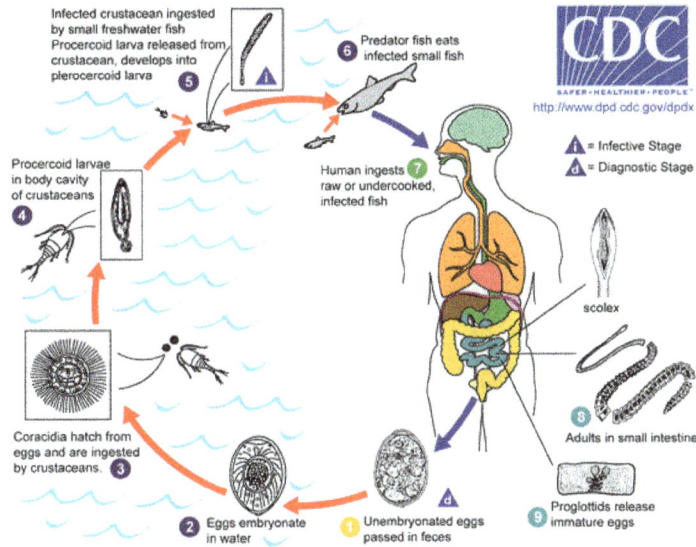

Life cycle of *D. latum*. Click the image to see full-size.

Because humans do not generally eat undercooked minnows and similar small freshwater fish, these do not represent an important source of infection. Nevertheless, these small second intermediate hosts can be eaten by larger predator species, for example trout, perch, walleye, and pike. In this case, the sparganum can migrate to the musculature of the larger predator fish and mammals can acquire the disease by eating these later intermediate infected host fish raw or undercooked. After ingestion of the infected fish, the plerocercoids develop into immature adults and then into mature adult tapeworms which will reside in the small intestine. The adults attach to the intestinal mucosa by means of the two bilateral grooves (bothria) of their scolex. The adults can reach more than 10 m (up to 30 ft) in length in some species such as *D. latum,* with more than 3,000 proglottids. One or several of the tape-like proglottid segments (hence the name tape-worm) regularly detach from the main body of the worm and release immature eggs in fresh water to start the cycle over again. Immature eggs are discharged from the proglottids (up to 1,000,000 eggs per day per worm) and are passed in the feces. The incubation period in humans, after which eggs begin to appear in the feces is typically 4–6 weeks, but can vary from as short as 2 weeks to as long as 2 years. The tapeworm can live up to 20 years.

Clinical Symptoms, Including Occasional Parasite-Induced B$_{12}$ Deficiency

Symptoms of diphyllobothriasis are generally mild, and can include diarrhea, abdominal pain, vomiting, weight loss, fatigue, constipation and discomfort. Approximately four out of five cases are asymptomatic and may go many years without being detected. In a small number of cases, this leads to severe vitamin B$_{12}$ deficiency due to the parasite absorbing 80% or more of the host's B$_{12}$ intake, and a megaloblastic anemia indistinguishable from pernicious anemia. The anemia can also lead to subtle demyelinative neurological symptoms (subacute combined degeneration of spinal cord). Infection for many years is ordinarily required to deplete the human body of vitamin B-12 to the point that neurological symptoms appear.

Diagnosis

Diagnosis is usually made by identifying proglottid segments, or characteristic eggs in the feces. These simple diagnostic techniques are able to identify the nature of the infection to the genus level, which is usually sufficient in a clinical setting. However, when the species needs to be determined (in epidemiological studies, for example), restriction fragment length polymorphisms can be effectively used. PCR can be performed on samples of purified eggs, or native fecal samples following sonication of the eggs to release their contents. Another interesting potential diagnostic tool and treatment is the contrast medium, Gastrografin, introduced into the duodenum, which allows both visualization of the parasite, and has also been shown to cause detachment and passing of the whole worm.

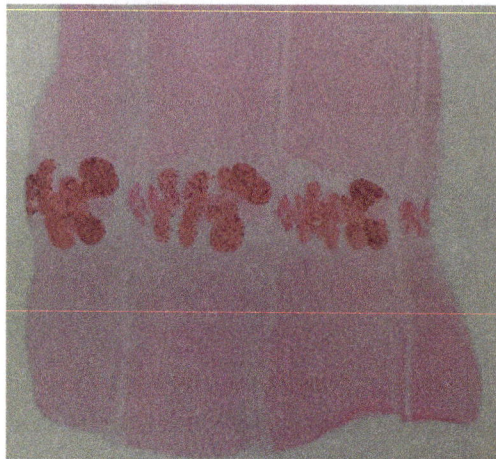

Diphyllobothrium latum proglottid

Treatment

The standard treatment for diphyllobothriasis, as well as many other tapeworm infections is a single dose of Praziquantel, 5–10 mg/kg PO once for both adults and children. An alternative treatment is Niclosamide, 2 g PO once for adults or 50 mg/kg PO once. One should note that Praziquantel is not FDA approved for this indication and Niclosamide is not available for human use in the United States.

Side Effects of Treatment

Praziquantel has few side effects, many of which are similar to the symptoms of diphyllobothriasis. They include malaise, headache, dizziness, abdominal discomfort, nausea, rise in temperature and occasionally allergic skin reactions. The side effects of Niclosamide are very rare, due to the fact that it is not absorbed in the gastrointestinal tract.

Epidemiology

People at high risk for infection have traditionally been those who regularly consume raw fish. Many regional cuisines include raw or undercooked food, including sushi and sashimi in Japanese cuisine, carpaccio di persico in Italian, tartare maison in French-speaking populations, ceviche in

Latin American cuisine and marinated herring in Scandinavia. With emigration and globalization, the practice of eating raw fish in these and other dishes has brought diphyllobothriasis to new parts of the world and created new endemic foci of disease.

Public Health Strategies

The most viable interventions include: prevention of water contamination both by raising public awareness of the dangers of defecating in recreational bodies of water and by implementation of basic sanitation measures; screening and successful treatment of people infected with the parasite; and prevention of infection of humans via consumption of raw, infected fish. The last of these can most easily be changed via education about proper preparation of fish. Fish that is thoroughly cooked, brined, or frozen at -10 °C for 24–48 hours can be consumed without risk of D. latum infection.

Pop Culture

In the 14th episode of season 3 of House entitled "Insensitive", the primary patient is a girl who has CIPA . She is ultimately diagnosed with *Diphyllobothrium* causing vitamin B12 deficiency

In the Mira Grant novel Parasite, the genetically engineered symbiotic tapeworm species at the core of the novel is engineered primarily based on *D. yonagoensis*.

Echinococcus Granulosus

Echinococcus granulosus, also called the hydatid worm, hyper tape-worm or dog tapeworm, is a cyclophyllid cestode that parasitizes the small intestine of canids as an adult, but which has important intermediate hosts such as livestock and humans, where it causes cystic echinococcosis, also known as hydatid disease. The adult tapeworm ranges in length from 3 mm to 6 mm and has three proglottids ("segments") when intact—an immature proglottid, mature proglottid and a gravid proglottid. The average number of eggs per gravid proglottid is 823. Like all cyclophyllideans, *E. granulosus* has four suckers on its scolex ("head"), and *E. granulosus* also has a rostellum with hooks. Several strains of *E. granulosus* have been identified, and all but two are noted to be infective in humans.

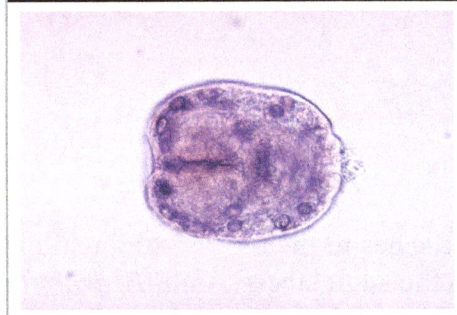

Echinococcus granulosus

The lifecycle of *E. granulosus* involves dogs and wild carnivores as a definitive host for the adult tapeworm. Definitive hosts are where parasites reach maturity and reproduce. Wild or domesticated ungulates, such as sheep, serve as an intermediate host. Transitions between life stages occur in intermediate hosts. The larval stage results in the formation of echinococcal cysts in intermediate hosts. Echinococcal cysts are slow growing, but can cause clinical symptoms in humans and be life-threatening. Cysts may not initially cause symptoms, in some cases for many years. Symptoms developed depend on location of the cyst, but most occur in the liver, lungs, or both.

E. granulosus was first documented in Alaska but is distributed world-wide. It is especially prevalent in parts of Eurasia, north and east Africa, Australia, and South America. Communities that practice sheep farming experience the highest risk to humans, but wild animals can also serve as an avenue for transmission. For example, dingoes serve as a definitive host before larvae infect sheep in the mainland of Australia. Sled dogs may expose moose or reindeer to *E. granulosus* in parts of North America and Eurasia.

Transmission

E. granulosus requires two host types, a definitive host and an intermediate host. The definitive host of this parasite are dogs and the intermediate host are most commonly sheep, however, cattle, horses, pigs, goats, and camels are also potential intermediate hosts. Humans can also be an intermediate host for *E. granulosus*, however this is uncommon and therefore humans are considered an aberrant intermediate host.

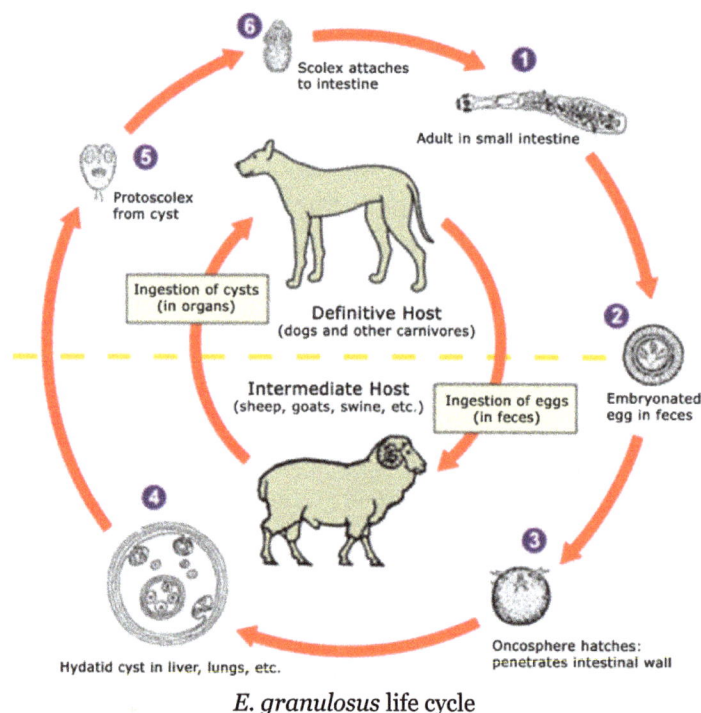

E. granulosus life cycle

E. granulosus is ingested and attaches to the mucosa of the intestines in the definitive host and there the parasite will grow into the adult stages. Adult *E. granulosus* release eggs within the intestine which will be transported out of the body via feces. When contaminated waste is excreted into the environment, intermediate host has the potential to contract the parasite by grazing in contaminated pasture, perpetuating the cycle.

E. granulosus is transmitted from the intermediate host (sheep) to the definitive host (dogs) by frequent feeding of offal, also referred to as "variety meat" or "organ meat". Consuming offal containing *E. granulosus* can lead to infection; however, infection is dependent on many factors.

The dog serves as the main definitive host for the dangerous parasite, with eggs being shed in its stool

The frequency of offal feedings, the prevalence of the parasites within the offal, and the age of the intermediate host are factors that affect infection pressure within the definitive host. The immunity of both the definitive and intermediate host plays a large role in the transmission of the parasite, as well as the contact rate between the intermediate and the definitive host (such as herding dogs and pasture animals being kept in close proximity where dogs can contaminate grazing areas with fecal matter).

The life expectancy of the parasite, coupled with the frequency of anthelminthic treatments, will also play a role in the rate of infection within a host. The temperature and humidity of the environment can affect the survival of *E. granulosus*.

Once sheep are infected, the infection typically remains within the sheep for life. However, in other hosts, such as dogs, treatment for annihilating the parasite is possible. However, the intermediate host is assumed to retain a greater life expectancy than the definitive host.

Diagnosis

Diagnosis in the definitive host, the dog, may be done by *post mortem* examination of the small intestine, or with some difficulty *ante mortem* by purging with arecoline hydrobromate. Detection of antigens in feces by ELISA is currently the best available technique. The prevalence of *Echinococcus granulosus* was found to be 4.35% in a 2008 study in Bangalore, India employing this coproantigen detection technique. Polymerase Chain Reaction(PCR) is also used to identify the parasite from DNA isolated from eggs or feces.

Risk in Humans

Humans should avoid handling stool of dogs and avoid eating infected animals and home slaughtering animals. If a human becomes infected there are a variety of methods for treatment. The most common treatment in the past years has been surgical removal of the hydatid cysts. Cyst manipulation should be performed with caution, as spilling of cyst contents can cause anaphylactic shock. However, in recent years, less invasive treatments have been developed such as cyst puncture, aspiration of the liquids, the injection of chemicals, and then re-aspiration. Benzimidazole-based chemotherapy is also a new treatment option for humans.

Prevention

Boiling offal containing hydatid cysts for 30 minutes kills the larvae of *E. granulosus* - a simple method for prevention in remote areas

In order to prevent transmission to dogs from intermediate hosts, dogs can be given anthelminthic vaccinations. In the case of intermediate hosts, especially sheep, these anthelminthic vaccinations do cause an antigenic response—meaning the body produces antibodi avinash response—however it does not prevent infection in the host. Clean slaughter and high surveillance of potential intermediate host during slaughter is key in preventing the spread this cestode to its definitive host. It is vital to keep dogs and potential intermediate host as separated as possible to avoid perpetuating infection. According to mathematical modeling, vaccination of intermediate hosts, coupled with dosing definitive hosts with anthelminths is the most effect method for intervening with infection rates.

Proper disposal of carcasses and offal after home slaughter is difficult in poor and remote communities and therefore dogs readily have access to offal from livestock, thus completing the parasite cycle of *Echinococcus granulosus* and putting communities at risk of cystic echinococcosis. Boiling livers and lungs which contain hydatid cysts for 30 minutes has been proposed as a simple, efficient and energy- and time-saving way to kill the infectious larvae.

Echinococcus Multilocularis

Echinococcus multilocularis is a cyclophyllid tapeworm that, along with some other members of the *Echinococcus* genus (especially *E. granulosus*), produces the disease known as echinococcosis in certain terrestrial mammals, including wolves, foxes, jackals, coyotes, domestic dogs and humans. Unlike *E. granulosus*, *E. multilocularis* produces many small cysts (also referred to as locules) that spread throughout the internal organs of the infected animal. Ingestion of these cysts, usually by a canid eating an infected rodent, results in a heavy infestation of tapeworms.

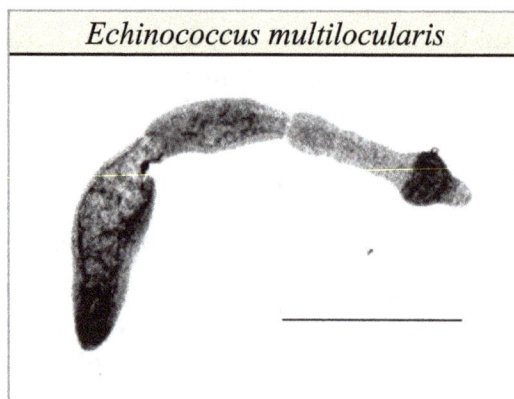

Echinococcus multilocularis

Signs and Symptoms

People infected with *E. multilocularis* may be asymptomatic for many years. Following the asymptomatic period of this disease, common symptoms are headache, nausea, vomiting, abdominal pain. Jaundice is rare, but hepatomegaly is a common physical finding.

Life Cycle

The life cycle of *E. multilocularis* involves a primary or definitive host and a secondary or intermediate host, each harboring different life stages of the parasite.

Foxes, coyotes, domestic dogs, and other canids are the definitive hosts for the adult stage of the parasite. The head of the tapeworm attaches to the intestinal mucosa by hooks and suckers. It then produces hundreds of microscopic eggs, which are dispersed through the feces.

Wild rodents such as mice serve as the intermediate host. Eggs ingested by rodents develop in the liver, lungs and other organs to form multilocular cysts. Humans could also become an intermediate host by handling infected animals or ingesting contaminated food, vegetable, and water. The life cycle is completed after a fox or canine consumes a rodent infected with cysts. Larvae within the cyst develop into adult tapeworms in the intestinal tract of the definitive host.

Except in rare cases where infected humans are eaten by canines, humans are a dead-end or incidental host (an intermediate host that does not allow transmission to the definitive host) for *E. multilocularis*.

Summary of the life cycle

1. adult worm present in intestine of definitive host

2. eggs passed in feces, ingested by humans or intermediate host

3. onchosphere penetrates intestinal wall, carried via blood vessels to lodge in organs

4. hydatid cysts develop in liver, lungs, brain, heart

5. protoscolices (hydatid sand) ingested by definitive host

6. ingested protoscolices attach to small intestine and develops into adult worm

Morphology

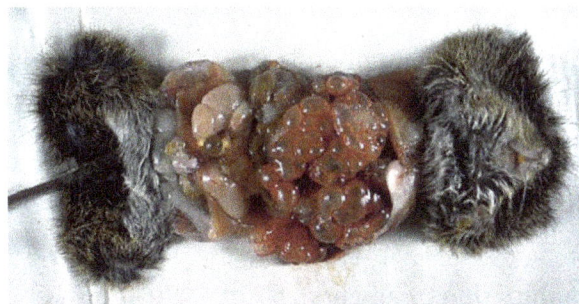

Cotton rat infected with *Echinococcus multilocularis*

The adult parasite is a small tapeworm that is 3- 6mm long, and lives in the small intestine of canines. The segmented worm contains a scolex with suckers and hooks that enable attachment to the mucosal wall, since tapeworms do not have a digestive tract. A short neck connects the head to three proglottids, the body segment of the worm which contains the eggs to be excreted in the feces.

Diagnosis

Serological and imaging tests are commonly used to diagnose this disease. Frequently used serological tests include antibody tests, ELISA and indirect hemaglutination (IHA). Also, an intradermal allergic reaction test (Casoni test) has also been used to diagnose patients. Imaging tests include: X-rays, cat scans, MRI, and ultrasound.

Disease Staging

Alveolar echinococcosis (AE) is a highly lethal helminthic disease in humans, caused by the larval form of the parasitic tapeworm *E. multilocularis*. The disease represents a serious public threat in China, Siberia, and central Europe. However, since the 1990s, the prevalence of the disease seems to be increasing in Europe, not only in the historically endemic areas but its neighboring regions. AE primarily affects the liver by inducing a hepatic disorder similar to liver cancer, therefore becoming extremely dangerous and difficult to diagnose. If the infection metastasizes, it may spread to any other organ and could be lethal if not treated. The most common treatment for AE is to surgically remove the parasite. Since it is difficult and not always possible to remove the entire parasite, medicine such as Albendazole is utilized to keep the cyst from growing back.

Guided by the Tumor-Node-Metastasis (TNM) system of liver cancer, the European Network for Concerted Surveillance of Alveolar Echinococcosis and the World Health Organization Informal Working Group on Echinococcosis, a clinical classification system has been proposed. This classification system has been designated as the "PNM" system (P = parasitic mass, N = involvement of neighboring organs, M = metastasis). The system was developed by a retrospective analysis of records from 97 patients treated in France and Germany (2 treatment centers). Amongst other characteristics, the system takes into consideration the localization of the parasite in the liver, the extent of lesion involvement, regional involvement, and metastasis.

Treatment

If no specific therapy is initiated, in 94% of patients the disease is fatal within 10–20 years following diagnosis.

- Currently, benzimidazoles (such as albendazole) are used to treat AE: only halt their proliferation and do not actually kill the parasites, side effects such as liver damage

- 2-ME2, a natural metabolite of estradiol, is tested with some results *in vitro*: decreased transcription of 14-3-3-pro-tumorogenic zeta-isoform, causes damage to germinal layer but does not kill parasite *in vivo*

- Treatment with a combination of albendazole/2-ME2 showed best results in reducing parasite burden

- Despite the improvements in the chemotherapy of echinococcosis with benzimidazole derivatives, complete elimination of the parasitic mass cannot be achieved in most infected patients, although studies indicate that long-term treatment with mebendazole may cause the death of the parasite.

Epidemiology

The incidence of human infestation with *E. multilocularis* and disease is increasing in urban areas, as wild foxes (an important reservoir species of the sylvatic cycle) are migrating to urban and suburban areas and gaining closer contact with human populations. Also, restocking fox enclosures for fox hunting with infected animals spreads the disease. Children, health care workers and domestic animals are at risk of ingesting the cysts after coming into contact with the feces of infected wild foxes. Even with the improvement of health in developed/industrialized countries, the prevalence of alveolar echinococcosis (AE) did not decrease. On the contrary, incidents of AE have now also been registered in eastern European countries and sporadic incidences in other European countries.

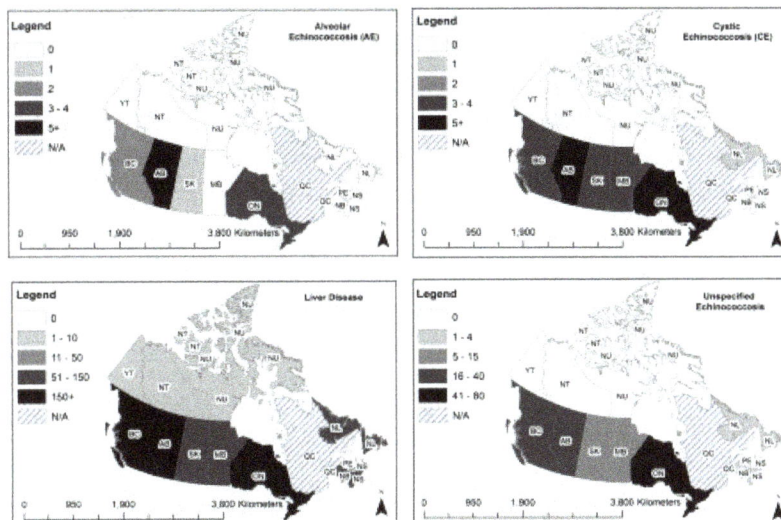

Epidemiology of *Echinococcus multilocularis* in Canada

The disease has extended its range in Europe in the last few decades. Still the infection is fairly rare. Between 1982 and 2000 a total of 559 cases were reported throughout Europe.

Recent findings indicate that *E. multilocularis* is likely expanding its range in the central region of the United States and Canada and that invasions of European strains might have occurred; the endemic presence of the parasite in urban areas and a recent human case in Alberta, Canada have been reported.

Hymenolepis Nana

Dwarf tapeworm (*Hymenolepis nana*, previously known as *Vampirolepis nana, Hymenolepis fraterna*, and *Taenia nana*) is a cosmopolitan species though most common in temperate zones, and is one of the most common cestodes (a type of intestinal worm or helminth) infecting humans, especially children.

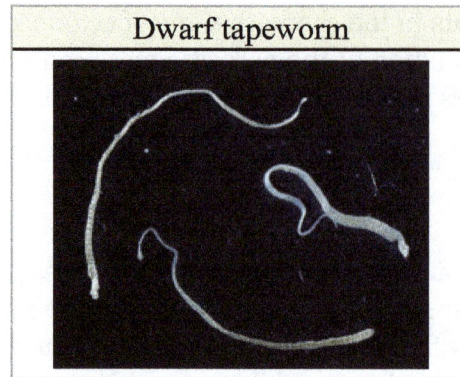

Dwarf tapeworm

Morphology

As its name implies (Greek: *nanos* – dwarf), it is a small species, seldom exceeding 40 mm long and 1 mm wide. The scolex bears a retractable rostellum armed with a single circle of 20 to 30 hooks. The scolex also has four suckers, or a tetrad. The neck is long and slender, and the segments are wider than long. Genital pores are unilateral, and each mature segment contains three testes. After apolysis, gravid segments disintegrate, releasing eggs, which measure 30 to 47 µm in diameter. The oncosphere is covered with a thin, hyaline, outer membrane and an inner, thick membrane with polar thickenings that bear several filaments. The heavy embryophores that give taeniid eggs their characteristic striated appearance are lacking in this and the other families of tapeworms infecting humans. The rostellum remains invaginated in the apex of the organ. Rostellar hooklets are shaped like tuning forks. The neck is long and slender, the region of growth. The strobila starts with short, narrow proglottids, followed with mature ones.

Lifecycle

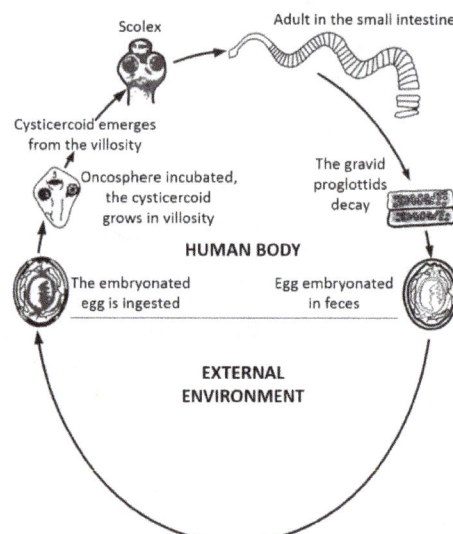

Hymenolepis nana **Life Cycle**, Cestode (Dwarf tapeworm)

Lifecycle of *H. nana* inside and outside of the human body

Infection is acquired most commonly from eggs in the feces of another infected individual, which are transferred in food, by contamination. Eggs hatch in the duodenum, releasing oncospheres, which penetrate the mucosa and come to lie in lymph channels of the villi. An oncosphere develops into a cysticercoid which has a tail and a well-formed scolex. It is made of longitudinal fibers and is spade-shaped with the rest of the worm still inside the cyst. In five to six days, cysticercoids emerge into the lumen of the small intestine, where they attach and mature.

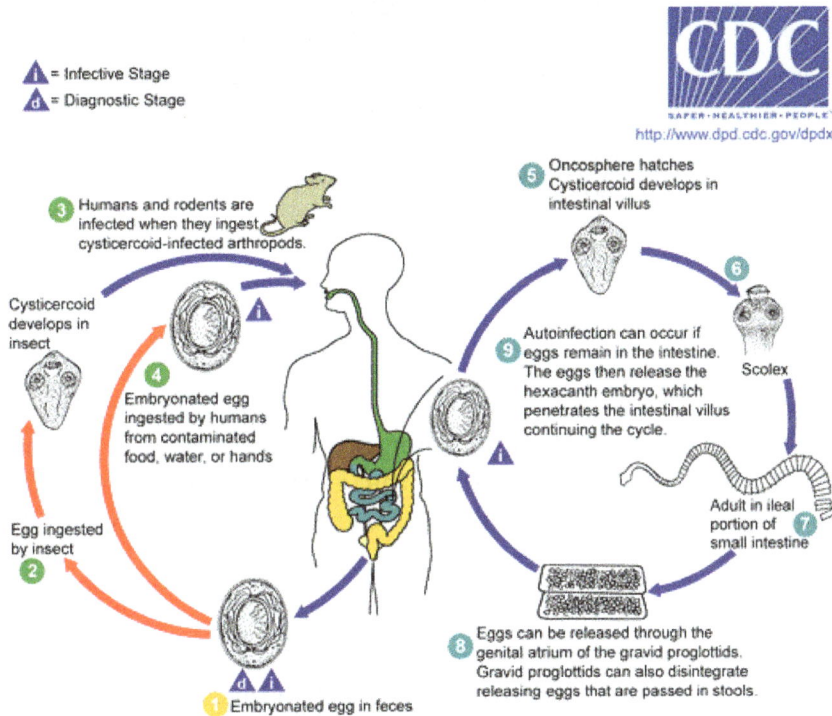

Hymenolepis nana lifecycle

The direct lifecycle is doubtless a recent modification of the ancestral two-host lifecycle found in other species of hymenolepidids, because cysticercoids of *H. nana* can still develop normally within larval fleas and beetles. One reason for facultative nature of the lifecycle is that *H. nana* cysticercoids can develop at higher temperatures than can those of the other hymenolepidids. Direct contaminative infection by eggs is probably the most common route in human cases, but accidental ingestion of an infected grain beetle or flea cannot be ruled out. The direct infectiousness of the eggs frees the parasite from its former dependence upon an insect intermediate host, making rapid infection and person-to-person spread possible. The short lifespan and rapid course of development also facilitate the spread and ready availability of this worm.

Reproduction

H. nana, like all tapeworms, contains both male and female reproductive structures in each proglottid. This means that the dwarf tapeworm, like other tapeworms is hermaphroditic. Each segment contains three testes and a single ovary. When a proglottid becomes old and unable to absorb any more nutrition, it is released and is passed through the host's digestive tract. This gravid proglottid contains the fertilized eggs, which are sometimes expelled with the feces. However, most of the time, the egg may also settle in the microvilli of the small intestine, hatch, and the larvae can develop to sexual maturity without ever leaving the host.

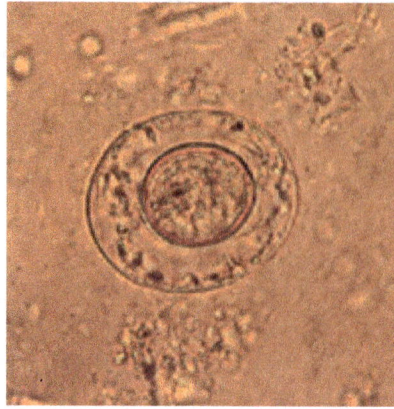

An egg of dwarf tapeworm

Behavior

The dwarf tapeworm, like all other tapeworms, lacks a digestive system and feeds by absorption on nutrients in the intestinal lumen. They have nonspecific carbohydrate requirements and they seem to absorb whatever is being passed through the intestine at that time. When it becomes an adult, it attaches to the intestinal walls with its suckers and toothed rostellum and has its segments reaching out into the intestinal space to absorb food.

History

In 1887, Grassi demonstrated that transmission from rat to rat did not require an intermediate host. Later, in 1921, Saeki demonstrated a direct cycle of transmission of *H. nana* in humans, transmission without an intermediate host. In addition to the direct cycle, Nicholl and Minchin demonstrated that fleas can serve as intermediate hosts between humans.

Hymenolepis Diminuta

Hymenolepis diminuta
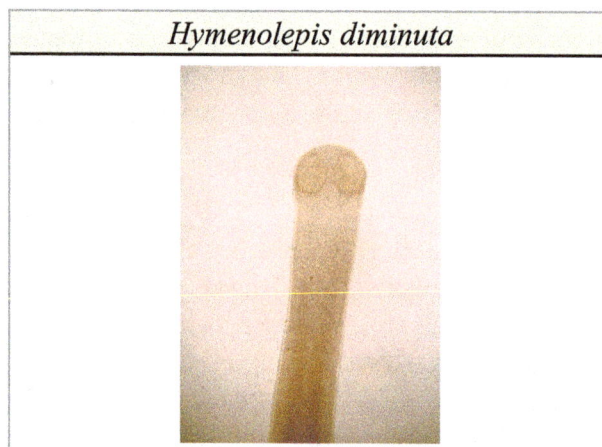

Hymenolepis diminuta, also known as rat tapeworm, is a species of *Hymenolepis* tapeworm that causes hymenolepiasis. It has slightly bigger eggs and proglottids than *H. nana* and infects mammals using insects as intermediate hosts. The adult structure is 20 to 60 cm long and the mature proglottid is similar to that of *H. nana*, except it is larger.

H. diminuta is prevalent worldwide, but only a few hundred human cases have been reported. Few cases have ever been reported in Australia, United States, Spain, and Italy. In countries such as Malaysia, Thailand, Jamaica, Indonesia, the prevalence is higher.

Life Cycle

The cycle begins as arthropods ingest the eggs. Arthropods are then able to act as the intermediate host. When ingested, the eggs develop into cysticercoids. As shown in the CDC life cycle, oncospheres hatch and then penetrate the intestinal wall. Rodents can become infected when they eat arthropods. Humans, especially children, can ingest the arthropods as well and therefore become infected via the same mechanism. Rodents, especially rats, are definitive hosts and natural reservoirs of *H. diminuta*. The intermediate hosts are the coprophilic arthropods (fleas, lepidoptera, and coleoptera). As the definitive host (rats) eats an infected arthropod, cysticercoids present in the body cavity transform into the adult worm. The resulting eggs are then passed through the stool. In recent findings, beetle-to-beetle transmission of *H. diminuta* can be seen via the feces. Additionally, more infections occur due to this mechanism of egg dispersal.

Prevalence

H. diminuta infection in humans is rare, typically occurring in isolated cases. As such, several studies of *H. diminuta* exist as case reports describing a single affected individual.

In rural Devghar, India, a place heavily infested with rodents and cockroaches, *H. diminuta* eggs were found in a 12-year-old girl living in a small village.

In an urban area of Rome, a 2-year-old boy was also infected by *H. diminuta*. However, in this instance, investigators found no evidence of rodent or other possible sources of infection in the places habitually occupied by the affected boy.

In 1989, a child from St. James Parish, Jamaica was the subject of the first documented case of *H. diminuta* occurring in Jamaica, West Indies.

Influence on Host Behavior

In a behavioral study of the beetle *Tenebrio molitor* with cysticercoids of the rat tapeworm *H. diminuta*, findings suggested that the parasite impairs a beetle's ability to conceal itself. The study followed a rat and a beetle infected with the parasite. Infected beetles were slower than the control group; however, they still maintained the same learning level. In the initial phase of infection, the beetle was in high stress. As time progressed, this did not worsen their ability to learn. Overall, the training experiment portrayed that infected beetles were unable to hide from the rat, illustrating the high impact the parasite had on its host, the beetle.

Role in Human Diseases

H. diminuta is often asymptomatic. However, abdominal pain, irritability, itching, and eosinophilia are among the existing symptoms in a few of the reported cases.

H. diminuta has been cited as a possible candidate species for helminthic therapy, i.e. the controlled use of live organism parasites for the prevention and control of diseases of modern living. These diseases include allergies, asthma, autoimmune diseases and autism spectrum disorders.

Treatment

Since data regarding praziquantel treatment of *H. diminuta* is sparse, scientists have recommended that every case and treatment of *H. diminuta* be reported for development of protocols and parasitological purposes.

A 2-year-old Italian boy affected by tuberous sclerosis was infected by *H. diminuta*. Due to concerns over his neurological condition, the boy was treated with niclosamide rather than praziquantel. In this case, niclosamide treatment proved to be successful.

Taenia Saginata

Taenia saginata (synonym *Taeniarhynchus saginata*), commonly known as the beef tapeworm, is a zoonotic tapeworm belonging to the order Cyclophyllidea and genus *Taenia*. It is an intestinal parasite in humans causing taeniasis (a type of helminthiasis) and cysticercosis in cattle. Cattle are the intermediate hosts, where larval development occurs, while humans are definitive hosts harbouring the adult worms. It is found globally and most prevalently where cattle are raised and beef is consumed. It is relatively common in Africa, Europe, Southeast Asia, South Asia, and Latin America. Humans are generally infected as a result of eating raw or undercooked beef which contains the infective larvae, called cysticerci. As hermaphrodites, each body segment called proglottid has complete sets of both male and female reproductive systems. Thus, reproduction is by self-fertilisation. From humans, embryonated eggs, called oncospheres, are released with faeces and are transmitted to cattle through contaminated fodder. Oncospheres develop inside muscle, liver, and lungs of cattle into infective cysticerci.

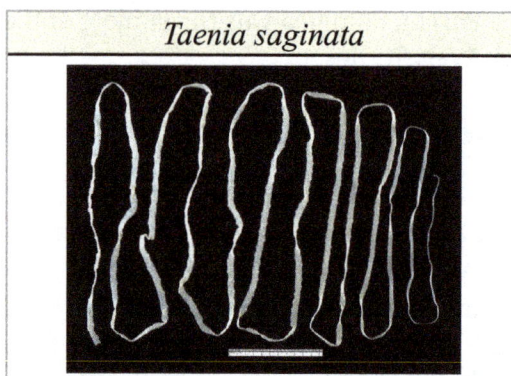

Taenia saginata

T. saginata has a strong resemblance to the other human tapeworms, such as *Taenia asiatica* and *Taenia solium*, in structure and biology, except for few details. It is typically larger and longer, with more proglottids, more testes, and higher branching of the uteri. It also lacks an armed scolex unlike other *Taenia*. Like the other tapeworms, it causes taeniasis inside the human intestine, but does not cause cysticercosis. Its infection is relatively harmless and clinically asymptomatic.

Description

T. saginata is the largest of species in the genus *Taenia*. An adult worm is normally 4 to 10 m in length, but can become very large; specimens over 22 m long are reported. Typical of cestodes, its body is flattened dorsoventrally and heavily segmented. It is entirely covered by a tegument. The body is white in colour and consists of three portions: scolex, neck, and strobila. The scolex has are four suckers, but they have no hooks. Lack of hooks and a rostellum is an identifying feature from other *Taenia* species. The rest of the body proper, the strobila, is basically a chain of numerous body segments called proglottids. The neck is the shortest part of the body, and consists of immature proglottids. The midstrobila is made of mature proglottids that eventually lead to the gravid proglottids, which are at the posterior end. An individual can have as many as 1000 to 2000 proglottids.

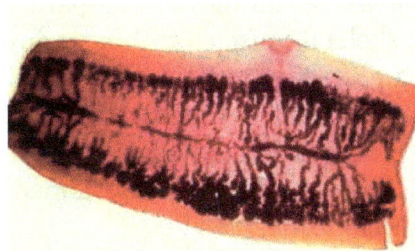

T. saginata proglottid stained to show uterine branches: The pore on the side identifies it as a cyclophyllid cestode.

T. saginata does not have a digestive system, mouth, anus, or digestive tract. It derives nutrients from the host through its tegument, as the tegument is completely covered with absorptive hair-like microtriches. It is also an acoelomate, having no body cavity. The inside of each mature proglottid is filled with muscular layers and complete male and female reproductive systems, including the tubular unbranched uterus, ovary, genital pore, testes, and vitelline gland. In the gravid proglottid, the uterus contains up to 15 side branches filled with eggs.

Life Cycle

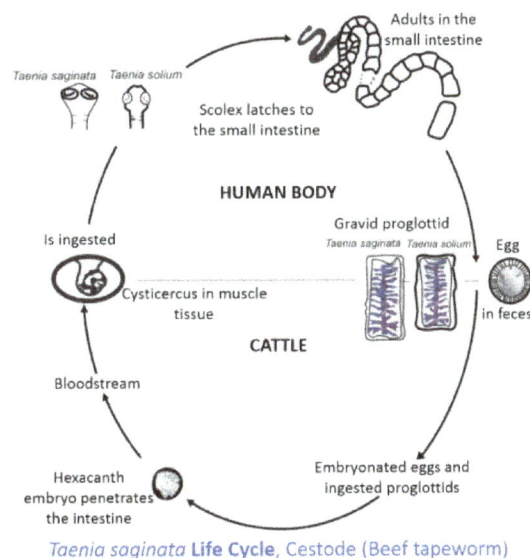

Taenia saginata **Life Cycle**, Cestode (Beef tapeworm)

Life cycle of Taenia Saginata inside and outside of the human body

The life cycle of *T. saginata* is indirect and digenetic, involving cattle and humans, with an interim of living in the environment. Humans as the definitive host harbour adult worms which release infective eggs into the environment through defecation. Cattle as the intermediate host pick up the viable eggs from contaminated vegetation.

Intermediate Host

Cattle acquire the embryonated eggs, the oncospheres, when they eat contaminated food. Oncospheres enter duodenum, the anterior portion of small intestine, and hatch there under the influence of gastric juices. The embryonic membranes are removed, liberating free hexacanth ("six-hooked") larvae. With their hooks, they attach to the intestinal wall and penetrate the intestinal mucosa into the blood vessels. The larvae can move to all parts of the body by the general circulatory system, and finally settle in skeletal muscles within 70 days. Inside the tissue, they cast off their hooks and instead develop a protective cuticular shell, called the cyst. Thus, they become fluid-filled cysticerci. Cysterci can also form in lungs and liver. The inner membrane of the cysticercus soon develops numerous protoscolices (small scolices) that are invertedly attached to the inner surface. The cysticercus of *T. saginata* is specifically named *cysticercus bovis* to differentiate from that of *T. solium, cysticercus cellulosae*.

Definitive Host

Humans contract infective cysticerci by eating raw or undercooked meat. Once reaching the jejunum, the inverted scolex becomes evaginated to the exterior under stimuli from the digestive enzymes of the host. Using the scolex, it attaches to the intestinal wall. The larva to mature into adults about 5 to 12 weeks later. Adult worms can live about 25 years in the host. Usually, only a single worm is present at time, but multiple worms are also reported. In each mature proglottid, self-fertilisation produces zygotes, which divide and differentiate into embryonated eggs called oncospheres. With thousands of oncospheres, the oldest gravid proglottids detach. Unlike in other *Taenia*, gravid proglottids are shed individually. In some cases, the proglottid ruptures inside the intestine, and the eggs are released. The free proglottids and liberated eggs are removed by peristalsis into the environment. On the ground, the proglottids are motile and shed eggs as they move. These oncospheres in an external environment can remain viable for several days to weeks in sewage, rivers, and pastures.

Epidemiology

The disease is relatively common in Africa, some parts of Eastern Europe, the Philippines, and Latin America. This parasite is found anywhere where beef is eaten, even in countries such as the United States, with strict federal sanitation policies. In the US, the incidence of infection is low, but 25% of cattle sold are still infected. The total global infection is estimated to be between 40 and 60 million. It is most prevalent in Sub-Saharan Africa and the Middle East.

Symptoms

T. saginata infection is usually asymptomatic, but heavy infection often results in weight loss, dizziness, abdominal pain, diarrhea, headaches, nausea, constipation, chronic indigestion, and loss of appetite. Intestinal obstruction in humans can be alleviated by surgery. The tapeworm can also expel antigens that can cause an allergic reaction in the individual. It is an also rare cause of ileus, pancreatitis, cholecystitis, and cholangitis.

Diagnosis

The basic diagnosis is done from a stool sample. Feces are examined to find parasite eggs. The eggs look like other eggs from the family Taeniidae, so it is only possible to identify the eggs to the family, not to the species level. Since it is difficult to diagnose using eggs alone, looking at the scolex or the gravid proglottids can help identify it as *Taenia saginata*. Proglottids sometimes trickle down the thighs of infected humans and are visible with unaided eye, so can aid with identification. Observation of scolex help distinguish between *T. saginata*, *T. solium* and *T. asiatica*. When the uterus is injected with India ink, its branches become visible. Counting the uterine branches enables some identification (*T. saginata* uteri have 12 or more branches on each side, while other species such as *T. solium* only have five to 10).

Differentiation of the species of *Taenia*, such as *T. solium* and *T. asiatica*, is notoriously difficult because of their close morphological resemblance, and their eggs are more or less identical. Identification often requires histological observation of the uterine branches and PCR detection of ribosomal 5.8S gene. The uuteri of *T. saginata* stem out from the center to form 12 to 20 branches, but in contrast to its closely related *Taenia* species, the branches are much less in number and comparatively thicker; in addition, the ovaries are bilobed and testes are twice as many.

Eosinophilia and elevated IgE levels are chief hematological findings. Also Ziehl–Neelsen stain can be used to differentiate between mature *T. saginata* and *T. solium*, in most cases *T. saginata* will stain while *T. solium* will not, but the method is not strictly reliable.

Treatment

Taenaisis is easily treated with praziquantel (5–10 mg/kg, single-administration) or niclosamide (adults and children over 6 years: 2 g, single-administration after a light breakfast, followed after 2 hours by a laxative; children aged 2–6 years: 1 g; children under 2 years: 500 mg). Albendazole is also highly effective for treatment of cattle infection.

Prevention

Adequate cooking (56°C for 5 minutes) of beef viscera destroys cysticerci. Refrigeration, freezing (-10°C for 9 days) or long periods of salting is lethal to cysticerci. Inspection of beef and proper disposal of human excreta are also important measures.

Taenia Solium

Taenia solium is the pork tapeworm belonging to cyclophyllid cestodes in the family Taeniidae. It is an intestinal zoonotic parasite found throughout the world, and is most prevalent in countries where pork is eaten. The adult worm is found in humans and has a flat, ribbon-like body, which is white in color and measures 2 to 3 m in length. Its distinct head, the scolex, contains suckers and a rostellum as organs of attachment. The main body, the strobila, consists of a chain of segments known as proglottids. Each proglottid is a complete reproductive unit; hence, the tapeworm is a hermaphrodite. It completes its life cycle in humans as the definitive host and pigs as intermediate host. It is transmitted to pigs through human faeces or contaminated fodder, and to humans through uncooked or undercooked pork. Pigs ingest embryonated eggs, morula, which develop into larvae, the oncospheres, and ultimately into infective larvae, cysticerci. A cysticercus grows into an adult worm in human small intestines. Infection is generally harmless and asymptomatic. However, accidental infection in humans by the larval stage causes cysticercosis. The most severe form is neurocysticercosis, which affects the brain and is a major cause of epilepsy.

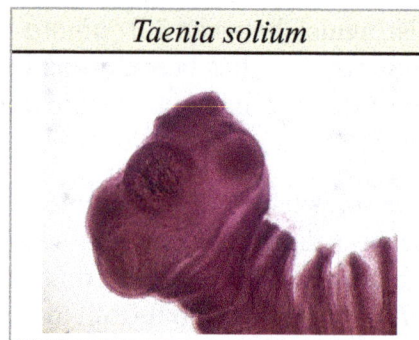

Taenia solium

Human infection is diagnosed by the parasite eggs in the faeces. For complicated cysticercosis, imaging techniques such as computed tomography and nuclear magnetic resonance are employed. Blood samples can also be tested using antibody reaction of enzyme-linked immunosorbent assay. Broad-spectrum anthelmintics such as praziquantel and albendazole are the most effective medications.

Description

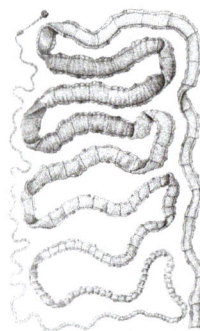

Taenia solium adult

Adult *T. solium* is a triploblastic acoelomate, having no body cavity. It is normally 2 to 3 m in length, but can become much larger, sometimes over 8 m long. It is white in colour and flattened into a ribbon-like body. The anterior end is a knob-like head called a scolex, which is 1 mm in diameter. The scolex bears four radially arranged suckers (acetabula) that surround the rostellum. These are the organs of attachment to the intestinal wall of the host. The rostellum is armed with two rows of spiny hooks, which are chitinous in nature. The 22 to 32 rotelllar hooks can be differentiated into short (130-μm) and long (180-μm) types. The elongated body is called the strobila, which is connected to the scolex through a short neck. The entire body is covered by a special covering called tegument, which is an absorptive layer consisting of a mat of minute hair-like microtriches. The strobila is divided into segments called proglottids, 800 to 900 in number. Body growth starts from the neck region, so the oldest proglottids are at the posterior end. Thus, the three distinct proglottids are immature proglottids towards the neck, mature proglottids in the middle, and gravid proglottids at the posterior end. A monoecious species, each mature proglottid contains a set of male and female reproductive systems. The numerous testes and a bilobed ovary open into a common genital pore. The oldest gravid proglottids are full of fertilised eggs,

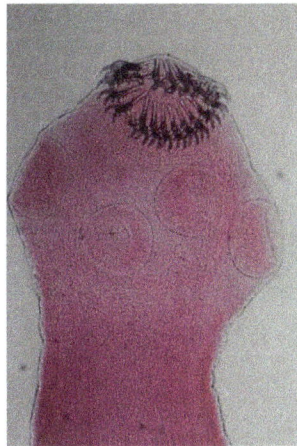

Taenia solium scolex (x400)

The infective larave, cysticerci, in humans, have three morphologically distinct types. The common one is the ordinary "cellulose" cysticercus, which has a fluid-filled bladder 0.5 to 1.5 cm in length and an invaginated scolex. The intermediate form has a scolex, while the "racemose" has no evident scolex, but is believed to be larger and much more dangerous. They are 20 cm in length and have 60 ml of fluid, and 13% of patients can have all three types in the brain.

Life Cycle

The life cycle of *T. solium* is indirect. It passes through pigs, as intermediate hosts, into humans, as definitive hosts. From humans, the eggs are released in the environment where they await ingestion by another host. Humans as the definitive hosts are directly infected from contaminated meat.

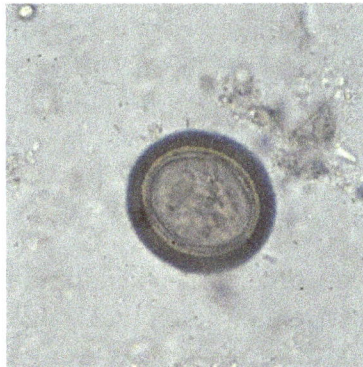

Egg of *T. solium*

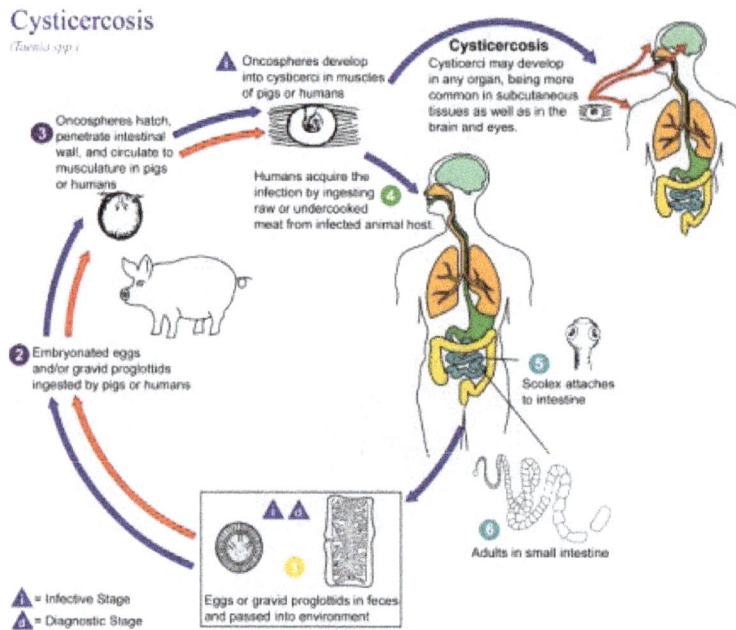

Lifecycle of *T. solium*

Definitive Host

Humans are infected by the larval stage, the cysticercus (*Cysticercus cellulosae*), from measly pork. A cysticercus is oval in shape, containing an inverted scolex (specifically "protoscolex"), which pops out externally once inside the small intestine. This process of evagination is stimulated by bile juice and digestive enzymes of the host. Using the scolex, it anchors to the intestinal wall. It grows in size using nutrients from the surroundings. Its strobila lengthens as new proglottids are formed at the neck. In 10–12 weeks after initial infection, it becomes an adult worm. As a hermaphrodite, it reproduces by self-fertilisation, or cross-fertilisation if gametes are exchanged between two different proglottids. Spermatozoa fuse with the ova in the fertilisation duct, where the zygotes are produced. The zygote undergoes holoblastic and unequal cleavage resulting in three cell types, small micromeres, medium mesomeres, and large megameres. Megameres develop into syncytial layer called outer embryonic membrane. Mesomeres develop into radially striated inner embryonic membrane or embryophore. Micromeres become the morula. The morula transforms into a six-hooked embryo known as oncosphere, or sometimes hexacanth ("six hooked") larva. A single gravid proglottid can contain more than 50,000 embryonated eggs. Gravid proglottids often rupture in the intestine, liberating the eggs in faeces. The intact gravid proglottids are shed off in groups of four or five. The free eggs and detached proglottids are released into the environment through peristalsis. Eggs can survive in the environment for up to two months.

Intermediate Host

Pigs ingest the eggs from human faeces or vegetation contaminated with human excreta. The embryonated eggs enter the intestine where they hatch into motile oncospheres. The embryonic and basement membranes are removed by the host's digestive enzymes (particularly pepsin). Then the free oncospheres get attached on the intestinal wall using their hooks. With the help of digestive enzymes from the penetration glands, they penetrate the intestinal mucosa to enter blood and lymphatic vessels. They move along the general circulatory system to various organs, and large numbers are cleared in the liver. The surviving oncospheres preferentially migrate to striated muscles, as well as the brain, liver, and other tissues, where they settle to form cysts called cysticerci. A single cysticercus is spherical, measuring 1–2 cm in diameter, and contains an invaginated protoscolex. The central space is filled with fluid like a bladder, hence it is also called bladder worm. Cysticerci are usually formed within 70 days and may continue to grow for a year.

Humans are also accidental secondary hosts when they are infected by embryonated eggs, either by autoinfection or ingestion of contaminated food. As in pigs, the oncospheres hatch, enter blood circulation, and have a predilection for brain tissue and other soft muscle tissues. When they settle to form cysts, clinical symptoms of cysticercosis appear. The cysticercus is often called the metacestode. If they localize in the brain, serious neurocysticercosis follows.

Pathogenesis

Intestinal infection of *T. solium* is called taeniasis which is quite asymptomatic. Only in severe cases, conditions of intestinal irritation, anaemia, and indigestion occur, which can lead to loss of appetite and emaciation. Cysticercus is clinically pathogenic. Ingestion of *T. solium* eggs or proglottids which rupture within the host intestines can cause larvae to migrate into host tissue to cause cysticercosis. This is the most frequent and severe disease caused by *T. solium*. In symptomatic cases, a wide spectrum of symptoms may be expressed, including headaches, dizziness, and occasional seizures. In more severe cases, dementia or hypertension can occur due to perturbation of the normal circulation of cerebrospinal fluid. (Any increase in intracranial pressure will result in a corresponding increase in arterial blood pressure, as the body seeks to maintain circulation to the brain.) The severity of cysticercosis depends on location, size and number of parasite larvae in tissues, as well as the host immune response. Other symptoms include sensory deficits, involuntary movements, and brain system dysfunction. In children, ocular location of cysts is more common than cystation in other locations of the body.

In many cases, cysticercosis in the brain can lead to epilepsy, seizures, lesions in the brain, blindness, tumor-like growths, and low eosinophil levels. It is the cause of major neurological problems, such as hydrocephalus, paraplegy, meningitis, convulsions, and even death.

Prevention and Control

The best way to avoid getting tapeworms is to not eat undercooked pork. Moreover, a high level of sanitation and prevention of faecal contamination of pig feeds also plays a major role in prevention. Infection can be prevented with proper disposal of human faeces around pigs, cooking meat thoroughly and/or freezing the meat at −10 °C for 5 days. For human cysticercosis, dirty hands are attributed to be the primary cause, and especially common among food handlers. Therefore, personal hygiene such as washing one's hands before eating is an effective measure.

Epidemiology

T. solium is found worldwide, but is more common in cosmopolitan areas. Because pigs are intermediate hosts of the parasite, completion of the life cycle occurs in regions where humans live in close contact with pigs and eat undercooked pork. Therefore, high prevalences are reported in Mexico, Latin America, West Africa, Russia, India, Pakistan, Manchuria, and Southeast Asia. In Europe it is most widespread among Slavic countries. Cysticercosis is often seen in areas where poor hygiene allows for contamination of food, soil, or water supplies. Prevalence rates in the United States have shown immigrants from Mexico, Central and South America, and Southeast Asia account for most of the domestic cases of cysticercosis. Taeniasis and cysticercosis are very rare in predominantly Muslim countries, as Islam forbids the consumption of pork. Human cysticercosis is acquired by ingesting *T. solium* eggs shed in the feces of a human tapeworm carrier by gravid proglottids, so can occur in populations that neither eat pork nor share environments with pigs, although the completion of the life cycle can occur only where humans live in close contact with pigs and eat pork.

In 1990 and 1991, four unrelated members of an Orthodox Jewish community in New York City developed recurrent seizures and brain lesions, which were found to have been caused by *T. solium*. All of the families had housekeepers from Latin American countries and were suspected to be source of the infections.

Spirometra Erinaceieuropaei

Spirometra erinaceieuropaei is a tapeworm that infects domestic animals and humans. In humans infection is called sparganosis. *Spirometra erinaceieuropaei's* distribution is cosmopolitan, meaning that it can be found nearly anywhere the parasite can complete its life cycle. This species is closely related to *Spirometra mansonoides*, and few morphological differences exist between the two. One difference is that the uterus of *S. mansonoides* is a "U" shape, but in *S. erinaceieuropaei* the uterus consists of two sections that resemble horns. The life cycle of both species is very similar.

In 2014 a British man was found to have been infected by the tapeworm from an unknown cause (possibly a traditional frog meat poultice) while in China. The parasitic worm was recorded on successive MRI scans of his brain, moving location by about 5 cm before doctors realized it was alive. The 50-year-old first visited doctors in 2008 suffering from headaches, seizures, memory loss, and complaining that his sense of smell had changed. The 1-cm ribbon-shaped larval worm was removed during a surgical procedure and the man recovered.

Genomics

The genome of *S. erinaceieuropaei* recovered from the patient's brain was sequenced in 2014 and is available through the WormBase ParaSite website.

Life Cycle

The worm has an interesting lifecycle. The adult worm lives in the small intestine of cats and dogs, where it may grow as long as 1.5 meters. Eggs from the worm are passed with the host feces, when they develop into a procercoid larva. This larva may be directly ingested by humans or may enter an intermediate host which include frogs, birds, snakes, rats and mice and become a plerocercoid larva. When cats, dogs, foxes or wolves eat the intermediate host, the worm completes its life cycle becoming an egg producing adult. Because humans would normally ingest the worm at the procercoid stage and are not usually eaten by cats and dogs, the human is a dead-end host.

Pathology

Although humans can get infected with this parasite, it should be understood that they cannot contract it from an infected cat or dog. People can't get infected by ingesting the eggs, which is what the pet would be shedding. They would have to eat the procercoid stage, which is found in the intermediate hosts. If the meat of an intermediate host, such as chicken, is undercooked and it happens to be contaminated by the parasite, the person can get infected.

Diagnoses and Treatment

An easy way to determine if an animal is infected with any type of tapeworm is seeing the proglottids in the feces. These are the white segments that break off from the parasite. To determine the type of species, a fecal sample under the microscope to see the eggs would be the best way. The eggs of any *Spirometra* species are oval shaped with a distinct operculum at one pole.The treatment for *Spirometra erinaceieuropaei* is the drug praziquantel, which is typical for tapeworm infections.

References

- Cheng TC (1986). General Parasitology (2nd edn). Academic Press, Division of Hardcourt Brace & Company, USA, pp. 402-416. ISBN 0-12-170755-5

- Scholar, Eric M.; Pratt, William B. (2000). "Treatment of Parasitic Infection". The Antimicrobial Drugs. Oxford University Press. pp. 465–466. ISBN 9780199759712.

- Eckert J. (2005). "Helminths". In Kayser, F.H.; Bienz, K.A.; Eckert, J.; Zinkernagel, R.M. Medical Microbiology. Stuttgart: Thieme. pp. 560–562. ISBN 9781588902450.

- Somers, Kenneth D.; Morse, Stephen A. (2010). Lange Microbiology and Infectious Diseases Flash Cards (2nd ed.). New York: Lange Medical Books/ McGraw-Hill. pp. 184–186. ISBN 9780071628792.

- Pawlowski, Z.S.; Prabhakar, Sudesh (2002). "Taenia solium: basic biology and transmission". In Gagandeep Singh, Sudesh Prabhakar. Taenia solium Cysticercosis from Basic to Clinical Science. Wallingford, Oxon, UK: CABI Pub. pp. 1–14. ISBN 9780851998398.

- Gutierrez, Yezid (2000). Diagnostic Pathology of Parasitic Infections with Clinical Correlations (2nd ed.). New York [u.a.]: Oxford University Press. pp. 635–652. ISBN 9780195121438.

- Mayta, Holger (2009). Cloning and Characterization of Two Novel Taenia Solium Antigenic Proteins and Applicability to the Diagnosis and Control of Taeniasis/cysticercosis. ProQuest. pp. 4–12. ISBN 9780549938996.

- Dworkin, Mark S. (2010). Outbreak Investigations Around the World: Case Studies in Infectious Disease. Jones and Bartlett Publishers. pp. 192–196. ISBN 978-0-7637-5143-2. Retrieved August 9, 2011.

- Tortora, Gerard J.; Funke, Berdell R; Case, Christine L. (2016, 2013, 2010). Microbiology: An Introduction 12th Edition. Benjamin-Cummings Publishing Company, Subs of Addison Wesley Longman, Inc. p. 347. ISBN 9780321929150.

- Reeder, P.E.S. Palmer, M.M. (2001). Imaging of Tropical Diseases : With Epidemiological, Pathological, and Clinical Correlation (2 (revised) ed.). Heidelberg, Germany: Springer-Verlag. pp. 641–642. ISBN 978-3-540-56028-9.

Flukes and Its Classification

Flukes have numerous classifications; some of these are liver flukes, paragonimus and schistosoma. Adult flukes that are found in the liver of mammals are known as liver flukes. They feed on blood and also reproduce into the intestine. This section helps the reader in understanding flukes and their characteristics.

Liver Fluke

Liver flukes are parasites. They are a polyphyletic group of trematodes (phylum Platyhelminthes). Adults of liver flukes are localized in the liver of various mammals, including humans. These flatworms can occur in bile ducts, gallbladder, and liver parenchyma. They feed on blood. Adult flukes produce eggs which are passed into the intestine. It depends on two intermediate hosts (a snail and a fish) to complete its life cycle.

Fasciola hepatica Egg of Dicrocoelium sp.

Examples include:

- *Clonorchis sinensis* (the "Chinese liver fluke" or the "Oriental liver fluke")
- *Dicrocoelium dendriticum* (lancet liver fluke)
- *Dicrocoelium hospes*
- *Fasciola hepatica* (the "sheep liver fluke")
- *Fascioloides magna* (the "giant liver fluke")
- *Fasciola gigantica*
- *Fasciola jacksoni*

- *Metorchis conjunctus*

- *Metorchis albidus*

- *Protofasciola robusta*

- *Parafasciolopsis fasciomorphae*

- *Opisthorchis viverrini* (Southeast Asian liver fluke)

- *Opisthorchis felineus* (cat liver fluke).

- *Opisthorchis guayaquilensis*

Clonorchis Sinensis

Clonorchis sinensis, the Chinese liver fluke, is a human liver fluke in the class Trematoda, phylum Platyhelminthes. This parasite lives in the liver of humans, and is found mainly in the common bile duct and gall bladder, feeding on bile. These animals, which are believed to be the third most prevalent worm parasite in the world, are endemic to Japan, China, Taiwan, and Southeast Asia, currently infecting an estimated 30,000,000 humans. 85% of cases are found in China.

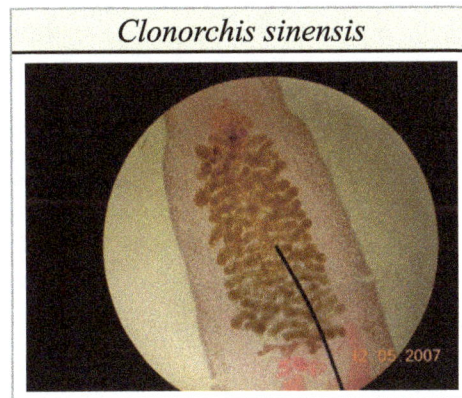

Clonorchis sinensis

It is the most prevalent human trematode in Asia, and still actively transmitted in Korea, China, Vietnam and also Russia, with 200 million people at constant risk. Recent studies have proved that it is capable of causing cancer of liver and bile duct, and in fact the International Agency for Research on Cancer has classed it as a group 1 biological carcinogen in 2009.

Life Cycle

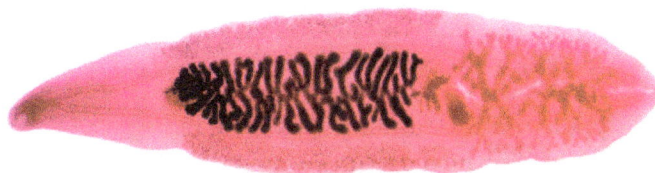

An adult *Clonorchis sinensis* has these main body parts: oral sucker, pharynx, caecum, ventral sucker, vitellaria, uterus, ovary, Mehlis' gland, testes, excretory bladder. (H&E stain)

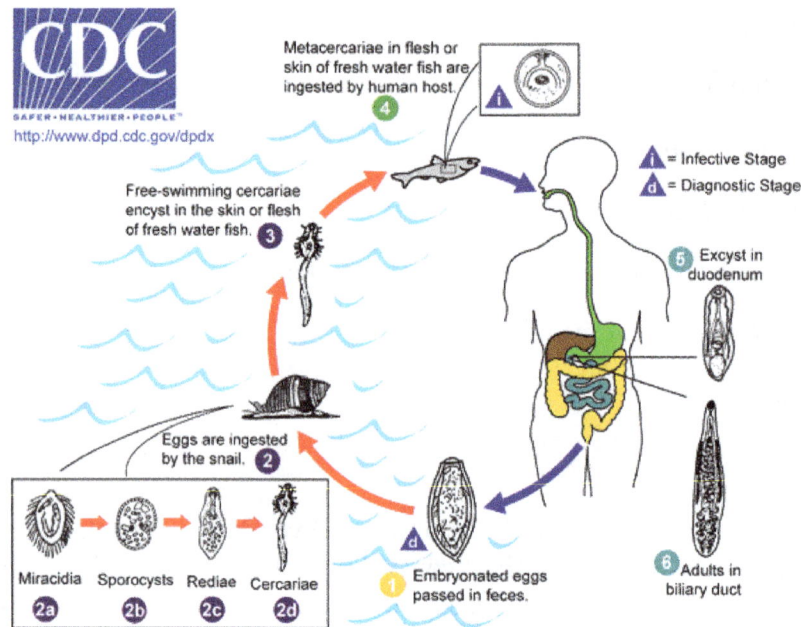

Life cycle of *Clonorchis sinensis*

The eggs of a *C. sinensis*, which contain the miracidium that develops into the adult form, float in fresh water until eaten by a snail.

First Intermediate Host

Freshwater snail *Parafossarulus manchouricus* - synonym: *Parafossarulus striatulus*, often serves as a first intermediate host for *C. sinensis* in China, Japan, Korea and Russia.

Other snail hosts include:

- *Bithynia longicornis* - synonym: *Alocinma longicornis* - in China

- *Bithynia fuchsiana* - in China

- *Bithynia misella* - in China

- *Parafossarulus anomalospiralis* - in China

- *Melanoides tuberculata* - in China

- *Semisulcospira libertina* - in China

- *Assiminea lutea* - in China

- *Tarebia granifera* - in Taiwan

Once inside of the snail body, the miracidium hatches from the egg, and parasitically grows inside of the snail. The miracidium develops into a sporocyst, which in turn house the asexual reproduction of redia, the next stage. The redia themselves house the asexual reproduction of free-swimming cercaria. This system of asexual reproduction allows for an exponential multiplication of cercaria individuals from one miracidium. This aids the *Clonorchis* in reproduction, because it enables the miracidium to capitalize on one chance occasion of passively being eaten by a snail before the egg dies.

Once the redia mature, having grown inside the snail body until this point, they actively bore out of the snail body into the freshwater environment.

Second Intermediate Host

There, instead of waiting to be consumed by a host (as is the case in their egg stage), they seek out a fish. Boring their way into the fish's body, they again become parasites of their new hosts.

Once inside of the fish muscle, the cercaria create a protective metacercarial cyst with which to encapsulate their bodies. This protective cyst proves useful when the fish muscle is consumed by a human.

The second intermediate hosts are freshwater fish: Common carp (Cyprinus carpio), grass carp (Cteno-pharyngodon idellus), crucian carp (Carassius carassius), goldfish (Carassius auratus), *Pseudorasbora parva, Abbottina rivularis, Hemiculter* spp., *Opsariichthys* spp., *Rhodeus* spp., *Sarcocheilichthys* spp., *Zacco platypus, Nipponocypris temminckii* , and pond smelt (Hypomesus olidus).

Definitive Host

The acid-resistant cyst enables the metacercaria to avoid being digested by the human gastric acids, and allows the metacercaria to reach the small intestine unharmed. Reaching the small intestines, the metacercaria navigate toward the human liver, which becomes its final habitat. *Clonorchis* feed on human blood. In the human liver, the mature *Clonorchis* reaches its stage of sexual reproduction, in which it produces eggs every 1–30 seconds.

The definitive hosts are fish-eating mammals such as dogs, cats, rats, pigs, badgers, weasels, camels, and buffaloes.

Effects on Human Health

Dwelling in the bile ducts, *Clonorchis* induces an inflammatory reaction, epithelial hyperplasia and sometimes even cholangiocarcinoma, the incidence of which is raised in fluke-infested areas.

One adverse effect of *Clonorchis* is the possibility for the adult metacercaria to consume all bile created in the liver, which would inhibit the host human from digesting, especially fats. Another possibility is obstruction of the bile duct by the parasite or its eggs, leading to biliary obstruction and cholangitis (specifically oriental cholangitis).

In a report by *Dr. John Chiao-nan Chang, M.D. and Dr. Yin-Ping Wang, M.D.* entitled, *Central Serous Retinopathy* (CSR), 80 cases were studied in Hong Kong. On page 125 of their report, it was observed that 19% of the cases of CSR in their sample tested positive for *Clonorchis sinensis*.

Symptoms

While normally asymptomatic most pathological manifestations result from inflammation and intermittent obstruction of the biliary ducts. The acute phase consists of abdominal pain with associated nausea and diarrhea. Long-standing infections consist of fatigue, abdominal discomfort, anorexia, weight loss, diarrhea, and jaundice. The pathology of long-standing infections consist of bile stasis, obstruction, bacterial infections, inflammation, periductal fibrosis, and hyperplasia. Development of cholangiocarcinoma is progressive.

Diagnosis and Treatment

Infection is detected mainly on identification of eggs by microscopic demonstration in faeces or in duodenal aspirate. But other sophisticated methods have been developed such as ELISA, which has become the most important clinical technique. Diagnosis by detecting DNAs from eggs in faeces are also developed using PCR, real-time PCR, and LAMP, which are highly sensitive and specific. Imaging diagnosis has been studied in depth and is now widely used. Drugs used to treat infestation include triclabendazole, praziquantel, bithionol, albendazole, levamisole and mebendazole. However, benzimidazoles are very weak as vermicide. As with other trematodes, praziquantel is the drug of choice. Lately, tribendimidine has been acknowledged as an effective and safe drug.

Fasciola Hepatica

Fasciola hepatica, also known as the common liver fluke or sheep liver fluke, is a parasitic trematode (fluke or flatworm, a type of helminth) of the class Trematoda, phylum Platyhelminthes. It infects the livers of various mammals, including humans. The disease caused by the fluke is called fasciolosis or fascioliasis, which is a type of helminthiasis and has been classified as a neglected tropical disease. Fasciolosis is currently classified as a plant/food-borne trematode infection, often acquired through eating the parasite metacercariae encysted on plants. *F. hepatica* which is distributed worldwide has been known as an important parasite of sheep and cattle for many years and causes great economic losses to these livestock species, up to £23 million in the UK alone. Because of its size and economic importance, it has been the subject of many scientific investigations and may be the best-known of any trematode species. *F. hepatica's* closest relative is *Fasciola gigantica*. These two flukes are sister species; they share many morphological features and can mate with each other.

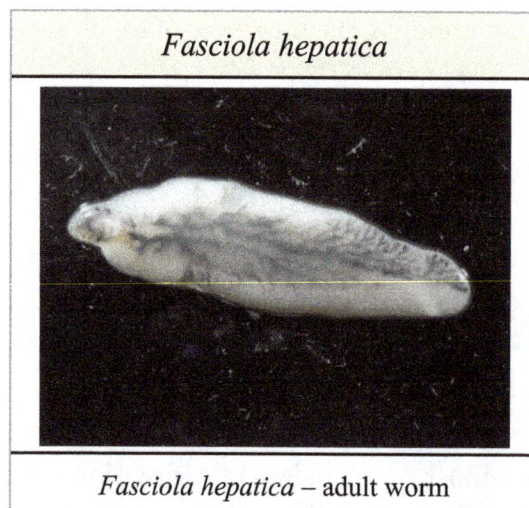

Fasciola hepatica

Fasciola hepatica – adult worm

Life Cycle

Galba truncatula, an amphibious freshwater lymnaeid snail that serves as the main intermediate host of *Fasciola hepatica* in Europe

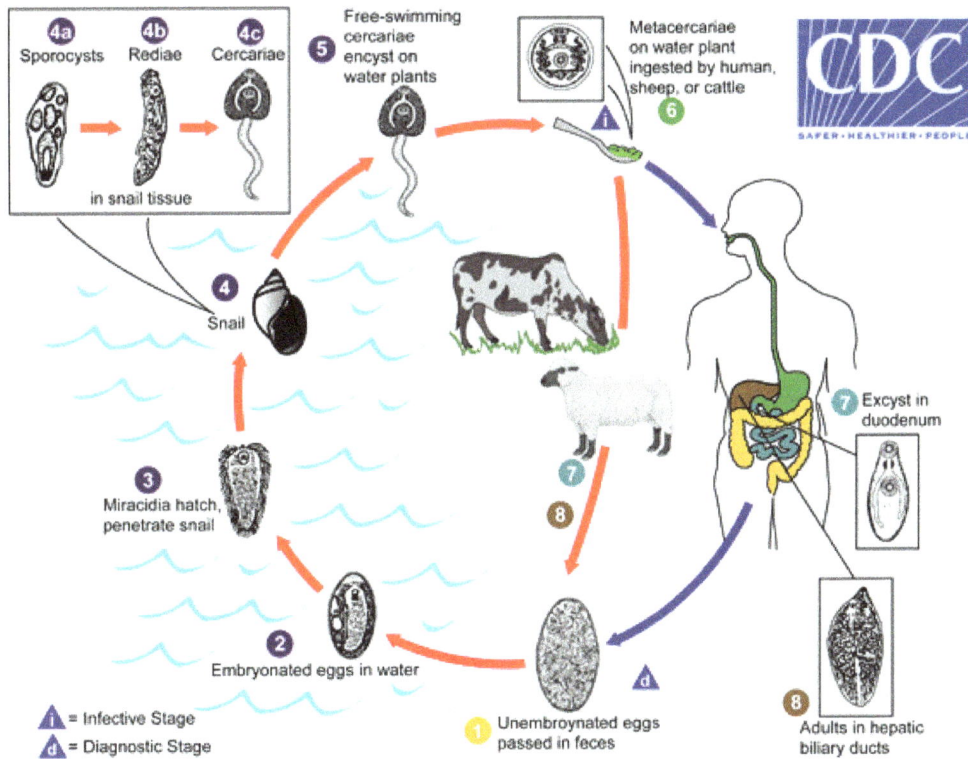

The life cycle of *Fasciola hepatica*

Fasciola hepatica occurs in liver of definitive host and its life cycle is indirect. Definitive hosts of the fluke are cattle, sheep and buffaloes. Wild ruminants and other mammals, including humans, can act as definitive hosts as well. The life cycle of *F. hepatica* goes through the intermediate host and several environmental larval stages. Intermediate hosts of *F. hepatica* are air-breathing freshwater snails from the family Lymnaeidae. Although several lymnaeid species susceptible to *F. hepatica* have been described, the parasite develops only in one or two major species on each continent. *Galba truncatula* is the main snail host in Europe, partly in Asia, Africa and South America. *Lymnaea viator, L. neotropica, Pseudosuccinea columella* and *L. cubensis* are most common intermediate hosts in Central and South America. Several other lymnaeid snails may be naturally or experimentally infected with *F. hepatica* but their role in transmission of the fluke is low. The list of lymnaeid snails that may serve as natural or experimental intermediate hosts of *F. hepatica* include:

- *Austropeplea ollula*
- *Austropeplea tomentosa*
- *Austropeplea viridis*
- *Fossaria bulimoides*
- *Galba truncatula*
- *Lymnaea cousini*
- *Lymnaea cubensis*
- *Lymnaea diaphana*
- *Lymnaea humilis*
- *Lymnaea neotropica*
- *Lymnaea occulta*
- *Lymnaea stagnalis*
- *Lymnaea viatrix*
- *Omphiscola glabra*
- *Pseudosuccinea columella*
- *Radix auricularia*
- *Radix lagotis*
- *Radix natalensis*
- *Radix peregra*
- *Radix rubiginosa*
- *Stagnicola caperata*
- *Stagnicola fuscus*
- *Stagnicola palustris*
- *Stagnicola turricula*

The metacercariae are released from the freshwater snail as cercariae, and form cysts on various surfaces including aquatic vegetation. The mammalian host then eats this vegetation and can become infected. Humans can often acquire these infections through eating freshwater plants such as watercress. Inside the duodenum of the mammalian host, the metacercariae are released from within their cysts. From the duodenum, they burrow through the lining of the intestine and into the peritoneal cavity. They then migrate through the intestines and liver, and into the bile ducts. Inside the bile ducts, they develop into an adult fluke. In humans, the time taken for *F. hepatica* to mature from metacercariae into an adult fluke is roughly 3 to 4 months. The adult flukes can then produce up to 25,000 eggs per fluke per day. These eggs are passed out via stools and into freshwater. Once in freshwater, the eggs become embryonated, allowing them to hatch as miracidia, which then find a suitable intermediate snail host of the Lymnaeidae family. Inside this snail, the miracidia develop into sporocysts, then to rediae, then to cercariae. The cercariae are released from the snail to form metacercariae and the life cycle begins again.

Morphology and Anatomy

Fasciola hepatica is one of the largest flukes of the world, reaching a length of 30 mm and a width of 13 mm (*Fasciola gigantica*, on the other hand, is even bigger and can reach up to 75 mm). It is leaf-shaped, pointed at the back (posteriorly) and wide in the front (anteriorly). The oral sucker is small but powerful and is located at the end of a cone-shape projection at the anterior end. The acetabulum is a larger sucker than the oral sucker and is located at the anterior end.

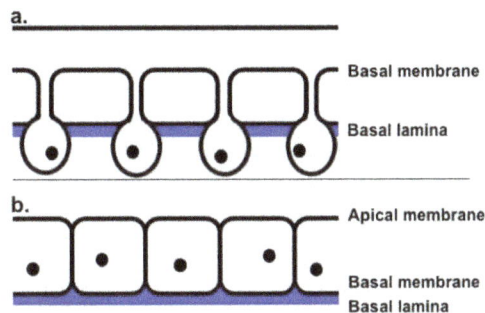

A simple diagram to show the difference between the teguments of free-living and parasitic flatworms: a. shows the syncytial epithelial tegument found in parasitic flatworms, such as *F. hepatica*. b. shows the multicellular, non-syncytial, epithelia, found in non-parasitic, free-living flatworms.

Tegument

The outer surface of the fluke is called the tegument. This is composed of scleroprotein and its primary function is to protect the fluke from the destructive digestive system of the host. Its also used for renewal of the surface plasma membrane and the active uptake of nutrients. On the surface of the tegument there are also small spines. Initially, these spines are single pointed, then, just prior to the fluke entering the bile ducts, they become multipointed. At the anterior end of the fluke the spines have between 10 and 15 points, whereas at the posterior end, they have up to 30 points. The tegument is a syncytial epithelium. This means it is made from the fusion of many cells, each containing one nucleus, to produce a multinucleated cell membrane. In the case of *F. hepatica*, there are no nuclei in the outer cytoplasm between the basal and apical membranes. Thus, this region is referred to as anucleate. Instead, the nuclei are found in the cell bodies, also known as tegumental cells, these connect to the outer cytoplasm via thin cytoplasmic strands. The tegumental cells contain the usual cytoplasmic organelles (mitochondria, Golgi bodies and endoplasmic reticulum).

The tegument plays a key role in the fluke's infection of the host. Studies have shown that certain parts of the tegument (in this case, the antigen named Teg) can actually suppress the immune response of the mammalian host. This means that the fluke is able to weaken the immune response, and increase its chances of a successful infection. A successful infection is needed in order for the fluke to have enough time to develop into an adult and continue its lifecycle.

Digestive System

The alimentary canal of *F. hepatica* has a single mouth which leads into the blind gut; it has no anus. The mouth is located within the anterior sucker on the ventral side of the fluke. This mouth leads to the pharynx, which is then followed by a narrow oesophagus. The oesophagus, which is lined with a thin layer of epithelial cells, then opens up into the large intestine. As there is no anus, the intestine branches, with each branch ending blindly near the posterior end of the body. It has been shown that flukes migrate into smaller capillaries and bile ducts when feeding within the host. They use their mouth suckers to pull off and suck up food, bile, lymph and tissue pieces from the walls of the bile ducts. *F. hepatica* relies on extracellular digestion which occurs within the intestine of the host. The waste materials are egested through the mouth. The non-waste matter is adsorbed back in through the tegument and the general surface of the fluke. The tegument facilitates this adsorption by containing many small folds to increase the surface area.

Respiratory System

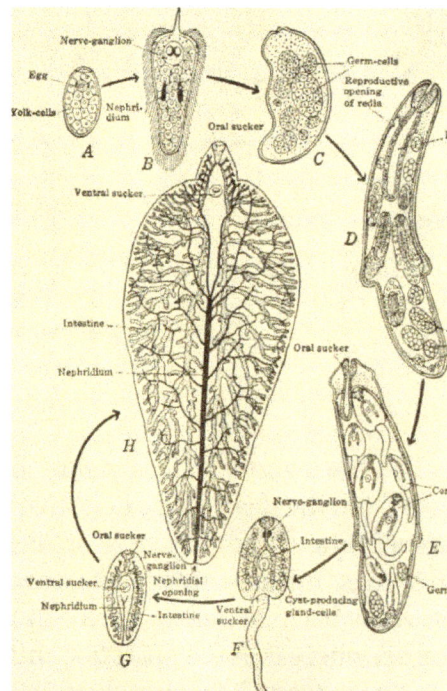

Diagram of the main organ systems of *Fasciola hepatica* throughout the progressive life stages of the fluke (1938). *A* - Egg; *B* - Miracidium; *C* - Sporocyst; *D* - Rediae, *E* - Immature cercaria, *F* - Cercaria, *G* - Encysted stage, *H* - Adult fluke (nervous and reproductive systems omitted)

F. hepatica has no respiratory organs: the adult flukes respire anaerobically (without oxygen). Glycogen, taken from within the host is broken down via glycolysis to produce carbon dioxide and fatty acids. This process provides the fluke with energy. In contrast, the free-living miracidia stages of the parasite generally develop within oxygen rich environments. It is therefore believed that the free-living stages of the parasite respire aerobically, to gain the most energy from their environment.

Excretory System

F. hepatica's excretory system contains a network of tubules surrounding one main excretory canal. This canal leads to the excretory pore at the posterior end of the fluke. This main canal branches into four sections within the dorsal and ventral regions of the body. The role of *F. hepatica's* excretory system is excretion and osmoregulation. Each tubule within the excretory system is connected to a flame cell, otherwise known as protonephridia. These cells are modified parenchyme cells. In *F. hepatica* their role is to perform excretory, but more importantly, osmoregulatory functions. Flame cells are therefore primarily used to remove excess water.

Nervous System and Sensory Organs

The nerve system of *Fasciola hepatica* consists of a pair of nerve ganglia, each one is located on either side of the oesophagus. Around the oesophagus is a nerve ring. This nerve ring connects the two nerve ganglia together. The nerves stem off from this ring, reaching all the way down to the posterior end of the body. At the posterior end, one pair of nerves become thicker than the others, these are known as the lateral nerve cords. From these lateral nerve cords, the other nerves branch. Sensory organs are absent from *F. hepatica*.

Reproductive System

F. hepatica adult flukes are hermaphrodite, this means each fluke contains both male and female reproductive organs. The male and female reproductive organs open up into the same chamber within the body, which is called the genital atrium. The genital atrium is an ectodermal sac which opens up to the outside of the fluke via a genital pore. The testes are formed of two branched tubules, these are located in the middle and posterior regions of the body. From the epithelium lining of the tubules sperm is produced. The sperm then passes into the vas deferens and then into the seminal vesicle. From the seminal vesicle projects the ejaculatory duct and this is what opens up into the genital atrium, many prostate glands surround this opening. On the right hand side of the anterior testis there is a branched, tubular ovary. From here, a short oviduct passes to the vitelline duct. This duct connects, via a junction, the ovaries, the uterus and the yolk reservoir. From this junction, the uterus opens into the genital atrium, this opening is surrounded by Mehlis glands. In some flukes, the terminal end of the uterus is strengthened with muscles and spines.

F. hepatica reproduces both sexually, via the hermaphrodite adult flukes, and also asexually. The miracidia can reproduce asexually within the intermediate snail host.

Prevalence

Currently, *F. hepatica* has one of the widest geographical spread of any parasitic and vector-borne disease. Originating in Europe, it has expanded to colonize over 50 countries, covering all continents except Antarctica. In contrast, *F. gigantica* is generally considered more geographically constricted to the tropical regions of Africa, Asia and the Middle East, there is some overlap between the two species.

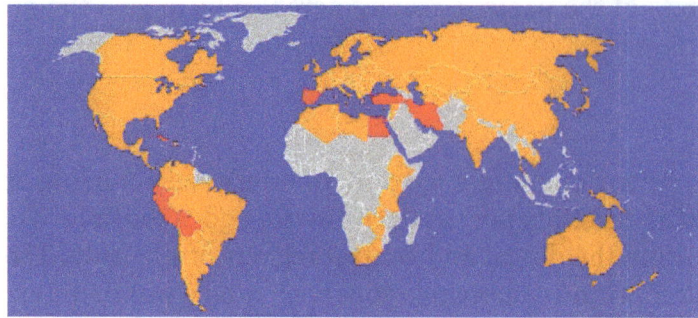

Fasciola hepatica prevalence. The countries in red are those with high prevalence, those in orange have low-medium prevalence.

Climate affects both *F. hepatica* itself and its definitive host, the snail. For example, the development of *F. hepatica* miracidia and larvae, and the reproduction of *Galba truncatula*, require a temperature range of 10-25 °C. In addition to this, they both require high levels of moisture in the air, as both are at risk of desiccation. Due to this, the prevalence, along with the intensity of infection, of *F. hepatica* is primarily dependent on rainfall levels and temperature.

Parasitic Adaptations

Fasciola hepatica's tegument protects it from the enzymes of the host's stomach, whilst still allowing water to pass through. Free-swimming larvae have cilia and the cercariae have a flagella-like tail to help them swim through the aquatic environment and also allow them to reach the plants on which they form a cyst. To attach within the host, *F. hepatica* has oral suckers and body spines. Their pharynx also helps it to suck onto the tissues within the body, particularly within the bile ducts. The adult fluke's respiration is anaerobic, this is ideal as there is no oxygen available in the liver. *F. hepatica* is adapted to produce a large number of eggs, this increases its chances of survival, as many eggs are destroyed on release into the environment. Also, *F. hepatica* is hermaphrodite, thus all flukes can produce eggs, increasing the number of offspring produced by the population.

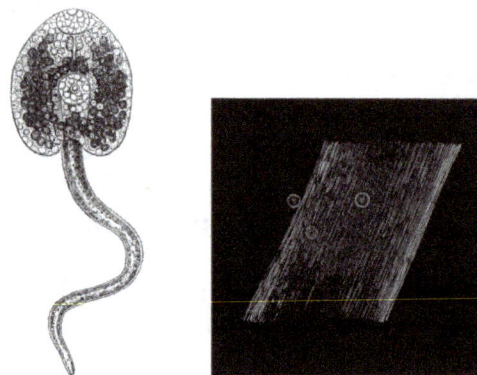

The left image shows the free-swimming cercariae, the flagella is clearly visible. The right-hand-side of the diagram shows the cysts attached to grass

The genome for *Fasciola hepatica* was published in 2015. *F. hepatica's* genome, at 1.3 Gb, is one of the largest known pathogen genomes. The genome contains many polymorphisms, and this represents the potential for the fluke to evolve and rapidly adapt to changes in the environment, such as host availability and drug or vaccine interventions.

Epidemiology

Infection begins when cyst-covered aquatic vegetation is eaten or when water containing metacercariae is drunk. In the United Kingdom, *F. hepatica* frequently causes disease in ruminants, most commonly between March and December.

Humans become infected by eating watercress or by drinking 'Emoliente', a Peruvian drink that uses drops of watercress juice. Cattle and sheep are infected when they consume the infectious stage of the parasite from low-lying, marshy pasture.

Human infections have been reported from more than 75 countries around the world. In Asia and Africa, people are infected both by *F. hepatica* and *F. gigantica* whereas human fasciolosis is caused only by *F. hepatica* in South and Central America and Europe.

The presence of *F. hepatica* can interfere with the detection of bovine tuberculosis in cattle. Cattle co-infected with *F. hepatica*, compared to those infected with *M. bovis* alone, react weakly to the single intradermal comparative cervical tuberculin (SICCT) test. Therefore, an infection from *F. hepatica* can make it difficult to detect bovine tuberculosis, this is, of course, a major problem in the farming industry.

Fasciolosis

Both *F. hepatica* and *F. gigantica* can cause fasciolosis. Human symptoms vary depending on if the disease is chronic or acute. During the acute phase, the immature worms begin penetrating the gut, causing symptoms of fever, nausea, swollen liver, skin rashes and extreme abdominal pain. The chronic phase occurs when the worms mature in the bile duct, and can cause symptoms of intermittent pain, jaundice and anemia. In cattle and sheep, classic signs of fasciolosis include persistent diarrhea, chronic weight loss, anemia and reduced milk production. Some remain asymptomatic. *F. hepatica* can cause sudden death in both sheep and cattle, due to internal hemorrhaging and liver damage.

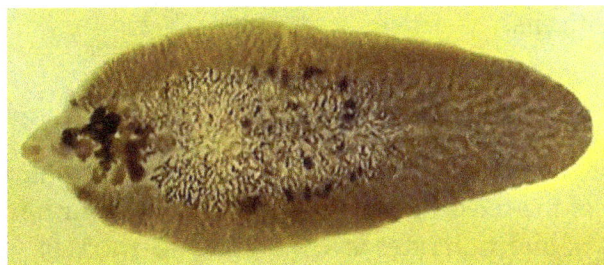

Slide showing *Fasciola hepatica's* internal organs

Fasciolosis is an important cause of both production and economic losses in the dairy and meat industry. Over the years, the prevalence has increased and it is likely to continue increasing in the future. Livestock are often treated with Flukicides, which are chemicals toxic to flukes. The two chemicals used are triclabendazole and bithionol. Ivermectin, which is widely used for many helminthic parasites, has low effectivity against *F. hepatica*, as does praziquantel. For humans, the type of control depends on the setting. One important method is through the strict control over the growth and sales of edible water plants such as watercress. This is particularly important in highly endemic areas. Some farms are irrigated with polluted water, hence, vegetables farmed from such land should be thoroughly washed and cooked before being eaten.

The best way to prevent Fasciolosis is by reducing the lymnaeid snail population or separating livestock from areas with these snails. These two methods are not always the most practical, so control by treating the herd before they are potentially infected is commonly practiced.

Diagnosis

A diagnosis may be made by finding yellow-brown eggs in the stool. They are indistinguishable from the eggs of *Fascioloides magna*, although the eggs of *F. magna* are very rarely passed in sheep, goats, or cattle. If a patient has eaten infected liver, and the eggs pass through the body and out via the faeces, a false positive result to the test can occur. Daily examination during a liver-free diet will unmask this false diagnosis.

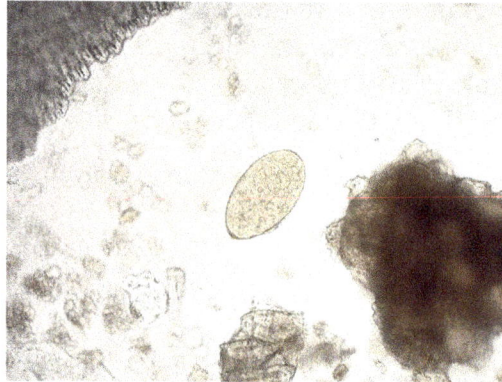

F. hepatica egg in stool sample.

An enzyme-linked immunosorbent assay (ELISA) test is the diagnostic test of choice. ELISA is available commercially and can detect anti-hepatica antibodies in serum and milk; new tests intended for use on faecal samples are being developed. Using ELISA is more specific than using a Western blot or Arc2 immunodiffusion. Proteases secreted by *F. hepatica* have been used experimentally in immunizing antigens.

Fasciola Gigantica

Fasciola gigantica is a parasitic flatworm of the class Trematoda, which causes tropical fascioliasis. It is regarded as one of the most important single platyhelminth infections of ruminants in Asia and Africa. Estimates of infection rates are as high as 80-100% in some countries. The infection is commonly called fasciolosis.

The prevalence of *F. gigantica* often overlaps with that of *Fasciola hepatica*, and the two species are so closely related in terms of genetics, behaviour, and morphological and anatomical structures that it is notoriously difficult to distinguish them. Therefore, sophisticated molecular techniques are required to correctly identify and diagnose the infection.

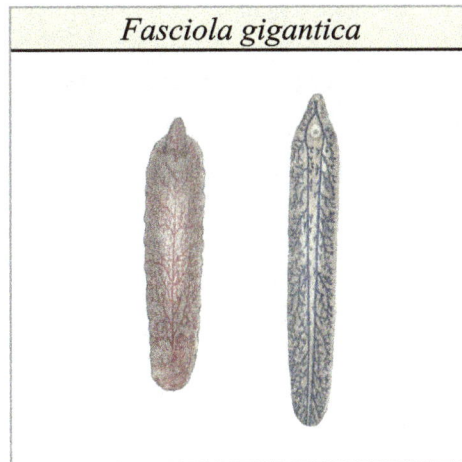

Fasciola gigantica

Distribution

Fasciola gigantica causes outbreaks in tropical areas of southern Asia, Southeast Asia, and Africa. The geographical distribution of *F. gigantica* overlaps with *Fasciola hepatica* in many African and Asian countries and sometimes in the same country, although in such cases the ecological requirement of the flukes and their snail host are distinct. Infection is most prevalent in regions with intensive sheep and cattle production. In Egypt *F. gigantica* has existed in domestic animals since the times of the pharaohs.

Life Cycle

The life cycle of *Fasciola gigantica* is as follows: eggs (transported with feces) → eggs hatch → miracidium → miracidium infect snail intermediate host → (parthenogenesis in 24 hours) sporocyst → redia → daughter redia → cercaria → (gets outside the snail) → metacercaria → infection of the host → adult stage produces eggs.

Intermediate Hosts

As with other trematodes, *Fasciola* develop in a molluscan intermediate host. Species of the freshwater snails from the family Lymnaeidae are well known for their role as intermediate hosts in the life cycle of *Fasciola gigantica*; however, throughout the years an increasing number of other molluscan intermediate hosts of *F. gigantica* have been reported. It has been reported that the Lymnaeid intermediate hosts of *F. gigantica* are distinguishable from those of *F. hepatica*, both morphologically and as to habitat requirement. The species of *Fasciola* can become adapted to new intermediate hosts under certain conditions at least based on laboratory trials. The most important intermediate host for *F. gigantica* is *Radix auricularia*. However, other species are also known to harbour the fluke including *Lymnaea rufescens* and *Lymnaea acuminata* in the Indian Subcontinent; *Radix rubiginosa* and *Radix natalensis* in Malaysia and in Africa respectively; and the synonymous *Lymnaea cailliaudi* in east Africa. Other snails also serve as natural or experimental intermediate such as *Austropeplea ollula*, *Austropeplea viridis*, *Radix peregra*, *Radix luteola*, *Pseudosuccinea columella* and *Galba truncatula*. The Australian *Lymnaea tomentosa* (host of *F. hepatica*) was shown to be receptive to miracidia of *F. gigantica* from East Africa, Malaysia and Indonesia.

Definitive Hosts

Fasciola gigantica is a causative agents (together with *Fasciola hepatica*) of fascioliasis in ruminants and in humans worldwide.

The parasite infects cattle and buffalo and can also be seen regionally in goats, sheep, and donkeys.

Infection and Pathogenicity

Infection with *Fasciola* spp. occurs when metacercariae are accidentally ingested on raw vegetation. The metacercariae exist in the small intestine, and move through the intestinal wall and peritoneal cavity to the liver where adults mature in the biliary ducts of the liver. Eggs are passed through the bile ducts into the intestine where they are then passed in the feces.

Diagnosis

Despite the importance to differentiate between the infection by either fasciolid species, due to their distinct epidemiological, pathological and control characteristics, there is, unfortunately, coprological (excretion-related) or immunological diagnosis are difficult. Especially in humans, specific detection by clinical, pathological, coprological or immunological methods are unreliable. Molecular assays are the only promising tools, such as PCR-RFLP assay, and the very rapid loop-mediated isothermal amplification (LAMP).

Treatment

Triclabendazole is the drug of choice in fasciolosis as it is highly effective against both mature and immature flukes. Artemether has been demonstrated *in vitro* to be equally effective. Though slightly less potent, artesunate is also useful in human fasciolosis.

Fasciolopsis

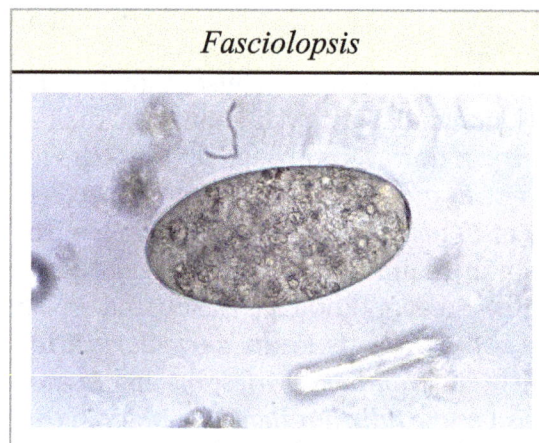

Fasciolopsis

Fasciolopsis of trematodes. Just one species is recognised: *Fasciolopsis buski*. It is a notable parasite of medical importance in humans and veterinary importance in pigs. It is prevalent in Southern and Eastern Asia. The term for infestation with *Fasciolopsis* is fasciolopsiasis.

Fasciolopsis buski

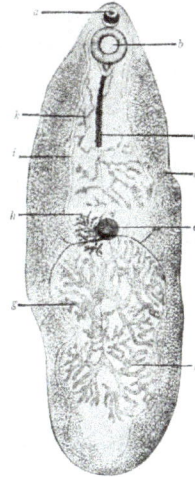

General anatomy

Fasciolopsis buski is commonly called the giant intestinal fluke, because it is an exceptionally large parasitic fluke, and the largest known to parasitise humans. Its size is variable and a mature specimen might be as little as 2 cm long, but the body may grow to a length of 7.5 cm and a width of 2.5 cm. It is a common parasite of humans and pigs and is most prevalent in Southern and Southeastern Asia. It is a member of the family Fasciolidae in the order Echinostomida. The Echinostomida are members of the class Trematoda, the flukes. The fluke differs from most species that parasitise large mammals, in that they inhabit the gut rather than the liver as *Fasciola* species do. *Fasciolopsis buski* generally occupies the upper region of the small intestine, but in heavy infestations can also be found in the stomach and lower regions of the intestine. *Fasciolopsis buski* is the cause of the pathological condition fasciolopsiasis.

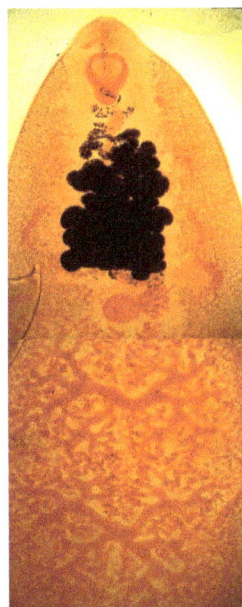

Foreparts of a cleared, stained specimen of *Fasciolopsis buski*.

In London, George Busk first described *Fasciolopsis buski* in 1843 after finding it in the duodenum of a sailor. After years of careful study and self experimentation, in 1925, Claude Heman Barlow determined its life cycle in humans.

Morphology

Fasciolopsis buski is a large, leaf-shaped, dorsoventrally flattened fluke characterized by a blunt anterior end, undulating, unbranched ceca (sac-like cavities with single openings), tandem dendritic testes, branched ovaries, and ventral suckers to attach itself to the host. The acetabulum is larger than the oral sucker. The fluke has extensive vitelline follicles. It can be distinguished from other fasciolids by a lack of cephalic cone or "shoulders" and the unbranched ceca.

Life Cycle

Adults produce over 25,000 eggs every day which take up to seven weeks to mature and hatch at 27–32 °C. Immature, unembryonated eggs are discharged into the intestine and stool. In two weeks, eggs become embryonated in water, and after about seven weeks, eggs release tiny parasitic organisms called miracidia, which invade a suitable snail intermediate host. Several species of genera *Segmentina* and *Hippeutis* serve as intermediate hosts. In the snail the parasite undergoes several developmental stages (sporocysts, rediae, and cercariae). The cercariae are released from the snail and encyst as metacercariae on aquatic plants such as water chestnut, water caltrop, lotus, bamboo, and other edible plants. The mammalian host, or the final host, becomes infected by ingesting metacercariae on the aquatic plants. After ingestion, the metacercariae excyst in the duodenum in about three months and attach to the intestinal wall. There they develop into adult flukes (20 to 75 mm by 8 to 20 mm) in approximately 3 months, attached to the intestinal wall of the mammalian hosts (humans and pigs). The adults have a life span of about one year.

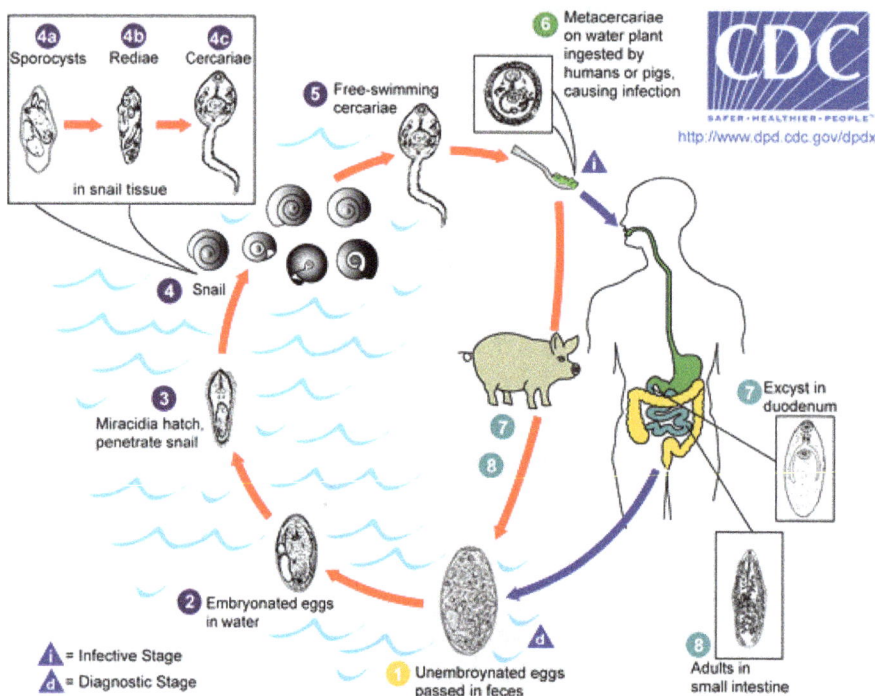

Metorchis Conjunctus

Metorchis conjunctus, common name Canadian liver fluke, is a species of trematode parasite in the family Opisthorchiidae. It can infect mammals that eat raw fish in North America. The first intermediate host is a freshwater snail and the second, a freshwater fish.

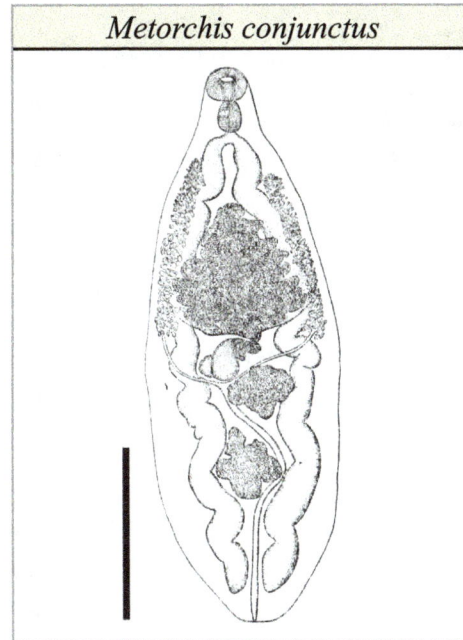

Metorchis conjunctus

Taxonomy

This species was discovered and described by Thomas Spencer Cobbold in 1860.

Distribution

The distribution of *Metorchis conjunctus* includes:

- east Greenland

- from Quebec to Saskatchewan, Canada

- Maine, Connecticut, South Carolina, US

Description

The body of *Metorchis conjunctus* is pear-shaped and flat. The body length is $\frac{1}{4}$–$\frac{3}{8}$ inch (6.4–9.5 mm). There is a weakly muscular terminal oral sucker. There is no prepharynx. The pharynx is strongly muscular. The esophagus is very short. The intestinal ceca vary from almost straight to sinuous. The acetabulum is slightly oval and weakly muscular. There is an anterior testis and a posterior testis. The testes vary from almost round to oval, and may be deeply lobed or slightly indented. There is no cirrus pouch. The seminal vesicle is slender. The ovary is trilobed. The receptaculum seminis is elongated or pyriform, and slightly twisted, and situated to the right and behind the ovary.

The Eggs are Oval and Yellowish Brown.

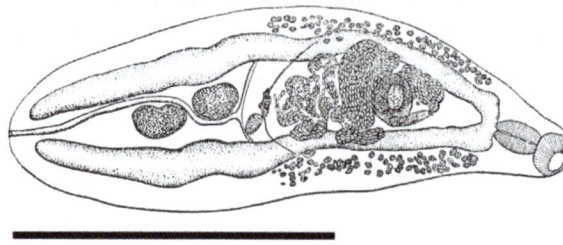

Drawing of dorsal view of *Metorchis conjunctus*. Scale bar is 1 mm.

Life Cycle

The first intermediate host of *Metorchis conjunctus* is a freshwater snail, *Amnicola limosus*.

The second intermediate host is a freshwater fish: *Catostomus catostomus*, *Salvelinus fontinalis*, *Perca flavescens*, or *Catostomus commersoni*. Metacercaria of *M. conjunctus* were also found in northern pike (*Esox lucius*).

The definitive hosts are fish-eating mammals such as domestic dogs (*Canis lupus familiars*), domestic cats (*Felis catus*), wolves (*Canis lupus*), red foxes (*Vulpes vulpes*), gray foxes (*Urocyon cinereoargenteus*), coyotes (*Canis latrans*), raccoons (*Procyon lotor*), muskrats (*Ondatra zibethicus*), American minks (*Neovison vision*), fishers (*Martes pennanti*), or bears. It can also infect humans. It lives in the bile duct and in the gallbladder.

Effects on Human Health

Metorchis conjunctus causes a disease called metorchiasis. It has been known to infect humans since 1946. Humans had eggs of *M. conjunctus* in their stools, but they were asymptomatic (they had no symptoms of the disease). Sashimi from raw *Catostomus commersoni* was identified as a source for an outbreak in Montreal in 1993. It was the first symptomatic disease in humans caused by *M. conjunctus*.

Symptoms

After ingestion of fish infected with *Metorchis conjunctus*, it takes about 1–15 days for symptoms to occur, namely for eggs to be detected in the stool (incubation period).

The acute phase consists of upper abdominal pain and low-grade fever. There are high concentrations of eosinophil granulocytes in blood. There are also higher concentrations of liver enzymes. When untreated, symptoms may last from three days to four weeks. Symptoms of chronic infection were not reported.

Diagnosis and Treatment

Eggs of *Metorchis conjunctus* can be found by stool analysis. Serologic analysis can be also used: ELISA test for IgG antibodies against antigens of *M. conjunctus*.

Drugs used to treat infestation include praziquantel: 75 mg/kg in 3 doses per day (the same dosage applies for adults and for children).

Effects on Animal Health

Watson and Croll (1981) studied symptoms of cats. Prevention include feeding with cooked fish (not raw fish).

Metorchis conjunctus was found to be a common infection of domestic dogs in Indian settlements in in 1973.

The prevalence of *M. conjunctus* in wolves in Canada is 1–3%. In wolves, *M. conjunctus* causes cholangiohepatitis with periductular fibrosis in the liver. It sometimes causes chronic inflammation and fibrosis of the pancreas in wolves.

Opisthorchis Viverrini

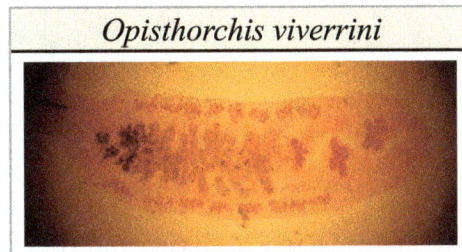

Opisthorchis viverrini, common name Southeast Asian liver fluke, is a trematode parasite from the family Opisthorchiidae that attacks the area of the bile duct. Infection is acquired when people ingest raw or undercooked fish. Infection with the parasite is called opisthorchiasis. *Opisthorchis viverrini* infection also predisposes the infected for cholangiocarcinoma, a cancer of the gall bladder and/or its ducts.

An adult *Opisthorchis viverrini* has these main body parts: oral sucker, pharynx, caecum, ventral sucker, vitellaria, uterus, ovary, Mehlis' gland, testes, exretory bladder.

Opisthorchis viverrini (together with *Clonorchis sinensis* and *Opisthorchis felineus*) is one of the three most medically important species in the family Opisthorchiidae. In fact *O. viverrini* and *C.*

sinensis are capable of causing cancer in humans, and are classified by the International Agency for Research on Cancer as a group 1 biological carcinogen in 2009. *O. viverrini* is endemic throughout Thailand, the Lao People's Democratic Republic, Vietnam and Cambodia. In Northern Thailand, it is widely distributed, with high prevalence in humans, while in Central Thailand there is low rate of prevalence. The disease opisthorchiasis (caused by *Opisthorchis viverrini*) does not occur in southern Thailand.

Description

The testes of an adult *Opisthorchis viverrini* are lobed in comparison of dendritic testes of *Clonorchis sinensis*.

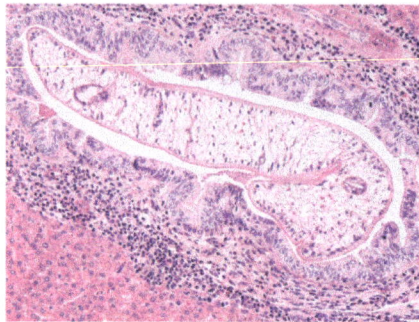

Photomicrograph of an adult *Opisthorchis viverrini* in bile ducts of experimentally infected hamster

The eggs of *Opisthorchis viverrini* are 30 × 12 μm in size and they are slightly narrower and more regularly ovoid than in *Clonorchis sinensis*. But eggs of *Opisthorchis viverrini* are visually indisgushiable in Kato technique smears from other eggs of flukes from other fluke family Heterophyidae.

An egg of *Opisthorchis viverrini*. 400× magnification.

The metacercariae of *Opisthorchis viverrini* are brownish, elliptical with two nearly equal-sized suckers: the oral sucker and the ventral sucker. They are 0.19–0.25 × 0.15–0.22 mm in size.

Life Cycle

Opisthorchis viverrini is a hermaphroditic liver fluke. Its life cycle is similar to the life cycle of *Clonorchis sinensis*. It involves a freshwater snail, in which asexual reproduction takes place, and freshwater cyprinid fishes (family Cyprinidae) as intermediate hosts. Fish–eating (= piscivorous) mammals, including humans, dogs and cats, act as definitive hosts, in which sexual reproduction occurs. As a result of poor sanitation practices and inadequate sewerage infrastructure, *Opisthorchis viverrini*-infected people pass the trematode's eggs in their feces into bodies of fresh water.

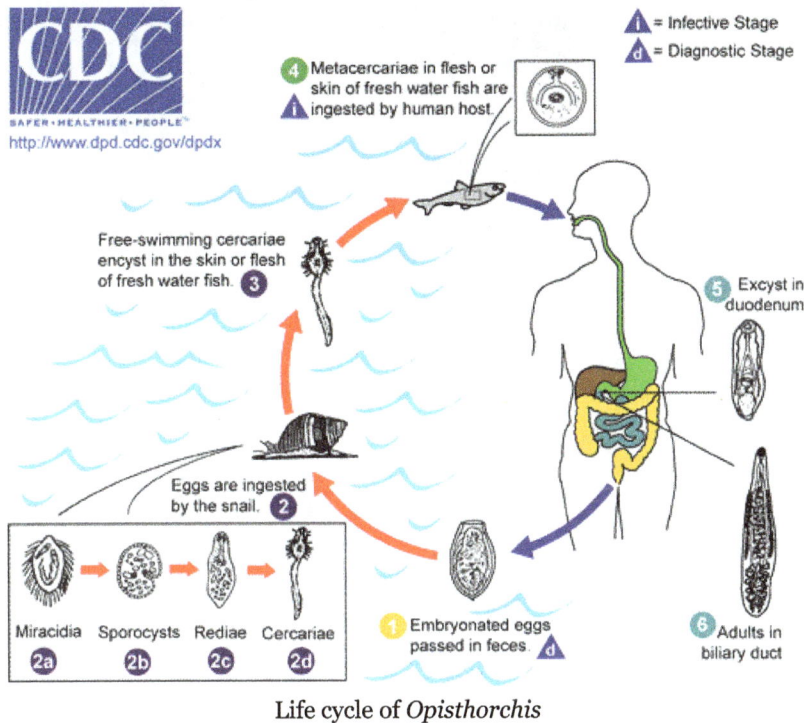

Life cycle of *Opisthorchis*

First Intermediate Host

The first intermediate hosts include freshwater snails of the genus *Bithynia*. The only known host is *Bithynia siamensis* (that include all its three subspecies). Aquatic snails, which represent the first intermediate hosts of *Opisthorchis viverrini*, ingest the eggs from which the miracidia undergo asexual reproduction before a population of the free swimming larval stage, called a cercaria, is shed from the infected snails.

Fluke-infected fish are plentiful in rivers such as the Chi River in Khon Kaen Province, Thailand.

Second Intermediate Host

The cercaria then locates a cyprinoid fish, encysts in the fins, skin and musculature of the fish, and becomes a metacercaria. Habitats of second intermediate hosts of *Opisthorchis viverrini* include freshwater habitats with stagnant or slow-moving waters (ponds, river, aquaculture, swamps, rice fields).

Thai fishermen catch fish (including infected ones) in nets and prepare fish-based meals with local herbs, spices, and condiments.

In 1965 there were known 9 fish hosts of *Opisthorchis viverrini*. Up to 2002 there were known 15 species of fishes from 7 genera of the family Cyprinidae, that serves as second intermediate host. Further research by Rim et al. (2008) showed additional five more host species:

- *Puntius brevis*
- *Puntius gonionotus* – synonym: *Barbonymus gonionotus*, Java barb
- *Puntius orphoides*
- *Puntius proctozysron* – synonym: *Puntioplites proctozystron*
- *Puntius viehoeveri* – synonym: *Barbonymus gonionotus*
- *Hampala dispar*
- *Hampala macrolepidota*
- *Cyclocheilichthys armatus* – synonym: *Cyclocheilichthys siaja*
- *Cyclocheilichthys repasson*
- *Labiobarbus lineatus*
- *Esomus metallicus*
- *Mystacoleucus marginatus*
- *Puntioplites falcifer*
- *Onychostoma elongatum*
- *Osteochilus hasseltii*
- *Hypsibarbus lagleri*
- *Barbodes gonionotus* – synonym: *Barbonymus gonionotus*

Definitive Host

The metacercarial stage is infective to humans and other fish-eating mammals including dogs, cats, rats, and pigs. The natural definitive host is the leopard cat (*Prionailurus bengalensis*). The young adult worm escapes from the metacercarial cyst in the upper small intestine and then migrates through the ampulla of Vater into the biliary tree, where it develops to sexual maturity over four to six weeks, thus completing the life cycle.

The finished dish of koi pla made of raw fish accompanied by rice and vegetables. This dish is a dietary staple of many northeastern Thai villagers and is a common source of infection with *Opisthorchis viverrini*.

Fish contain more metacercaria from September to February, before the dry season and this is the period, when humans are usually infected. Infection is acquired when people ingest raw or undercooked fish. Dishes of raw fish are common in the cuisine of Laos and the cuisine of Thailand: *koi-pla*, raw fish in spicy salad *larb-pla*, salted semi-fermented fish dishes called *pla-ra* (pla ra), *pla som* and Som fak.

The adult worms, which are hermaphrodites, can live for many years in the liver, even decades, shedding as many as 200 eggs per day which pass out via bile into the chyme and feces. The lifespan of *Opisthorchis viverrini* is over 10 years.

Opisthorchis viverrini secretes a granulin-like growth protein especially in its gut and integument.

Effect on Human Health

Medical care and loss of wages caused by *Opisthorchis viverrini* in Laos and in Thailand costs about $120 million annually or $120 million per year can cost Northeast Thailand only.

Infections with *Opisthorchis viverrini* and of other liver flukes in Asia affect the poor and poorest people. Opisthorchiasis has received less attention in comparison to other diseases, and it is a neglected disease in Asia.

Genetics

Currently, a total of only ~5,000 expressed sequence tags (ESTs) are publicly available for *Opisthorchis viverrini*, a dataset far too small to give sufficient insights into transcriptomes for the purpose of supporting genomic and other fundamental molecular research.

Although the genome size of *Opisthorchis viverrini* has not yet been reported, it is known to have six pairs of chromosomes, i.e. 2n = 12.

Opisthorchis Felineus

Opisthorchis felineus, or cat liver fluke is a trematode parasite that infects the liver in mammals. It was first discovered in 1884 in a cat's liver by Sebastiano Rivolta of Italy. In 1891, Russian scientist K.N. Vinogradov found it in a human, and named the parasite a "Siberian liver fluke". In the 1930s, helminthologist Hans Vogel of Hamburg published an article describing the life cycle of *Opisthorchis felineus*.

Distribution

Distribution of *Opisthorchis felineus* include: Spain, Italy, Albania, Greece, France, Macedonia, Switzerland, Germany, Poland, Russia, Turkey, and Caucasus.

Life Cycle

The first intermediate hosts of the parasite are freshwater snails:

- *Bithynia inflata* (synonym: *Codiella inflata*)
- *Bithynia troschelii*
- *Bithynia leachii*
- *Bithynia tentaculata*

The second intermediate hosts are freshwater fish: *Leuciscus idus, Tinca tinca, Abramis brama,* white-eye bream *Ballerus sapa, Barbus barbus,* common carp *Cyprinus carpio, Blicca bjoerkna, Leuciscus idus, Alburnus alburnus, Aspius aspius,* and common rudd *Scardinius erythropthalmus.*

The definitive hosts are fish-eating mammals such as dogs, foxes, cats, rats, pigs, rabbits, seals, lions, wolverines, martens, polecats and humans.

Effect on Human Health

It is estimated that 1.5 million people in Russia are infected with the parasite. Inhabitants of Siberia acquire the infection by consuming raw, slightly salted and frozen fish.

Opisthorchiasis, the disease caused by *Opisthorchis felineus*, ranges in severity from asymptomatic infection to severe illness. Patient outcome is dependent on early detection and treatment.

Human cases of opisthorchiasis may affect the liver, pancreas, and gall bladder. If not treated in the early stages, opisthorchiasis may cause cirrhosis of the liver and increased risk of liver cancer, but may be asymptomatic in children.

Two weeks after flukes enter the body, the parasites infect the biliary tract. Symptoms of infection include fever, general felling of tiredness, skin rash, and gastrointestinal disturbances. Severe anemia and liver damage may also incapacitate the infected person for 1–2 months. Treatment of opisthorchiasis is generally with a single dose of praziquantel.

Paragonimus

Paragonimus is a genus of flatworms (platyhelminths). Some tens of species have been described, but they are difficult to distinguish, so it is not clear how many of the named species may be synonyms. Several of the species are known as lung flukes. In humans some of the species occur as zoonoses; the term for the condition is paragonimiasis. The first intermediate hosts of *Paragonimus* include at least 54 species of freshwater snails from superfamilies Cerithioidea and Rissooidea.

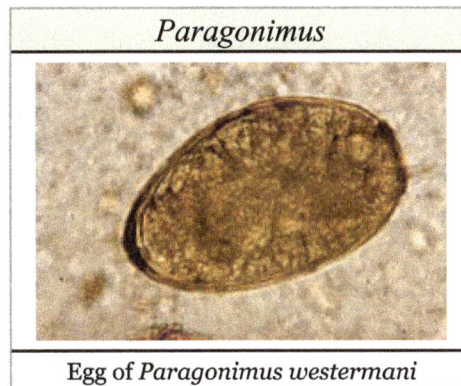

Egg of *Paragonimus westermani*

The most prominent species of Paragonimus in human medicine is *Paragonimus westermani*, an infectious lung fluke originating in eastern Asia. Worldwide, about nine species of *Paragonimus* are known to cause human paragonimiasis in which many of the species reside in East Asia, West Africa, and in North and South America.

Morphology

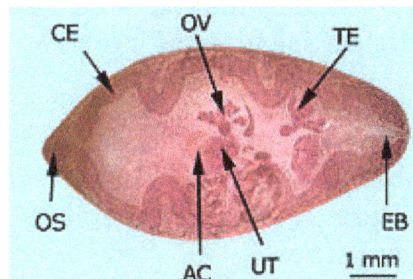

Morphology of typical Paragonimus:
AC: acetabulum (ventral sucker)
CE: cecum, EB: excretory bladder
OS: oral sucker, OV: ovary
TE: testes, UT: uterus

Species of *Paragonimus* vary in size; the adult stage might attain a length of up to 15 millimetres (0.59 in) and a width of up to 8 mm (0.31 in). The adult flatworm has an oval shape body with spines covering its thick tegument. Both the oral sucker and acetabulum are round and muscular. The acetabulum is slightly bigger than the oral sucker – 0.19 mm and 0.12 mm, respectively. Ovaries are located behind the acetabulum and posterior to the ovary are the testes. The seminal receptacle, the uterus and its metraterm, the thick-walled terminal part, lie between the acetabulum and the ovary.

Life Cycle

The parasite passes through two intermediate hosts, an aquatic snail and a crustacean. It enters its mammalian definitive hosts when they eat infected freshwater crustaceans. Typical hosts include dogs, cats, and humans. Humans usually contract paragonimiasis when they eat undercooked freshwater crabs or crayfish, that contain live metacercariae. In the intestine, the parasite will move into the abdomen and commonly into the lungs. In the lung, the parasites encyst and cross fertilize each other. The cyst eventually ruptures in the lungs and the eggs may be coughed up or swallowed and excreted in the feces. An egg landing in fresh water hatches and releases a ciliated miracidium. A successful miracidium swims about until it finds an intermediate host, usually an aquatic snail. A crustacean in turn becomes infected by eating infected snails. The definitive host completes the cycle if it eats infected crustaceans.

Epidemiology

Worldwide roughly 20 million people are infected with Paragonimus. Human infections are commonest in regions with many human and animal reservoir hosts plus an abundance of intermediate hosts, such as snails, crabs, or crayfish, and where in addition consumption of raw or undercooked seafood is common. Consumption of insufficiently cooked meat from infected land animal hosts, such as wild boar, commonly transmits the infection.

Symptoms

Symptoms of paragonimiasis may include abdominal pain, diarrhea, fever, and hives. If the infection remains untreated, the symptoms may peter out after only few months, but sometimes they last for decades. Paragonimiasis is caused by the body's natural immune response to the worms and eggs that are present and also migrating from the intestines to the lungs.

As a rule, the parasites begin to cause the symptoms about three weeks after ingesting live metacercariae. After about eight weeks, they begin to produce eggs in the lungs. Some patients develop brain damage if parasites establish in the brain and produce eggs. The brain damage commonly causes headache, vomiting, and seizures. Untreated cerebral paragonimaisis commonly results in death from increased intracranial pressure.

Treatment

Praziquantel has been used to effectively treat paragonimiasis by separating the tegument. An effectively complete rate of cure may be expected after three days of treatment if there has not been too much permanent damage, such as from intracranial effects. Other medications can also be used such as bithionol, niclofan, and triclabendazole with high cure rates.

Prevention

Thorough cooking of an infected crustacean kills all stages of the parasite. Crab meat should not be eaten raw, even if pickled, because the pickling solution often fails to kill all the parasites. Utensils and cutlery boards should be cleaned thoroughly before and after food preparation.

Paragonimus Westermani

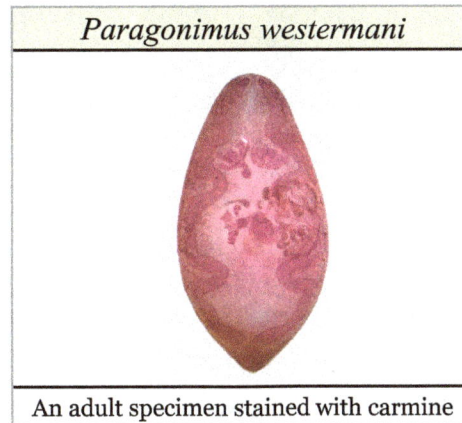

Paragonimus westermani

An adult specimen stained with carmine

Paragonimus westermani is the major species of lung fluke to infects humans, causing parago-nimiasis. The species sometimes is called the Japanese Lung fluke or Oriental Lung fluke. Human infections are most common in eastern Asia and in South America. *Paragonimus westermani* was discovered when two Bengal tigers died of paragonimiasis in zoos in Europe in 1878. Several years later Infections in humans were recognised in Formosa.

Introduction

Paragonimiasis is a food-borne parasitic infection caused by the lung fluke. It may cause a sub-acute to chronic inflammatory disease of the lung. It is one of the most familiar lung flukes with the widest geographical range. It was discovered by Coenraad Kerbert (1849-1927) in 1878.

Causative Agent

More than 30 species of trematodes (flukes) of the genus *Paragonimus* have been reported to infect animals and humans. Among the more than 10 species reported to infect humans, the most common is *Paragonimus westermani*, the oriental lung fluke.

Morphology

In size, shape, and color, *Paragonimus westermani* resembles a coffee bean when alive. Adult worms are 7.5 mm to 12 mm long and 4 mm to 6 mm wide. The thickness ranges from 3.5 mm to 5 mm. The skin of the worm (tegument) is thickly covered with scalelike spines. The oral and ventral suckers are similar in size, with the latter placed slightly pre-equatorially. The excretory bladder extends from the posterior end to the pharynx. The lobed testes are adjacent from each other located at the posterior end, and the lobed ovaries are off-centered near the center of the worm (slightly postacetabular). The uterus is located in a tight coil to the right of the acetabulum, which is connected, to the vas deferens. The vitelline glands, which produce the yolk for the eggs, are widespread in the lateral field from the pharynx to the posterior end. Inspection of the tegumental spines and shape of the metacercariae may distinguish between the 30-odd species of *Paragonimus spp.* but the distinction is sufficiently difficult to justify suspicion that many of the described species are synonyms.

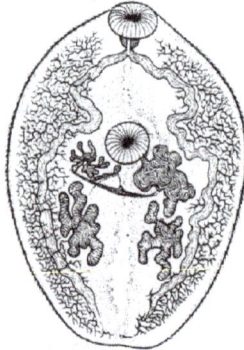

Paragonimus westermani. An adult of the hermaphroditic generation.

Egg of *Paragonimus westermani*

- Eggs: *Paragonimus westermani* eggs range from 80 to 120 μm long by 45 to 70 μm wide. They are yellow-brown, ovoid or elongate, with a thick shell, and often asymmetrical with one end slightly flattened. At the large end, the operculum is clearly visible. The opposite (abopercular) end is thickened. The eggs are unembryonated when passed in sputum or feces.

- Cercaria (not shown): Cercariae are often indistinguishable between species. There is a large posterior sucker, and the exterior is spined.

- Metacercaria: Metacercariae are usually encysted in tissue. The exterior is spined and has two suckers

- Adults: Adult flukes are typically reddish brown and ovoid, measuring 7 to 16 mm by 4 to 8 mm, similar in size and appearance to a coffee bean.They are hermaphroditic, with a lobed ovary located anterior to two branching testes. Like all members of the Trematoda, they possess oral and ventral suckers.

History of Discovery

P. westermani was discovered in the lungs of a human by Ringer in 1879 and eggs in the sputum were recognized independently by Manson and Erwin von Baelz in 1880. Manson proposed the snail as an intermediate host and various Japanese workers detailed the whole life cycle in the snail between 1916 and 1922. The species name *P. westermani* was named after Pieter Westerman (1859-1925) a zookeeper who noted the trematode in a Bengal tiger in an Amsterdam Zoo.

Life Cycle

Unembryonated eggs are passed in the sputum of a human or feline. Two weeks later, miracidia develop in the egg and hatches. The miracidia penetrate its first intermediate host (snail). Within the snail mother sporocyst form and produce many mother rediae, which subsequently produce many daughter rediae which shed crawling cercariae into fresh water. The crawling cercariae penetrate fresh water crabs and encyst in its muscles becoming metacercaria. Humans or felines then eat the infected crabs raw. Once eaten, the metacercaria excysts and penetrates the gut, diaphragm and lung where it becomes an adult worm in pairs.

The first intermediate hosts of the *Paragonimus westermani* are freshwater snails:

- *Semisulcospira amurensis*

- *Semisulcospira calculus*

- *Semisulcospira cancellata*

- *Semisulcospira extensa*

- *Semisulcospira gottschei*

- *Semisulcospira libertina* - synonym: *Semisulcospira toucheana*

- *Semisulcospira mandarina* - synonym: *Semisulcospira wegckiangensis*

- *Semisulcospira multicincta*

- *Semisulcospira nodiperda*

- *Semisulcospira nodiperda quinaria*

- *Semisulcospira paucincta*

- *Semisulcospira peregrinomum*

For many years *Tarebia granifera* was believed to be an intermediate host for the *Paragonimus westermani*, but Michelson showed in 1992 that this was erroneous.

Paragonimus has a quite complex life-cycle that involves two intermediate hosts as well as humans. Eggs first develop in water after being expelled by coughing (unembryonated) or being passed in human feces. In the external environment, the eggs become embryonated. In the next stage, the parasite miracidia hatch and invades the first intermediate host such as a species of freshwater snail. Miracidia penetrate its soft tissues and go through several developmental stages inside the snail but mature into cercariae in 3 to 5 months. Cercariae next invade the second intermediate host such as crabs or crayfish and encyst to develop into metacercariae within 2 months. Infection of humans or other mammals (definitive hosts) occurs via consumption of raw or undercooked crustaceans. Human infection with P. westermani occurs by eating inadequately cooked or pickled crab or crayfish that harbor metacercariae of the parasite. The metacercariae excyst in the duodenum, penetrate through the intestinal wall into the peritoneal cavity, then through the abdominal wall and diaphragm into the lungs, where they become encapsulated and develop into adults. The worms can also reach other organs and tissues, such as the brain and striated muscles, respectively. However, when this takes place completion of the life cycles is not achieved, because the eggs laid cannot exit these sites.

Epidemiology

Reservoir hosts of *Paragonimus* spp. include numerous species of carnivores including felids, canids, viverrids, mustelids, some rodents and pigs. Humans become infected after eating raw freshwater crabs or crayfish that have been encysted with the metacerciaria. Southeast Asia is more predominately more infected because of lifestyles. Raw seafood is popular in these countries. Crab collectors string raw crabs together and bring them miles inland to sell in Taiwan markets. These raw crabs are then marinated or pickled in vinegar or wine to coagulate the crustacean muscle. This method of preparation does not kill the metacercariae, consequently infecting the host. Smashing rice-eating crabs in rice paddies, splashing juices containing metacercariae, can also transmit the parasite, or using juices strained from fresh crabs for medicinal uses. This parasite is easily spread because it is able to infect other animals (zoonosis). An assortment of mammals and birds can be infected and act as paratenic hosts. Ingestion of the paratenic host can lead to infection of this parasite.

Paragonimus westermani is distributed in southeast Asia and Japan. Other species of Paragonimus are common in parts of Asia, Africa and South and Central America. *P. westermani* has been increasingly recognized in the United States during the past 15 years because of the increase of immigrants from endemic areas such as Southeast Asia. Estimated to infect 22 million people worldwide.

Transmission

Transmission of the parasite *P. westermani* to humans and mammals primarily occurs through the consumption of raw or undercooked seafood. In Asia, an estimated 80% of freshwater crabs carry *P. westermani*. In preparation, live crabs are crushed and metacercariae may contaminate the fingers/utensils of the person preparing the meal. Accidental transfer of infective cysts can occur via food preparers who handle raw seafood and subsequently contaminate cooking utensils and other foods. Consumption of animals which feed on crustaceans can also transmit the parasite, for cases have been cited in Japan where raw boar meat was the source of human infection. Food preparation techniques such as pickling and salting do not exterminate the causative agent. For example, in a Chinese study eating "drunken crabs" was shown to be particularly risky because the infection rate was 100% when crabs are immersed in wine for 3–5 minutes and fed to cats/dog.

Reservoir

Animals such as pigs, dogs, and a variety of feline species can also harbor *P. westermani*.

Vector

There is no vector, but various snail and crab species serve as intermediate hosts. In Japan and Korea, the crab species *Eriocheir* is an important item of food as well as a notable second intermediate host of the parasite.

Incubation Period

Time from infection to oviposition (laying eggs) is 65 to 90 days. Infections may persist for 20 years in humans.

Pathology

Once in the lung or ectopic site, the worm stimulates an inflammatory response that allows it to cover itself in granulation tissue forming a capsule. These capsules can ulcerate and heal over time. The eggs in the surrounding tissue become pseudotubercles. If the worm becomes disseminated and gets into the spinal cord, it can cause paralysis; capsules in the heart can cause death. The symptoms are localized in the pulmonary system, which include a bad cough, bronchitis, and blood in sputum (hemoptysis).

Diagnosis

Diagnosis is based on microscopic demonstration of eggs in stool or sputum, but these are not present until 2 to 3 months after infection. However, eggs are also occasionally encountered in effusion fluid or biopsy material. Furthermore, you can use morphologic comparisons with other intestinal parasites to diagnose potential causative agents. Finally, antibody detection is useful in light infections and in the diagnosis of extrapulmonary paragonimiasis. In the United States, detection of antibodies to Paragonimus westermani has helped physicians differentiate paragonimiasis from tuberculosis in Indochinese immigrants.

Additionally, radiological methods can be used to X-ray the chest cavity and look for worms. This method is easily misdiagnosed, because pulmonary infections look like tuberculosis, pneumonia, or spirochaetosis. A lung biopsy can also be used to diagnose this parasite.

Management and Treatment

According to the CDC, praziquantel is the drug of choice to treat paragonimiasis. The recommended dosage of 75 mg/kg per day, divided into 3 doses over 3 days has proven to eliminate P. westermani. Bithionol is an alternative drug for treatment of this disease but is associated with skin rashes and urticaria. For additional information, see the recommendations in The Medical Letter (Drugs for Parasitic Infections).

Clinical Presentation in Humans

Case study:

An 11½-year-old Hmong Laotian boy was brought into the emergency room by his parents with a 2- to 3-month history of decreasing stamina and increasing dyspnea [shortness of breath] on exertion. He described an intermittent nonproductive cough and decreased appetite and was thought to have lost weight. He denied fever, chills, night sweats, headache, palpitations, hemoptysis [coughing up blood], chest pain, vomiting, diarrhea or urticaria [skin rash notable for dark red, raised, itchy bumps]. There were no pets at home. At the time of immigration to the United States 16 months earlier, all family members had negative purified protein derivative intradermal tests except one brother, who was positive but had a normal chest radiograph and subsequently received isoniazid for 12 months… a left lateral thoracotomy was performed during which 1800 ml of an odorless, cloudy, pea soup-like fluid containing a pale yellow, cottage cheese-like, proteinaceous material was removed, along with a solitary, 6-mm-long, reddish brown fluke subsequently identified as Paragonimus westermani

Human infection with Paragonimus may cause acute or chronic symptoms, and manifestations may be either pulmonary or extrapulmonary.

Acute symptoms: The acute phase (invasion and migration) may be marked by diarrhea, abdominal pain, fever, cough, urticaria, hepatosplenomegaly, pulmonary abnormalities, and eosinophilia. The acute stage corresponds to the period of invasion and migration of flukes and consists of abdominal pain, diarrhea and urticaria, followed roughly 1 to 2 weeks later by fever, pleuritic chest pain, cough and/or dyspnea. Chronic Symptoms: During the chronic phase, pulmonary manifestations include cough, expectoration of discolored sputum, hemoptysis, and chest radiographic abnormalities. Chronic pulmonary paragonimiasis, the most common clinical pattern, is frequently mild, with chronic cough, brown-tinged sputum (the color being caused by expectorated clusters of reddish brown eggs rather than by blood) and true hemoptysis.

Confusion With Tuberculosis

Practitioners should always consider the possibility of tuberculosis in patients with fevers, cough, weight loss. However, in endemic areas it is prudent to consider paragonimiasis as well. Flukes occasionally cause confusion when they invade the pleural space without entering the lung parenchyma.

"In contrast to tuberculosis, pulmonary paragonimiasis is only rarely accompanied by rales or other adventitious breath sounds. Many patients are asymptomatic, and symptomatic patients frequently look well despite a prolonged course."

In pleural paragonimiasis, symptoms may be minimal and diagnosis complicated, since ova are not coughed or spit out or swallowed and there is frequently no cough. Such patients may develop pleural effusions and, because of the coendemicity with Mycobacterium tuberculosis (and co-infection in some patients), such effusions are often misdiagnosed as isolated tuberculosis.

- Adapted from Heath, Harley W & Susan G Marshall. "Pleural Paragonimiasis In A Laotian Child.*

Extra-pulmonary locations of the adult worms result in more severe manifestations, especially when the brain is involved. Extra-pulmonary paragonimiasis is rarely seen in humans, as the worms nearly exclusively migrate to the lungs. Despite this, cysts can develop in the brain and abdominal adhesions resulting from infection have been reported. Cysts may contain living or dead worms; a yellow-brownish thick fluid (occasionally hemorrhagic). When the worm dies or escapes, the cysts gradually shrink, leaving nodules of fibrous tissues and eggs which can calcify.

Worldwide the most common cause of hemoptysis is paragonimiasis.

Other case studies:

- *Pachucki CT, Levandowski RA, Brown VA, Sonnenkalb BH, Vruno MJ (1984). "American paragonimiasis treated with Praziquantel". N Engl J Med. 311: 582–3. doi:10.1056/nejm198408303110906.*

- *Procop GW, Marty AM, Scheck DN, Mease DR, Maw GM (2000). "North American Paragonimiasis: A case report". Acta Cytol. 44: 75–80. doi:10.1159/000326230.*

Public Health and Prevention Strategies

Prevention programs should promote more hygienic food preparation by encouraging safer cooking techniques and more sanitary handling of potentially contaminated seafood. The elimination of the first intermediate host, the snail, is not tenable due to the nature of the organisms habits. A key component to prevention is research, more specifically the research of everyday behaviors. This recent study was conducted as a part of a broader effort to determine the status of Paragonimus species infection in Laos. An epidemiological survey was conducted on villagers and schoolchildren in Namback District between 2003 and 2005. Among 308 villagers and 633 primary and secondary schoolchildren, 156 villagers and 92 children had a positive reaction on a Paragonimus skin test. Consequently, several types of crabs were collected from markets and streams in a paragonimiasis endemic area for the inspection of metacercariae and were identified as the second intermediate host of the Paragonimus species. In this case study, we see how high prevalence of paragonimiasis is explained by dietary habits of the population. Amongst schoolchildren, many students reported numerous experiences of eating roast crabs in the field. Adult villagers reported frequent consumption of seasoned crabs (Tan Cheoy Koung) and papaya salad (Tammack Koung) with crushed raw crab. In addition to this characteristic feature of the villagers' food culture, the denizens of this area drink fresh crab juice as a traditional cure for measles, and this was also thought to constitute a route for infection.

Paragonimus Skrjabini

Paragonimus skrjabini is classified as a species in the genus *Paragonimus*, which consists of many species of lung flukes that result in the food-borne parasitic disease paragonimiasis.

Introduction

Scientists have identified *P. skrjabini,* along with several other species including *P. westermani* and *P. miyazakii,* to be key pathogens in causing paragonimiasis in humans, primarily in Asian regions of the world. *P. skrjabini* is especially prevalent in 26 provinces in China with cases appearing more recently in India and Vietnam as well. From a morphological and genetic standpoint, *P. skrjabini* is most closely related to the species *P. miyazakii,* so much so that two sub-species have been classified separately within the *P. skrjabini* complex: *P. skrjabini skrjabini* and *P. skrjabini miyazakii.* Doanh PN (2007) establishes the importance of learning more about *P. skrjabini,* asserting that "among *Paragonimus* species, *P. westermani* followed by *P. skrjabini* complex are the major pathogens for human paragonimiasis in Asia."

History

Max Braun in 1899 first defined the *Paragonimus* genus, which initially only included the species *P. westermani.*

In Vietnam, since paragonimiasis was first reported there in 1906, it was presumed for 89 years that only one species of *Paragonimus* lung fluke, *P. westermani*, caused paragonmiasis in humans. However, scientists have been conducting many studies of crabs and humans infected with paragonimiasis, leading to the discovery of several previously unknown species. The few cases reported in North America of paragonimiasis are most likely the result of diseased individuals traveling to the area from a different country or people consuming infected, imported food.

"P. skrjabini" is endemic in China and was first described in the Guangdong Province in 1959, originating from a viverrid's lungs.

Today, according to the World Health Organization, estimates put the number of persons afflicted with paragonimiasis at 20.7 million and the number at risk of contracting the disease at 293 million. For *P. skrjabini,* it stands as a public health threat in certain areas of the world, such as the Three Gorges Reservoir where it is the primary paragonimiasis causing parasite. Changes in the environment such as pollution and the persistence of individuals' consumption habits of raw crab puts paragonimiasis epidemics at high risk.

Life Cycle

The life cycle of *P. skrjabini* involves three hosts. The first intermediate host is a mollusk (typically a snail), the second intermediate host is a crustacean (typically a crab), and the definitive host is a mammal such as a dog, cat, or a human. The mammal is the definitive host because it is the site where sexual reproduction occurs and adult *P. skrjabini* flukes develop. Infection begins when humans consume raw or uncooked crustaceans such as crabs that contain metacercariae of *P. skrjabini*. *P. skrjabini* metacercarie are typically located in the muscles of the crabs's bodies (Zhang et al. 2012). Next, in the animal or human's small intestine, the metacercarie excyst (emerge from a cyst) and travel to the abdomen before ultimately moving into the lungs. There, adult worms begin to grow and develop. *P. skrjabini* in humans, however, are known to often fail to make it to the lungs and thus don't reach the stage of adult development. Rather, immature *P. skrjabini* parasites stay undeveloped and enter the human host's brain, muscles, and various other subcutaneous tissues, leading to extrapulmonary neurologic and abdominal paragonimiasis. *P. skrjabini* trematodes in the mammal produce and fertilize eggs that then exit the host, typically through feces. In water, the eggs hatch and release miracidium that in turn infect a snail. A sporocyst that contains germinal cells forms in the snail's body cavity, and, following asexual reproduction, produces rediae. Rediae produce cercariae (the larval form of the parasite). The cercariae migrate from the snail to a crab, entering either through direct penetration or by the consumption of the snail by the crab. Often, multiple species of *Paragonimus* can be found coexisting in one crustacean, suggesting that metacercariae of different species do not compete with each other within the host. The life cycle of *P. skrjabini* starts over again as mammals or humans eat the crab.

Morphology

The encysted metacercaria of *P. skrjabini* tend to have a round and circular shape with a cyst wall that consists of a fragile outer later and thicker middle and inner layers. Excysted metacercariae of *P. skrjabini* have an oral sucker and a ventral sucker measuring 80-120 μm and 120-60 μm in diameter, respectively. The morphological characteristics of adult worms of *P. skrjabini* are an elongated body with scattered, singularly arranged cuticular spines, branched testes and ovaries, and a ventral sucker that is larger than the oral sucker. Due to the presence of both ovaries and testes in *P. skrjabini* parasites, they are hermaphroditic. An esophagus and truncated pharnyx make up the adult digestive system.

Symptoms and Diagnosis

In humans, *P. skrjabini* infections can result in a wide variety of symptoms, rendering it difficult to diagnose and challenging to quickly enact proper treatment. Diagnosis typically requires first a general recognition of the symptoms followed by laboratory and radiologic tests. Radiology, for example, can pick up on *P. skrjabini's* migration to ectopic places like the brain. Serological or immunological tests including intradermal test, immunodiffusion, indirect haemagglutination test, enzyme-linked immunosorbent assay, and Western blot also play a key role in not only diagnosing the overall presence of *Paragonimus* parasites in the body but also distinguishing the various different species invading. The ELISA test detects the antibody used to combat the *P. skrjabini* infection. Although cases of *P. skrjabini* infection can exhibit early symptoms such as abdominal pain, a lack of appetite, and high fever, it can also come on slowly and fail to exhibit identifiable symptoms for a period of latency between 20 days to as long as 3 months. Later on in the infection, the type of symptoms that appear primarily depends on which organs the *P. skrjabini* parasites enter. Infections and the resulting set of symptoms can be classified under five different types including the subcutaneous mass type, cerebral type, pericarditis type, abdominal type, and pleurisy type. Infections of the pleurisy type, for instance, exhibit symptoms like chest pain and cough while the cerebral type tends to result in vomiting and headaches. The pericarditis type is associated with shortness of breath and palpitation while the abdominal type infections lead to diarrhea and abdominal pain. *P. skrjabini* has shown no effect on mammalian reproductive abilities. Despite occasional cases resulting in death, *P. skrjabini* has not been widely known to cause morbidity and death. Surgery may be occasionally necessary for patients with cerebal or pleural paragonimiasis.

Children and young adults tend to show a higher rate of infection than adults.

Genetic Sequencing and Studies

The ITS2 (nuclear ribosomal second internal transcribed spacer region) and CO1 (partial mitochondrial cytochrome oxidase subunit 1 gene) sequences of *P. skrjabini* have been registered in GenBank, which is used to determine similarities and phylogenetic relationships between various species of *Paragonimus* in different parts of the world. ITS2 sequences are utilized when studying inter-species variation, and CO1 sequences are important when studying intra-species variation. For instance, Doanh et al. (2012) determined that, after analyzing the CO1 sequences of groups of *P. skrjabini* in India, China, and Vietnam, a fair amount of genetic differences exist between the groups. Doanh et al. (2012) also used ITS2 and CO1 sequences to prove that populations of *P.*

skrjabini in both Vietnam and China are the same genetically. Blair et al. (2005) studied CO1 sequences and found molecular similarities and differences in the *P. skrjabini* complex, concluding that there should be two sub-species of *P. skrjabini*.

Methods to study *P. skrjabini* consist of collecting samples of crabs to analyze how metacercarie is distributed in the body, purposefully infecting dogs in order to extract worms from it to later examine under a microscope, and various other tests and assays.

Prevention and Treatment

First and foremost, health education is necessary in order to teach people the ways in which to evade infections of *P. skrjabini*. Through a survey of people in the Three Gorges Reservoir in China where *P. skrjabini* cases have appeared, Zhang et al. (2012) reported that out of a sampling of 724 people, the consumption rate of raw crab was 68.09%. And out of the 213 individuals that tested positive for *P. skrjabini* infection, every single one of them had consumed raw crab. Reasons behind this eating habit can be explained by the various culinary preferences and unique customs in the local area where *Paragonimus* species are endemic. But, people must be educated in ways that encourage fully cooking meals that contain crab and routinely washing their hands after handling any crustacean. In Korea, for instance, education proved to be successful in reducing the prevalence of paragonimiasis cases after people learned about proper eating habits, ways to reduce pollution, and the importance of not using crayfish juice in medicine. The same method can be applied in areas where *P. skrjabini* remains a health concern, for all it takes is cooking the crabs to kill the *P. skrjabini* parasite.

Praziquantel can be used to treat individuals infected with *P. skrjabini* parasites. Prazinquatal acts quickly to cure paragonimiasis, showing results after even one day, unlike the previously used drug bithionol, which required extended treatment and involved more severe side effects. The latest drug being investigated to treat paragonimiasis is triclabendazole.

Schistosoma

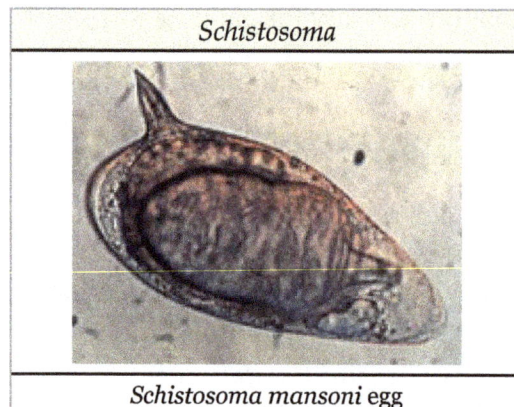

Schistosoma

Schistosoma mansoni egg

A genus of trematodes, *Schistosoma*, commonly known as blood-flukes, are parasitic flatworms responsible for a highly significant group of infections in humans termed schistosomiasis. Schistosomiasis is considered by the World Health Organization as the second most socioeconomically devastating parasitic disease, (after malaria), with hundreds of millions infected worldwide.

Adult flatworms parasitize blood capillaries of either the mesenteries or plexus of the bladder, depending on the infecting species. They are unique among trematodes and any other flatworms in that they are dioecious with distinct sexual dimorphism between male and female. Thousands of eggs are released and reach either the bladder or the intestine (according to the infecting species), and these are then excreted in urine or feces to fresh water. Larvae must then pass through an intermediate snail host, before the next larval stage of the parasite emerges that can infect a new mammalian host by directly penetrating the skin.

Evolution

The origins of this genus remain unclear. For many years it was believed that this genus had an African origin, but DNA sequencing suggests that the species (*S. edwardiense* and *S. hippopotami*) that infect the hippo (*Hippopotamus amphibius*) could be basal. Since hippos were present in both Africa and Asia during the Cenozoic era the genus might have originated as parasites of hippos. The original hosts for the South East Asian species were probably rodents.

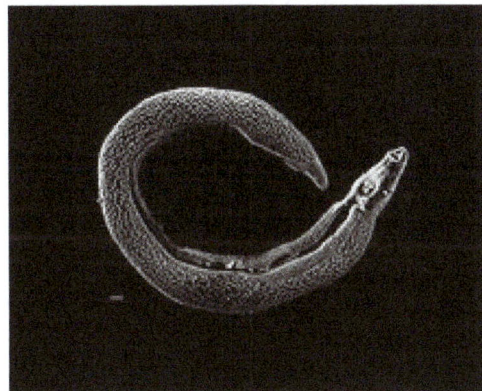

Electron micrograph of an adult male *Schistosoma* parasite worm. The bar (bottom left) represents a length of 500 μm.

Based on the phylogenetics of the host snails it seems likely that the genus evolved in Gondwana between 70 million years ago and 120 million years ago.

The sister group to *Schistosoma* is a genus of elephant-infecting schistosomes — *Bivitellobilharzia*. The cattle, sheep, goat and cashmere goat parasite *Orientobilharzia turkestanicum* appears to be related to the African schistosomes. This latter species has since been transferred to the genus *Schistosoma*.

Within the *haematobium* group *S. bovis* and *S. curassoni* appear to be closely related as do *S. leiperi* and *S. mattheei*.

S. mansoni appears to have evolved in East Africa 0.43–0.30 million years ago.

S. incognitum and *S. nasale* are more closely related to the African species rather than the *japonicum* group.

S. sinensium appears to have radiated during the Pliocene.

S. mekongi appears to have invaded South East Asia in the mid-Pleistocene.

Estimated speciation dates for the *japonicum* group: ~3.8 million years ago for *S. japonicum*/ South East Asian schistosoma and ~2.5 million years ago for *S. malayensis*/*S. mekongi*.

Schistosoma turkestanicum is found infecting red deer in Hungary. These strains appear to have diverged from those found in China and Iran. The date of divergence appears to be 270,000 years before present.

Taxonomy

The genus *Schistosoma* as currently defined is paraphyletic, so revisions are likely. Over twenty species are recognised within this genus.

The genus has been divided into four groups — *indicum*, *japonicum*, *haematobium* and *mansoni*. The affinities of the remaining species are still being clarified.

Thirteen species are found in Africa. Twelve of these are divided into two groups — those with a lateral spine on the egg (*mansoni* group) and those with a terminal spine (*haematobium* group).

The Mansoni Group

The four *mansoni* group species are: *S. edwardiense*, *S. hippotami*, *S. mansoni* and *S. rodhaini*.

The Haematobium Group

The nine *haematobium* group species are: *S. bovis*, *S. curassoni*, *S. guineensis*, *S. haematobium*, *S. intercalatum*, *S. kisumuensis*, *S. leiperi*, *S. margrebowiei* and *S. matthei*.

S. leiperi and *S. matthei* appear to be related. *S. margrebowiei* is basal in this group. *S. guineensis* is the sister species to the *S. bovis* and *S. curassoni* grouping. *S. intercalatum* may actually be a species complex of at least two species.

The Indicum Group

The *indicum* group has three species: *S. indicum*, *S. nasale* and *S. spindale*. This group appears to have evolved during the Pleistocene. All use pulmonate snails as hosts. *S. spindale* is widely distributed in Asia, but is also found in Africa. They occur in Asia and India.

S. indicum is found in India and Thailand.

The indicum group appears to be the sister clade to the African species.

The Japonicum Group

The *japonicum* group has three species: *S. japonicum*, *S. malayensis* and *S. mekongi*.

S. sinensium is a sister clade to the *S. japonicum* group and is found in China.

S. ovuncatum forms a clade with *S. sinensium* and is found in northern Thailand. The definitive host is the black rat (*Rattus rattus*) and the intermediate host is the snail *Tricula bollingi*. This species is known to use snails of the family Pomatiopsidae as hosts.

S. incognitum appears to be basal in this genus. It may be more closely related to the African/ Indian species than to the Southeast Asian group. This species uses pulmonate snails as hosts. Examination of the mitochondria suggests that *Schistosoma incognitum* may be a species complex.

New Species

As of 2012, four additional species have been transferred to this genus., previously classified as species in the genus *Orientobilharzia*. Orientobilharzia differs from Schistosoma morphologically only on the basis of the number of testes. A review of the morphological and molecular data has shown that the differences between these genera are too small to justify their separation. The four species are

- *Schistosoma bomfordi*

- *Schistosoma datta*

- *Schistosoma harinasutai*

- *Schistosoma turkestanicum*

Hybrids

The hybrid *S. haematobium-S.guineenis* was observed in Cameroon in 1996. *S. haematobium* could establish itself only after deforestation of the tropical rainforest in Loum next to the endemic *S. guineensis*; hybridization led to competitive exclusion of S. guineensis.

In 2003, a *S. mansoni-S. rodhaini* hybrid was found in snails in western Kenya, As of 2009, it had not been found in humans.

In 2009, *S. haematobium–Schistosoma bovis* hybrids were described in in northern Senegalese children. The Senegal River Basin had changed very much since the 1980s after the Diama Dam in Senegal and the Manantali Dam in Mali had been built. The Diama dam prevented ocean water to enter and allowed new forms of agriculture. Human migration, increasing number of livestock and sites where human and cattle both contaminate the water facilitated mixing between the different schistosomes in Nder e.g. The same hybrid was identified during the 2015 investigation of a schistosomiasis outbreak on Corsica, traced to the Cavu river.

Geographical Distribution

Schistosoma species have been found in tropical areas of Africa, the Middle East and Asia as well as the Caribbean and South America. There had been no cases in Europe since 1965, until an outbreak occurred on Corsica.

Schistosomiasis

The parasitic flatworms of *Schistosoma* cause a group of chronic infections called schistosomiasis known also as bilharziasis. An anti-schistosome drug is a schistosomicide.

Species Infecting Humans

Parasitism of humans by *Schistosoma* appears to have evolved at least three occasions in both Asia and Africa.

- *S. guineensis*, a recently described species, is found in West Africa. Known snail intermediate hosts include *Bulinus forskalii*.

- *S. haematobium*, commonly referred to as the *bladder fluke*, originally found in Africa, the Near East, and the Mediterranean basin, was introduced into India during World War II. Freshwater snails of the *Bulinus* genus are an important intermediate host for this parasite. Among final hosts humans are most important. Other final hosts are rarely baboons and monkeys.

- *S. intercalatum*. The usual final hosts are humans. Other animals can be infected experimentally.

- *S. japonicum*, whose common name is simply *blood fluke*, is widespread in East Asia and the southwestern Pacific region. In Taiwan this species only affects animals, not humans. Freshwater snails of the *Oncomelania* genus are an important intermediate host for *S. japonicum*. Final hosts are humans and other mammals including cats, dogs, goats, horses, pigs, rats and water buffalo.

- *S. malayensis* This species appears to be a rare infection in humans and is considered to be a zoonosis. The natural vertebrate host is von Muller's rat (*Rattus muelleri*). The snail host(s) are Robertsiella species (*R. gismanni*, *R. kaporensis* and *R. silvicola* (see Attwood et al. 2005 Journal of Molluscan Studies Volume 71, Issue 4 pp. 379–391).

- *S. mansoni*, found in Africa, Brazil, Venezuela, Suriname, the lesser Antilles, Puerto Rico, and the Dominican Republic. It is also known as *Manson's blood fluke* or *swamp fever*. Freshwater snails of the *Biomphalaria* genus are an important intermediate host for this trematode. Among final hosts humans are most important. Other final hosts are baboons, rodents and raccoons.

- *S. mekongi* is related to *S. japonicum* and affects both the superior and inferior mesenteric veins. *S. mekongi* differs in that it has smaller eggs, a different intermediate host (*Neotricula aperta*) and longer prepatent period in the mammalian host. Final hosts are humans and dogs. The snail *Tricula aperta* can also be experimentally infected with this species.

Human Schistosomes		
Scientific Name	**First Intermediate Host**	**Endemic Area**
Schistosoma guineensis	*Bulinus forskalii*	West Africa
Schistosoma intercalatum	*Bulinus* spp	Africa
Schistosoma haematobium	*Bulinus* spp.	Africa, Middle East
Schistosoma japonicum	*Oncomelania* spp.	China, East Asia, Philippines
Schistosoma malayensis	*Robertsiella* spp.	Southeast Asia

Schistosoma mansoni	Biomphalaria spp.	Africa, South America, Caribbean, Middle East
Schistosoma mekongi	Neotricula aperta	Southeast Asia

Species Infecting Other Animals

Schistosoma indicum, Schistosoma nasale, Schistosoma spindale,Schistosoma leiperi are all parasites of ruminants.

Schistosoma edwardiense and *Schistosoma hippopotami* are parasites of the hippo.

Schistosoma ovuncatum and *Schistosoma sinensium* are parasites of rodents.

Morphology

Adult schistosomes share all the fundamental features of the digenea. They have a basic bilateral symmetry, oral and ventral suckers, a body covering of a syncytial tegument, a blind-ending digestive system consisting of mouth, esophagus and bifurcated caeca; the area between the tegument and alimentary canal filled with a loose network of mesoderm cells, and an excretory or osmoregulatory system based on flame cells. Adult worms tend to be 10–20 mm (0.39–0.79 in) long and use globins from their hosts' hemoglobin for their own circulatory system.

Reproduction

Unlike other trematodes, the schistosomes are dioecious, *i.e.*, the sexes are separate. The two sexes display a strong degree of sexual dimorphism, and the male is considerably larger than the female. The male surrounds the female and encloses her within his *gynacophoric canal* for the entire adult lives of the worms. As the male feeds on the host's blood, he passes some of it to the female. The male also passes on chemicals which complete the female's development, whereupon they will reproduce sexually. Although rare, sometimes mated schistosomes will "divorce", wherein the female will leave the male for another male. The exact reason is not understood, although it is thought that females will leave their partners to mate with more genetically distant males. Such a biological mechanism would serve to decrease inbreeding, and may be a factor behind the unusually high genetic diversity of schistosomes.

Genome

The genomes of *Schistosoma haematobium*, *S. japonicum* and *S. mansoni* have been reported.

History

The eggs of these parasites were first seen by Theodor Maximilian Bilharz, a German pathologist working in Egypt in 1851 who found the eggs of *Schistosoma haematobium* during the course of a post mortem. He wrote two letters to his former teacher von Siebold in May and August 1851 describing his findings. Von Siebold published a paper in 1852 summarizing Bilharz's findings. Bilharz wrote a paper in 1856 describing the worms more fully and he named them *Distoma haematobium*. Their unusual morphology meant that they could not be comfortably included in *Distoma*. So in 1856 Meckel von Helmsback created the genus *Bilharzia* for them. In 1858 David Friedrich Weinland proposed the name *Schistosoma* (Greek: "split body") after the male worms' morphology. Despite *Bilharzia* having precedence, the genus name *Schistosoma* was officially adopted by the International Commission on Zoological Nomenclature. The term *Bilharzia* to describe infection with these parasites is still in use in medical circles.

Bilharz also described *Schistosoma mansoni*, but this species was redescribed by Louis Westenra Sambon in 1907 at the London School of Tropical Medicine who named it after his teacher Patrick Manson.

In 1898, all then known species were placed in a subfamily by Stiles and Hassel. This was elevated to family status by Looss in 1899. Poche in 1907 corrected a grammatical error in the family name. The life cycle was determined by the Brazilian parasitologist Pirajá da Silva (1873-1961) in 1908.

In 2009, the genomes of *Schistosoma mansoni* and *Schistosoma japonicum* were decoded opening the way for new targeted treatments. In particular, the study discovered that the genome of *S. mansoni* contained 11,809 genes, including many that produce enzymes for breaking down proteins, enabling the parasite to bore through tissue. Also, *S. mansoni* does not have an enzyme to make certain fats, so it must rely on its host to produce these

Schistosoma Mansoni

Schistosoma mansoni is a significant parasite of humans, a trematode that is one of the major agents of the disease schistosomiasis which is one type of helminthiasis, a neglected tropical disease. The schistosomiasis caused by *Schistosoma mansoni* is intestinal schistosomiasis.

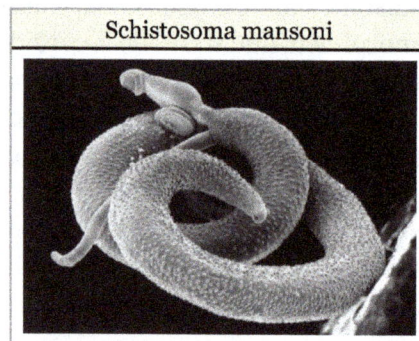

Schistosoma mansoni

Schistosomes are atypical trematodes in that the adult stages have two sexes (dioecious) and are located in blood vessels of the definitive host. Most other trematodes are hermaphroditic and are found in the intestinal tract or in organs, such as the liver. The lifecycle of schistosomes includes two hosts: a definitive host (i.e. human) where the parasite undergoes sexual reproduction, and a single intermediate snail host where there are a number of asexual reproductive stages. *S. mansoni* is named after Sir Patrick Manson, who first identified it in Formosa (now Taiwan).

Morphology of Adult Schistosomes

Schistosomes, unlike other trematodes, are long and slim worms. The male *S. mansoni* is approximately 1 cm long (0.6–1.1 cm) and is 0.1 cm wide. It is white, and it has a funnel-shaped oral sucker at its anterior end followed by a second pediculated sucker. The external part of the worm is composed of a double bilayer, which is continuously renewed as the outer layer, known as the membranocalyx, and is shed continuously. The tegument bears a large number of small tubercules. The suckers have small thorns in their inner part as well as in the buttons around them. The male genital apparatus is composed of 6 to 9 testicular masses, situated dorsally. There is one deferent canal beginning at each testicle, which is connected to a single deferent that dilates into a reservatory, the seminal vesicle, located at the beginning of the gynacophoric canal. The copula happens through the coaptation of the male and female genital orifices.

The female has a cylindrical body, longer and thinner than the male's (1.2 to 1.6 cm long by 0.016 cm wide). The female parasite is darker, and it looks gray. The darker color is due to the presence of a pigment (hemozoin) in its digestive tube. This pigment is derived from the digestion of blood. The ovary is elongated and slightly lobulated and is located on the anterior half of the body. A short oviduct conducts to the ootype, which continues with the uterine tube. In this tube it is possible to find 1 to 2 eggs (rarely 3 to 4) but only 1 egg is observed in the ootype at any one time. The genital pore opens ventrally. The posterior two-thirds of the body contain the vittelogenic glands and their winding canal, which unites with the oviduct a little before it reaches the ootype.

The digestive tube begins at the anterior extremity of the worm, at the bottom of the oral sucker. The digestive tube is composed of an esophagus, which divides in two branches (right and left) and that reunite in a single cecum. The intestines end blindly, meaning that there is no anus.

Physiology

Feeding and Nutrition

Developing *Schistosoma mansoni* worms that have infected their definitive hosts, prior to the sexual pairing of males and females, require a nutrient source in order to properly develop from cercariae to adults. The developing parasites lyse host red blood cells to gain access to nutrients; the hemoglobin and amino acids the blood cells contain can be used by the worm to form proteins. While hemoglobin is digested intracellularly, initiated by salivary gland enzymes, iron waste products cannot be used by the worms, and are typically discarded via regurgitation.

Kasschau et al. (1995) tested the effect of temperature and pH on the ability of developing *S. mansoni* to lyse red blood cells. The researchers found that the parasites were best able to destroy red blood cells for their nutrients at a pH of 5.1 and a temperature of 37 °C.

Locomotion

S. mansoni is locomotive in primarily two stages of its life cycle: as cercariae swimming freely through a body of freshwater to locate the epidermis of their human hosts, and as developing and fully-fledged adults, migrating throughout their primary host upon infection. Cercariae are attracted to the presence of fatty acids on the skin of their definitive host, and the parasite responds to changes in light and temperature in their freshwater medium to navigate towards the skin. Ressurreicao et al. (2015) tested the roles of various protein kinases in the ability of the parasite to navigate its medium and locate a penetrable host surface. Extracellular signal-regulated kinase and protein kinase C both respond to changes in medium temperature and light levels, and the stimulation of p38 mitogen-activated protein kinase, associated with recognition of parasite host surface, results in a glandular secretion that deteriorates the host epidermis, and allows the parasite to burrow into its host.

The parasite's nervous system contains bilobed ganglia and several nerve cords which splay out to every surface of the body; serotonin is a transmitter distributed widely throughout the nervous system and plays an important role in nervous reception, and stimulating mobility.

Life Cycle

After the eggs of the human-dwelling parasite are emitted in the faeces and into the water, the ripe miracidium hatches out of the egg. The hatching happens in response to temperature, light and dilution of faeces with water. The miracidium searches for a suitable freshwater snail (*Biomphalaria glabrata*, *Biomphalaria straminea*, *Biomphalaria tenagophila* or *Biomphalaria sudanica*) to act as an intermediate host and penetrates it. Following this, the parasite develops via a so-called mother-sporocyst and daughter-sporocyst generation to the cercaria. The purpose of the growth in the snail is the numerical multiplication of the parasite. From a single miracidium result a few thousand cercaria, every one of which capable of infecting a human.

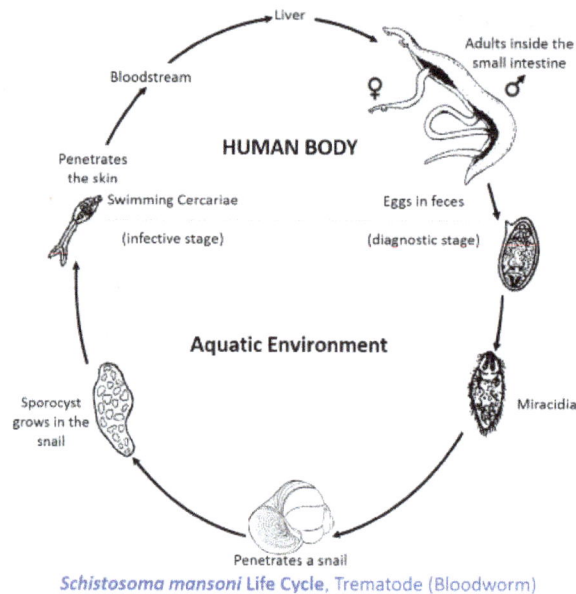

Schistosoma mansoni Life Cycle, Trematode (Bloodworm)

Schistosoma mansoni life cycle

Libora et al. (2010) have detected in Venezuela, that a land snail *Achatina fulica* can also serve as a host of *Schistosoma mansoni*.

The cercaria emerge from the snail during daylight and they propel themselves in water with the aid of their bifurcated tail, actively seeking out their final host. When they recognise human skin, they penetrate it within a very short time. This occurs in three stages, an initial attachment to the skin, followed by the creeping over the skin searching for a suitable penetration site, often a hair follicle, and finally penetration of the skin into the epidermis using cytolytic secretions from the cercarial post-acetabular, then pre-acetabular glands. On penetration, the head of the cercaria transforms into an endoparasitic larva, the schistosomule. Each schistosomule spends a few days in the skin and then enters the circulation starting at the dermal lymphatics and venules. Here, they feed on blood, regurgitating the haem as hemozoin. The schistosomule migrates to the lungs (5–7 days post-penetration) and then moves via circulation through the left side of the heart to the hepatoportal circulation (>15 days) where, if it meets a partner of the opposite sex, it develops into a sexually mature adult and the pair migrate to the mesenteric veins. Such pairings are monogamous.

Male schistosomes undergo normal maturation and morphological development in the presence or absence of a female, although behavioural, physiological and antigenic differences between males from single-sex, as opposed to bisex, infections have been reported. On the other hand, female schistosomes do not mature without a male. Female schistosomes from single-sex infections are underdeveloped and exhibit an immature reproductive system. Although the maturation of the female worm seems to be dependent on the presence of the mature male, the stimuli for female growth and for reproductive development seem to be independent from each other.

The adult female worm resides within the adult male worm's gynaecophoric canal, which is a modification of the ventral surface of the male, forming a groove. The paired worms move against the flow of blood to their final niche in the mesenteric circulation, where they begin egg production (>32 days). The *S. mansoni* parasites are found predominantly in the small inferior mesenteric blood vessels surrounding the large intestine and caecal region of the host. Each female lays approximately 300 eggs a day (one egg every 4.8 minutes), which are deposited on the endothelial lining of the venous capillary walls. Most of the body mass of female schistosomes is devoted to the reproductive system. The female converts the equivalent of almost her own body dry weight into eggs each day. The eggs move into the lumen of the host's intestines and are released into the environment with the faeces.

Genome

Schistosoma mansoni has 8 pairs of chromosomes (2n = 16)—7 autosomal pairs and 1 sex pair. The female schistosome is heterogametic, or ZW, and the male is homogametic, or ZZ. Sex is determined in the zygote by a chromosomal mechanism. The Schistosoma genome is approximately 270 MB with a GC content of 34%, 4–8% highly repetitive sequence, 32–36% middle repetitive sequence and 60% single copy sequence. Numerous highly or moderately repetitive elements have been identified, and their frequency in genomic sequence data also suggests at least 30% repetitive DNA. Chromosomes range in size from 18 to 73 MB and can be distinguished by size, shape, and C banding. There are estimated to be 15–20 thousand expressed genes.

In 2000, the first BAC library of Schistosome was constructed. In June 2003, a ~5x whole genome shotgun sequencing project was initiated at the Sanger Institute. Together with the shotgun data being generated by TIGR, an ~8x coverage of the genome will be obtained, assembled and annotated. Also in 2003, 163,000 ESTs (expressed sequence tags) were generated (by a consortium headed by the University of São Paulo) from six selected developmental stages of this parasite, resulting in 31,000 assembled sequences and an estimated 92% of the 14,000-gene complement.

In 2009 the genomes of both *S. mansoni* and *S. japonicum* were published, with each describing 11,809 and 13,469 genes, respectively. Analysis of the *S. mansoni* genome highlighted expansions in protease families and deficiencies in lipid anabolism; both observations can be directly related to S. mansoni's parasitic lifestyle. The former included the invadolysin (host penetration) and cathepsin (blood-feeding) gene families, while the latter encompassed several enzymes required for the de novo synthesis of fatty acids and sterols (so the worm must rely on its host for these products). The results open the way for research on new targeted treatments.

In 2012, an improved version of the *S. mansoni* genome was published, with only 885 scaffolds and more than 81% of the bases organised into chromosomes. In the same study, the authors have also used transcriptome sequencing (RNA-seq) from four time points in the parasite's lifecycle to refine 45% gene predictions and profile their expression levels.

Pathology

Schistosome eggs, which may become lodged within the hosts tissues, are the major cause of pathology in schistosomiasis. Some of the deposited eggs reach the outside environment by passing through the wall of the intestine; the rest are swept into the circulation and are filtered out in the periportal tracts of the liver, resulting in periportal fibrosis. Onset of egg laying in humans is sometimes associated with an onset of fever (Katayama fever). This "acute schistosomiasis" is not, however, as important as the chronic forms of the disease. For *S. mansoni* and *S. japonicum*, these are "intestinal" and "hepatic schistosomiasis", associated with formation of granulomas around trapped eggs lodged in the intestinal wall or in the liver, respectively. The hepatic form of the disease is the most important, granulomas here giving rise to fibrosis of the liver and hepatosplenomegaly in severe cases. Symptoms and signs depend on the number and location of eggs trapped in the tissues. Initially, the inflammatory reaction is readily reversible. In the latter stages of the disease, the pathology is associated with collagen deposition and fibrosis, resulting in organ damage that may be only partially reversible.

Granuloma formation is initiated by antigens secreted by the miracidium through microscopic pores within the rigid egg shell, and there is strong evidence that the vigorous granulomatous response, rather than the direct action of parasite egg antigens, is responsible for the pathologic tissue manifestations in schistosomiasis. The granulomas formed around the eggs impair blood flow in the liver and, as a consequence, induce portal hypertension. With time, collateral circulation is formed and the eggs disseminate into the lungs, where they cause more granulomas, pulmonary arteritis and, later, cor pulmonale. A contributory factor to portal hypertension is Symmers' fibrosis, which develops around branches of the portal veins. This fibrosis occurs only many years after the infection and is presumed to be caused in part by soluble egg antigens and various immune cells that react to them.

Recent research has shown that granuloma size is consistent with levels of IL-13, which plays a prominent role in granuloma formation and granuloma size. IL-13 receptor α 2 (IL-13Rα2) binds IL-13 with high affinity and blocks the effects of IL-13. Thus, this receptor is essential in preventing the progression of schistosomiasis from the acute to the chronic (and deadly) stage of disease. Synthetic IL-13Rα2 given to mice has resulted in significant decreases in granuloma size, implicating IL-13Rα2 as an important target in schistosomiasis.

Evasion of Host Immunity

Adult and larval worms migrate through the host's blood circulation avoiding the host's immune system. The worms have many tools that help in this evasion, including the tegument, antioxidant proteins, and defenses against host membrane attack complex (MAC).

Tegument

The tegument coats the worm and acts as a physical barrier to host antibodies and complement.

Antioxidant proteins

Host immune defenses are capable of producing superoxide, which has a tremendous detrimental effect on the worm. However, they are able to produce a number of antioxidant proteins that block the effect of superoxide. Schistosomes have four superoxide dismutases, and levels of these proteins increase as the schistosome develops and matures.

Antioxidant pathways were first recognised as a chokepoints for Schistosomes and later extended to other trematodes and cestodes. Targeting of this pathway with different inhibitors of the central antioxidant enzyme Thioredoxin Glutathione Reductase (TGR) results in reduced viability of worms

Defense against host MAC

Schistosomes have evolved ways to block host complement proteins. Immunocytochemistry techniques have found decay accelerating factor (DAF) protein on the tegument. DAF is found on host cells and protects host cells by blocking formation of MAC. It has also been found that the schistosome genome consists of human CD59 homologs. CD59 inhibits MAC.

Epidemiology

Schistosoma mansoni infects about 83 million people worldwide (data from 1999), causing the disease intestinal schistosomiasis (schistosomiasis caused by all the *Schistosoma* species infects over 200 million people.)

S. mansoni is the most widespread of the human-infecting schistosomes, and is present in 54 countries. These countries are predominantly in South America and the Caribbean, Africa including Madagascar, and the Middle East.

S. mansoni is commonly found in places with poor sanitation. Because of the parasite's fecal-oral transmission, bodies of water that contain human waste can be infectious. Water that contains large populations of the intermediate host snail species is more likely to cause infection. Young children living in these areas are at greatest risk because of their tendency to swim and bathe in cercaria-infected waters longer than adults . Any one travelling to the areas described above, and who is exposed to contaminated water, is at risk of schistosomiasis.

History

Schistosoma mansoni reached Egypt via infected slaves and baboons from the Land of Punt through migrations that occurred possibly as early as the Vth Dynasty.

Schistosoma Intercalatum

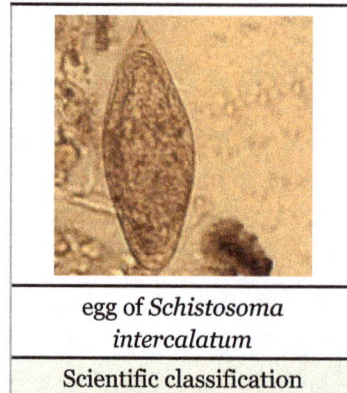

egg of *Schistosoma intercalatum*

Scientific classification

Schistosoma intercalatum is a parasitic worm found in parts of western and central Africa. There are two strains: the Lower Guinea strain and the Zaire strain. *S. intercalatum* is one of the major agents of the rectal form of schistosomiasis, also called bilharzia. It is a trematode, and being part of the *Schistosoma* genus, it is commonly referred to as a blood-fluke since the adult resides in blood vessels.

Humans are the definitive host and two species of freshwater snail make up the intermediate host, *Bulinus forskalii* for the Lower Guinea strain and *Bulinus africanus* for the Zaire strain.

Morphology

The clinically defining characteristic of most schistosome species are their eggs' size and shape. The eggs of *Schistosoma intercalatum* have a terminal spine and tend to be moderately larger than those of *S. haematobium* (approximately 130 × 75 μm). The origin of the name 'intercalatum' is from the observation that their eggs are of an intermediate range between the smaller *S. haematobium* and larger *S. bovis*. These eggs are unique because they will stain red when exposed to the Ziehl-Neelsen technique, aiding in identification. When viewed using scanning electron microscopy, it can be observed that the *S. intercalatum's* surface has a much lower amount of integumental elevations, or bosses, than *S. mansoni*. This feature is consistent with the tegument appearance of other terminally spined schistosomes.

Life Cycle

Schistosoma intercalatum's life cycle is very similar to that of *S. haematobium*, except for some key differences. To start the life cycle, the human host releases eggs with its feces. In water, the eggs hatch to become miracidia, which penetrate the freshwater snail intermediate host. *S. intercalatum* has two major strains, each with its own preferred bulinid host. The Zaire strain will use *Bulinus africanus*, while the Lower Guinea strain will use the extremely common *B. forskalii* as its intermediate host. The miracidia penetrate the snail tissue, and inside they become sporocysts and multiply. The sporocysts then mature into cercariae inside the snail host and are ready to leave. The cercariae are free-swimming in the surrounding water until they find their definitive host: a human. If there is a small temperature change, the cercariae of *S. intercalatum* will form concentrated aggregates near the surface of the water. This mechanism for body heat detection of a potential host restricts the formation of viable cercariae to small streams and slow moving bodies of water because of their high sensitivity.

The cercariae penetrate through the human's skin and lose their tail, becoming schistosomulae. The schistosomulae then migrate to the hepatic portal system of the liver to mature into adults. As adults, they make their way to the inferior mesenteric vein and mate, producing thousands of eggs. These eggs migrate down to the mesenteric venules of the colon and form polyps as the eggs attempt to cross into the lumen. *S. intercalatum's* eggs are specific to the colon, making them unique among the infectious African schistosomes.

Epidemiology

S. intercalatum is at risk of endangerment in large part due to the introduction of invasive species into its native habitat. Since 1973, both *S. mansoni* and *S. haematobium* have been found in places that have been traditionally inhabited by *S. intercalatum*. This is thought to be because of the increase in transportation accessibility and the increase in forestry jobs in these habitats. Male *S. mansoni* and *S. haematobium* will both take priority over *S. intercalatum* when it comes to mate selection, leading to a smaller proportion of female *S. intercalatum* available for mating. While crosses with *S. mansoni* give no viable offspring, the pairing with a male *S. haematobium* will result in a hybrid organism. Most hybrids will have a diluted genome that is more closely related to *S. haematobium*, helping to bring about a decline in *S. intercalatum* populations. The other obstacle restricting the parasite's population growth is its selective distribution. The cercariae are very particular over where they develop, needing small, forested areas with streams to infect their human host. There are only a few of these regions in Africa, and they decrease in size every day due to deforestation.

Prevalence

In 2009, there were an estimated 200 million human infections of schistosomiasis. In 1999, the noted number of *S. intercalatum* infections was 1.73 million.

Distribution

There are two major strains of *S. intercalatum*, both living in forested areas of Africa. One strain lives in the Congo area, particularly Zaire, and the other strain lives in the Lower Guinea area, mainly in Cameroon. Cameroon is a place of scientific interest because it is where all three species of human schistosomes live. Most relevant research conducted on *S. intercalatum* was performed in, or around, the Loum area in Cameroon.

Pathology

Symptoms

Symptoms of all forms of schistosomiasis are caused by the immune system's reaction to the eggs, rather than the adult worms themselves. A few hours to days after cercariae invade the skin, some people experience pruritus and raised papules at the site of penetration. This is called cercarial dermatitis, also known as swimmer's itch. It can last up to a few weeks, although, this stage is usually asymptomatic in local populations. *S. intercalatum* is associated with lower morbidity than the other schistosomes that infect humans. In a study done on schoolchildren in the Republic of São Tomé and Principe in western Africa–where *S. intercalatum* and *S. haematobium* are endemic—the only schistosome present in the sample was *S. intercalatum*, an overall prevalence of 10.9 percent in stool specimens.

Unlike the more pathogenic species, infection with *S. intercalatum* is usually only associated with bloody stool, and sometimes splenomegaly. Blood in the stool is caused by "inflammation, hypertrophy, and ulceration of the mucosa" of the intestine. These signs can be difficult to interpret because effected populations are often infected with multiple intestinal parasites. Clinical presentation of an established *S. intercalatum* infection can be different in the local population and non-immune tourists. The majority of infections of foreign travelers are asymptomatic and go unnoticed. Chronic schistosomiasis results in granulomata forming around eggs in the mesenteric vessels.

Diagnosis

Diagnosis is usually made using clinical and epidemiological information. Infection with *S. intercalatum* can be distinguished from that of *S. mansoni* or *S. haematobium* based on where eggs manifest outside the body and the morphology of the eggs. In Africa, the only species of schistosome are *S. intercalatum*, *S. mansoni*, and *S. haematobium*. *S. haematobium* causes urinary schistosomiasis, so eggs will be shed in the urine; *S. mansoni* and *S. intercalatum* reside in the mesenteric venous plexus, so eggs will be shed in the feces. Looking at the stool specimen under a microscope, the species can be distinguished; *S. intercalatum* eggs have a terminal spine (as seen in the figure above) and *S. mansoni* eggs have a lateral spine.

Serologic testing looks for the presence of antibodies against the adult schistosome in the blood. This can only take place 6 to 8 weeks after initial infection in order for the parasite to reach the adult stage and the immune system to produce antibodies against it. However, serologic testing is not useful for patients with previous infections.

Treatment

Praziquantel is an effective treatment against all species of *Schistosoma* that infect humans. Administering treatment at the correct time is important since the drug only works against the adult worm and there must be a strong antibody response from the immune system. Thus, it should be administered 6 to 8 weeks after suspected infection (contact with infested freshwater). There has been limited evidence on possible drug resistance among the schistosomes due to reports of low cure rates. Oxaminiquine is another treatment for schistosomiasis, but it is not widely available, nor is it routinely used.

Schistosoma Haematobium

Adults are found in the venous plexuses around the urinary bladder and the released eggs travels to the wall of the urine bladder causing haematuria and fibrosis of the bladder. The bladder becomes calcified, and there is increased pressure on ureters and kidneys otherwise known as hydronephrosis. Inflammation of the genitals due to *S. haematobium* may contribute to the propagation of HIV. Studies have shown the relationship between S. haematobium infection and the development of squamous cell carcinoma of the bladder.

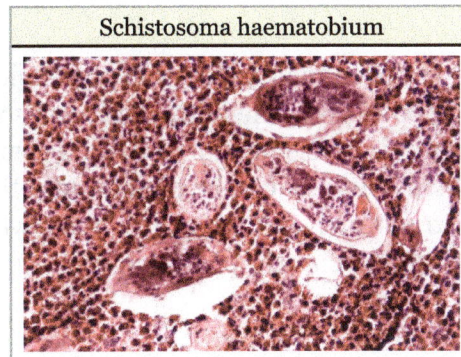

Schistosoma haematobium is an important digenetic trematode, and is found in Africa and the Middle East. It is a major agent of schistosomiasis; more specifically, it is associated with urinary schistosomiasis.

Life Cycle

The free swimming infective larval cercariae burrow into human skin when it comes into contact with contaminated water. The cercariae enter the blood stream of the host where they travel to the liver to mature into adult flukes. In order to avoid detection by the immune system inside the host, the adults have the ability to coat themselves with host antigen After a period of about three weeks the young flukes migrate to the urinary bladder veins to copulate. The female fluke lays as many as 30 eggs per day which migrate to the lumen of the urinary bladder and ureters. The eggs are eliminated from the host into the water supply with micturition. In fresh water, the eggs hatch forming free swimming miracidia which penetrate into the intermediate snail host *(Bulinus* spp., e.g. *B. globosus, B. forskalii, B. nyassanus* and *B. truncatus)*. Inside the snail, the miracidium sheds it epithelium and develops into a mother sporocyst. After two weeks the mother begins forming daughter sporocysts. Four weeks after the initial penetration of the miracidium into the snail furcocercous cercariae begin to be released. The cercariae cycle from the top of the water to the bottom for three days in the search of a human host. Within half an hour the cercariae enter the host epithelium.

Geographical Distribution

S. hematobium infects snails in Africa and the Middle East.

Diagnosis

Traditionally, diagnoses has been made by examination of the urine for eggs. In chronic infections, or if eggs are difficult to find, an intradermal injection of schistosome antigen to form a wheal is effective in determining infection. Alternatively diagnosis can be made by complement fixation tests, As of 2012 commercial serologic tests included ELISA and an Indirect immunofluorescence test, hampered by a low sensitivity ranging from 21% to 71%.

Prevention

The main cause of schistomiasis is the dumping of human waste into water supplies. Hygienic disposal of waste would be sufficient to eliminate the disease.

Immunopathology

The immune system responds to eggs in liver causing hypersensitivity; an immune response is necessary to prevent damage to hepatocytes. The hosts' antibodies which bind to the tegument of the Schistosome don't bind for long since the tegument is shed every few hours. The schistosome can also take on host proteins. Schistomiasis can be divided into three phases: (1) the migratory phase lasting from penetration to maturity,(2) the acute phase which occurs when the schistosomes begin producing eggs, and (3) the chronic phase which occurs mainly in endemic areas.

Pathology

The ova are initially deposited in the muscularis propria which leads to ulceration of the overlaying tissue. Infections are characterized by pronounced acute inflammation, squamous metaplasia, blood and reactive epithelial changes. Granulomas and multinucleated giant cells may be seen.

Treatment

The drug of choice is praziquantel, a quinolone derivative.

Schistosoma Japonicum

Schistosoma japonicum is an important parasite and one of the major infectious agents of schistosomiasis.This parasite has a very wide host range, infecting at least 31 species of wild mammals, including 9 carnivores, 16 rodents, one primate (Human), two insectivores and three artiodactyls and therefore it can be considered a true zoonosis.

Disease

Schistosoma japonicum is the only human blood fluke that occurs in China and Philippines. It is the cause of schistosomiasis japonica, a disease that still remains a significant health problem especially in lake and marshland regions. Schistosomiasis is an infection caused mainly by three schistosome species; *Schistosoma mansoni, Schistosoma japonicum* and *Schistosoma haematobium. S. japonicum* being the most infectious of the three species. Infection by schistosomes is followed by an acute Katayama fever. Historical accounts of Katayama disease dates back to the discovery of *S. Japonicum* in Japan in 1904. The disease was named after an area it was endemic to, Katayama district, Hiroshima, Japan. If left untreated, it will develop into a chronic condition characterized by hepatosplenic disease and impaired physical and cognitive development. The severity of *S. japonicum* arises in 60% of all neurological diseases in schistosomes due to the migration of schistosome eggs to the brain.

Morphology

The *S. japonicum* worms are yellow or yellow-brown. The males of this species are slightly larger than the other Schistosomes and they measure ~ 1.2 cm by 0.5 mm. The females measure 2 cm by 0.4 mm. The adult worms are longer and narrower than the related *S. mansoni* worms.

By electron microscopy there are no bosses or spines on the dorsal surface of the male, which is ridged and presents a spongy appearance. Many spines cover the inner surface of the oral sucker and extend to the pharyngeal opening. The oral sucker shows a rim with spines of variable size and sharpness inward and outward from the rim. The ventral sucker possesses many spines which are smaller than in the oral sucker. The lining of the gynecophoric canal is roughened by minute spines. The integument of the female is ridged and pitted and possesses fewer spines than in the oral sucker, the ventral sucker, and the gynecophoric canal of the male. Anterior to the acetabulum, the integumental surfaces are devoid of spines. However, in the other areas, spines are equally distributed except for the vicinity of the excretory pore.

The ova are about 55 - 85 µm by 40 - 60 µm, oval with a minute lateral spine or knob.

Life Cycle

The life cycles of *Schistosoma japonicum* and *Schistosoma mansoni* are very similar. In brief, eggs of the parasite are released in the feces and if they come in contact with water they hatch into free-swimming larva, called miracidia. The larva then has to infect a snail of the genus *Oncomelania* such as species of *Oncomelania hupensis* within one or two days. Inside the snail, the larva undergo asexual reproduction through a series of stages called sporocysts. After the asexual reproduction stage cercaria (another free-swimming larva) are generated in large quantities, which then leave (shed into the environment) the snail and must infect a suitable vertebrate host. Once the cercaria penetrates the skin of the host it loses its tail and becomes a schistosomule. The worms then migrate through the circulation ending at the mesenteric veins where they mate and start laying eggs. Each pair deposits around 1500 – 3500 eggs per day in the vessels of the intestinal wall. The eggs infiltrate through the tissues and are passed in the feces.

Pathology

Once the parasite has entered the body and begun to produce eggs, it uses the hosts' immune system (granulomas) for transportation of eggs into the gut. The eggs stimulate formation of granuloma around them. The granulomas, consisting of motile cells, carry the eggs to the intestinal lumen. When in the lumen, granuloma cells disperse leaving the eggs to be excreted within feces. Unfortunately, about two-thirds of eggs are not excreted, instead they build up in the gut. Chronic infection can lead to characteristic Symmer's fibrosis (also known as "clay pipe stem" fibroses, these occur due to intrahepatic portal vein calcification which assume the shape of a clay pipe in cross section). *S. japonicum* is the most pathogenic of the schistosoma species because it produces up to 3,000 eggs per day, ten times greater than that of *S. mansoni*..

As a chronic disease, *S. japonicum* can lead to Katayama fever, liver fibrosis, liver cirrhosis, liver portal hypertension, splenomegaly, and ascites. Some eggs may pass the liver and enter lungs, nervous system and other organs where they can adversely affect the health infected individual.

Diagnosis

Microscopic identification of eggs in stool or urine is the most practical method for diagnosis. Stool examination should be performed when infection with *S. mansoni* or *S. japonicum* is suspected, and urine examination should be performed if *S. haematobium* is suspected.

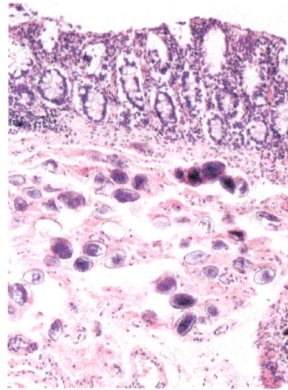

Histopathological image of old state of schistosomiasis incidentally found at autopsy. The deposition of calcified eggs in the colonic submucosa suggests prior infection of Schistosoma japonicum.

Eggs can be present in the stool in infections with all Schistosoma species. The examination can be performed on a simple smear (1 to 2 mg of fecal material). Since eggs may be passed intermittently or in small amounts, their detection will be enhanced by repeated examinations and/or concentration procedures (such as the formalin - ethyl acetate technique). In addition, for field surveys and investigational purposes, the egg output can be quantified by using the Kato-Katz technique (20 to 50 mg of fecal material) or the Ritchie technique.

Eggs can be found in the urine in infections with *S. haematobium* (recommended time for collection: between noon and 3 PM) and with *S. japonicum*. Detection will be enhanced by centrifugation and examination of the sediment. Quantification is possible by using filtration through a Nucleopore membrane of a standard volume of urine followed by egg counts on the membrane. Tissue biopsy (rectal biopsy for all species and biopsy of the bladder for *S. haematobium*) may demonstrate eggs when stool or urine examinations are negative.

Since the eggs of *S. japonicum* are small, concentration techniques may be required. Biopsies are mostly performed to test for chronic schistomiasis with no eggs. An ELISA test can be performed to test for antibodies specific to schistosomes. A positive result indicates a present or recent infection (within the past two years). Ultrasonographic examination can be performed to assess the extent of hepatic and spleen-related morbidity. The problems with immunodiagnostic methods are that 1) It is only positive a certain time after infection 2) They can cross interact with other helminthes infections.

Treatment

The therapy of choice is praziquantel, a quinolone derivative. Praziquantel is generally administered in an oral form in one or two doses from 40–60 mg/kg body weight.

Combination treatment may prevent morbidity due to schistosomiasis. Praziquantel is most active against adult worms. However, it has been found that artemether prevents the development of adult worms, thus decreasing egg production in the host. If both praziquantel and artemether can be used together, the entire lifespan of *S. japonicum* would be covered in the vertebrate host.

Prevention

Human waste should be hygienically disposed of. Human waste in water with the *Oncomelania* snail intermediate host is a major cause to the perpetuation of schistosomiasis. To prevent this from occurring, human waste should never be used for nightsoiling (fertilization of crops with human waste) and unsanitary conditions should be improved. To avoid infection, individuals should avoid contact with water that is contaminated by human or animal waste, especially water sources that are endemic to *Oncomelania* snails.

If necessary to enter potentially infected water, cercarial repellants and cercaricidal ointments can be applied to the skin before entering the water. Barrier cream with a dimethicone base offered high levels of protection for at least 48 hours.

The search for a practical vaccine continues and could greatly benefit affected areas.

Control

Control against infection of *S. japonicum* requires multiple efforts consisting of education, eliminating the disease from infected individuals, controlling the vector, and providing a protective vaccine.

Education can be highly effective, but difficult with lack of resources. Also, asking people to change customs, traditions and behaviors can prove a difficult task.

Controlling *S. japonicum* with molluscicide has proved ineffective because *Oncomelania* snails are amphibious and only frequent water to lay their eggs.

Social Impacts

Individuals at risk to infection from *S. japonicum* are farmers who often wade in their irrigation water, fisherman that wade in streams and lakes, children that play in water, and people who wash clothes in streams.

Ablution is a religious requirement in some Moslem countries to achieve cleanliness by washing of the anal or urethral orifices after urination or defecation. However, this act leads to the transmission of schistosomiasis. The water source typically used for ablution is a contaminated river or canal from previously deposited human waste, thus furthering the contamination in the population.

Important factors to influence transmission are age, sex of an individual, as well as the economic and educational level of a population. Males show the highest rates of infection, as well as the most intense infections. This may be due to occupational risk. As was the case of Suriname, the highest prevalence occurs in both sexes where both male and females work in fields.

Climate change may have potential impact on the transmission of schistosomiasis in China. The development of *S. japonicum* in the intermediate host *Oncomelania hupensis* occurred at the threshold of 15.4 °C. Previously, *O. hupensis* has been restricted to areas where the mean January temperature has been over 0 °C. With rising climate change, it is predicted that by 2050, *O. hupensis* will be able to cover 8.1% of the surface area of China, thus leading to greater concern to new populations being at risk to schistosomiasis.

Echinostoma

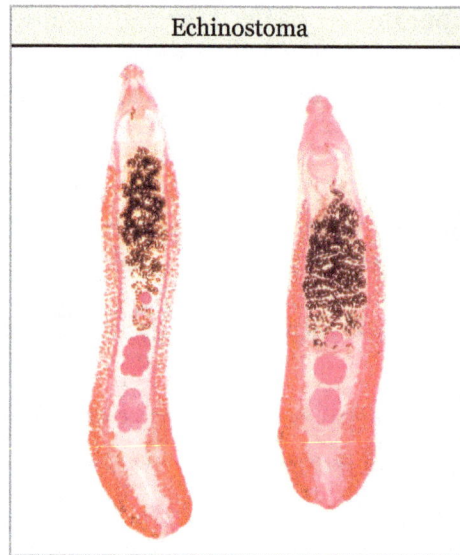

Echinostoma

Echinostoma is a genus of trematode parasites, which can infect both humans and other animals. These intestinal flukes have a three-host life cycle with snails or aquatic organisms as intermediate hosts, and a variety of animals, including humans, as their definitive hosts.

Echinostoma infect the gastrointestinal tract of humans, and can cause a disease known as echinostomiasis. The parasites are spread when humans or animals eat infected raw or undercooked food, such as bivalve molluscs or fish

Taxonomy

It has been estimated that there are between 61 and 114 species of *Echinostoma*. *Echinostoma* are difficult to classify and are known as a cryptic species (different lineages are considered to be the same species, due to high morphological similarity between them). Many species of *Echinostoma* have been re-classified several times. For example, the species now known as *Echinostoma caproni*, was previously known by a variety of names including *E. liei*, *E. parasensei* and *E. togoensis*.

Methods for classifying *Echinostoma* species, such as the *Echinostoma revolutum* group, were devised by Kanev. The *Echinostoma* species in this group are now classified according to their shared morphological and biological characteristics, such as the presence of 37 collar spines.

Molecular methods, such as sequencing mitochondrial DNA and ribosomal DNA, are also used to distinguish between species of *Echinostoma* as an alternative to morphological classification methods.

Morphology

Echinostoma are internal digenean trematode parasites which infect the intestines and bile duct of their hosts.

The length and width of adult *Echinostoma* varies between species, but they tend to be approximately 2-10mm × 1-2mm in size.

Adult *Echinostoma* have two suckers: an anterior oral sucker and a ventral sucker. They also have a characteristic head collar with spines surrounding their oral sucker. The number of collar spines varies between *Echinostoma* species, but there are usually between 27 and 51. These spines can be arranged in one or two circles around the sucker, and their arrangement may be a characteristic feature of an *Echinostoma* species.

Echinostoma have a digestive system consisting of a pharynx, oesophagus and an excretory pore.

Echinostoma are hermaphrodites, and have both male and female reproductive organs. The testes are found in the posterior part of the fluke's body, in the area furthest from the mouth. The ovary is also found in this location, close to the testes.

The eggs (ova) of *Echinostoma* are operculate and vary in size, but are typically in the range of 80-135μm × 55-80μm.

Geographic Distribution

The genus *Echinostoma* has a global distribution. These parasites are particularly common in South East Asia, in countries such as South Korea and the Philippines. However, they are also found in some European countries, and species such as *Echinostoma trivolvis* are found in North America.

Life Cycle

Echinostoma have three hosts in their life cycle: a first intermediate host, a second intermediate host and a definitive host. Snail species such as *Lymnaea* spp. are common intermediate hosts for *Echinostoma*, although fish and other bivalve molluscs can be also be intermediate hosts for these parasites.

Echinostoma species have low specificity for their definitive hosts, and can infect a variety of different species of animal, including amphibians, aquatic birds, mammals and humans. A definitive host which is infected with *Echinostoma* will shed unembryonated *Echinostoma* eggs in their faeces. When the eggs are in contact with fresh water they may become embryonated, and will then hatch and release miracidia. The miracidia stage of *Echinostoma* is free-swimming, and actively penetrates the first intermediate snail host, which then becomes infected.

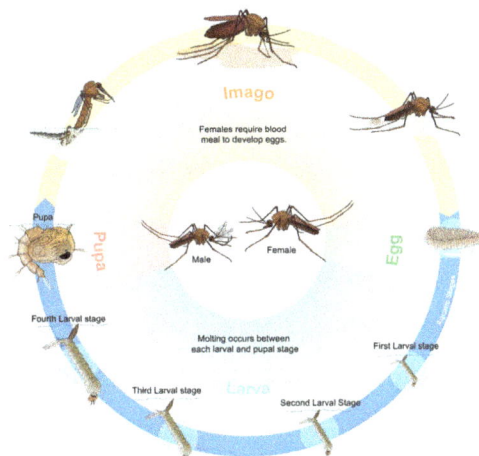

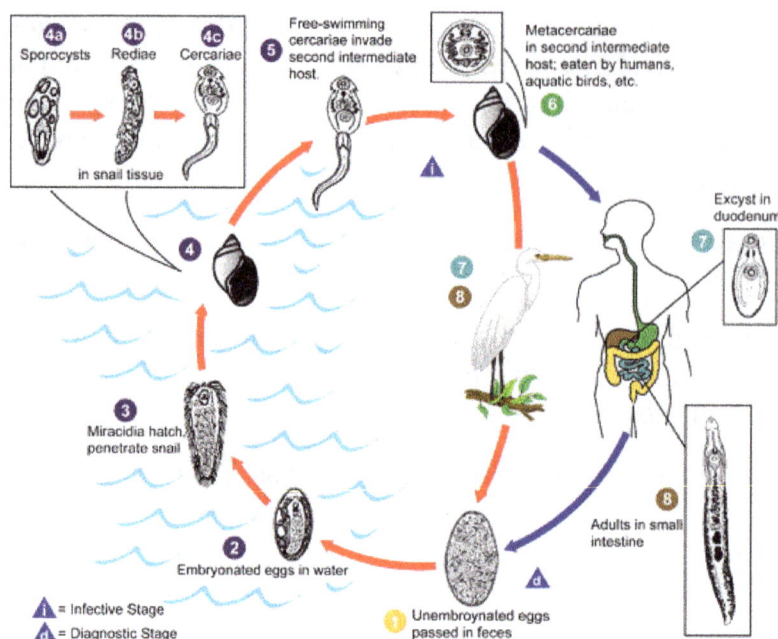

Life cycle of *Echinostoma*.

In the first intermediate host, the miracidium undergoes asexual reproduction for several weeks, which includes sporocyst formation, a few generations of rediae and the production of cercariae. The cercariae are released from the snail host into water and are also free-swimming. The cercariae penetrate a second intermediate host, or they remain in the first intermediate host, where they form metacercariae. Definitive hosts become infected by eating secondary hosts which are infected with metacercariae. Once the metacercariae have been eaten, they excyst in the intestine of the definitive host where the parasite then develops into an adult.

Echinostoma are hermaphrodites. A single adult individual has both male and female reproductive organs, and is capable of self-fertilization. Sexual reproduction of adult *Echinostoma* in the definitive host leads to the production of unembryonated eggs. The life cycle of *Echinostoma* is temperature dependent, and occurs quicker at higher temperatures. *Echinostoma* eggs can survive for about 5 months and still have the ability to hatch and develop into the next life cycle stage.

Echinostomiasis

Infection of humans with members of the family Echinostomatidae, including *Echinostoma*, can lead to a disease called echinostomiasis. *E. revolutum*, *E. echinatum*, *E. malaynum* and *E. hortense* are particularly common causes of *Echinostoma* infections in humans. Humans can become infected with *Echinostoma* by eating infected raw or undercooked food, particularly fish, clams and snails. Infection with these parasites tends to be common in regions where cultural dishes require the use of raw or undercooked food that may be infected with *Echinostoma*. A mild infection may not have any symptoms. If symptoms are present they can include abdominal pain, diarrhoea, tiredness and weight loss.

Epidemiology of Echinostomiasis

Echinostomiasis is endemic in South East Asia and the Far East, in countries including China, Korea, Taiwan, Philippines, Malaysia, Indonesia and India. Echinostomiasis has also been reported in Japan, Singapore, Romania, Hungary and Italy. The prevalence of echinostomiasis varies between countries but there tend to be foci of infection in areas where raw or undercooked hosts of *Echinostoma*, such as snails or fish, are widely consumed.

Pathogenesis

Echinostoma are not highly pathogenic. Symptoms of greater severity tend to be seen in an echinostomiasis infection where there is a higher number of flukes. The flukes cause damage to the intestinal mucosa, which leads to ulceration and inflammation.

Diagnosis

An *Echinostoma* infection can be diagnosed by observing the parasite eggs in the faeces of an infected individual, under a microscope. Methods such as the Kato-Katz procedure can be used to do this. The eggs typically have a yellow-brown appearance, and are ellipsoid in shape. To confirm which species is causing the infection adult worms must be recovered from the infected individual, such as with anthelmintic treatment.

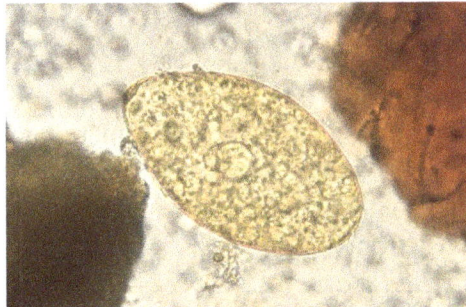

Unstained *Echinostoma* egg.

Treatment and Prevention

Echinostomiasis can be treated with the anthelmintic drug praziquantel, as for other intestinal trematode infections. Side effects of anthelmintic drug treatment may include nausea, abdominal pain, headaches or dizziness.

Echinostomiasis can be controlled at the same time as other food-borne parasite infections, using existing control programmes. Interrupting the parasite's lifecycle by efficient diagnosis and subsequent treatment of infected individuals, and preventing reinfection, may help to control this disease. As echinostomiasis is acquired through the consumption of raw or undercooked infected food, cooking food thoroughly will prevent infection.

Trichobilharzia Regenti

Trichobilharzia regenti is a neuropathogenic parasitic flatworm of birds which also causes cercarial dermatitis in humans. The species was originally described in 1998 in the Czech Republic and afterwards it was detected also in other European countries, e.g. Denmark, France, Iceland or Russia, and even in Iran. For its unique neurotropic behaviour in vertebrate hosts, the host-parasite interactions are extensively studied in terms of molecular biology, biochemistry and immunology.

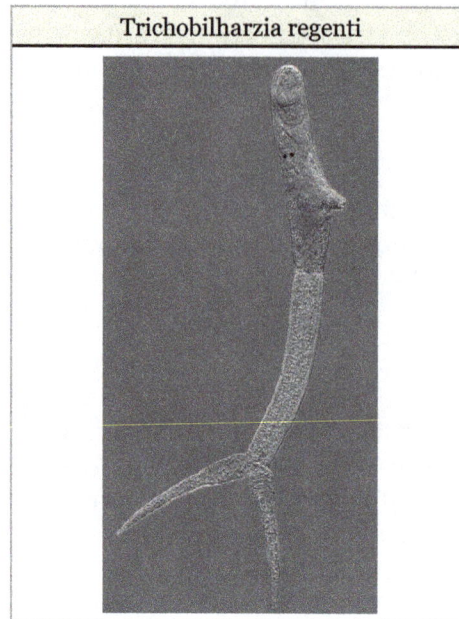

Trichobilharzia regenti

Life Cycle

The life cycle of *T. regenti* is analogous to that of human schistosomes. Adult flukes mate in a nasal mucosa of anatid birds (e.g. *Anas platyrhynchos* and *Cairina moschata*) and produce eggs with miracidia which hatch directly in the host tissue and leak outside when the bird is drinking/feeding. Once in water, the miracidia swim using their cilia and actively search for a proper molluscan intermediate host (*Radix lagotis, Radix labiata, Radix peregra*). In the snail, the miracidia develop into a primary sporocyst in which secondary sporocysts are formed and give rise to cercariae later on.

Cercariae, infective larvae, exit the snail and penetrate the skin of an avian host. After penetration, they transform to schistosomula (subadult stage) and look for peripheral nerves to use them to get to the spinal cord. Through it they continue their migration to the brain and, finally, the nasal tissue in a bill. Here, they mature, copulate and lay eggs while causing pathology (inflammatory infiltration, haemorrhages).

If mammals are infected by cercariae (instead of birds), the parasites die in the skin being entrapped by immune response. The clinical manifestation of such infection is known as an neglected allergic disease called cercarial dermatitis (or swimmer's itch). In mice, especially in immunodeficient ones, migration of the parasite to the spinal cord was observed.

Migration in Vertebrate Hosts

When cercariae of *T. regenti* find either avian or mammalian host, they penetrate its skin. For this purpose, they are equipped with cysteine peptidases present in their excretory/secretory products, which are capable of keratin and collagen degradation. Experiments with laboratory prepared recombinant form of the cysteine peptidase cathepsin B2 of *T. regenti* (TrCB2) confirmed its ability to cleave skin proteins (collagen, keratin and elastin).

After penetration the skin, cercariae transform to schistosomula and start a migration through the host's body. They avoid penetration into blood capillaries and rather prefer entering peripheral nerves in host's limbs. Schistosomula are found in peripheral nerves of ducks and mice as soon as 1.5 and 1 day post infection (DPI), respectively. In both types of hosts, schistosomula exhibit a high affinity to the central nervous system which they enter *via* spinal roots. Based on recent observation by 3D imaging techniques (ultramicroscopy and micro-CT), schistosomula appear to migrate preferably through the white matter of the spinal cord in both birds and mammals.

The next course of the infection differs in final and accidental hosts. In ducks, schistosomula are observed in synsacral segments of a spinal cord 3 DPI and 7–8 days latter (10–11 DPI) they reach the brain. In their final localisation (the nasal tissue), they occur 13–14 DPI and laying eggs starts 15 DPI. In mice, the first schistosomula are found in a lumbar spinal cord as early as 2 DPI and *medulla oblongata* is invaded the day after, but only in some individuals. Most of schistosomula stay localised in the thoracic and cervical spinal cord and only exceptionally migrate to the brain. Neither the presence of worms has been detected in a nasal cavity nor has their maturation been noticed in the nervous tissue. Schistosomula development in mice is suppressed likely due to the host immune response and/or the presence/absence of some essential (nutritional, stimulatory) host factors.

Pathology in Vertebrate Hosts

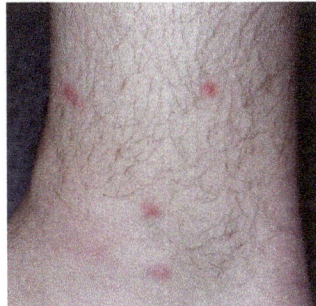

Cercarial dermatitis.

In vertebrate hosts infected by *T. regenti*, pathological states might be caused by:

- penetrating cercariae transforming to schistosomula in the skin,

- schistosomula migrating through the central nervous system,

- adults laying eggs in nasal mucosa (only in avian hosts).

Schistosomulum of *T. regenti*.

Although mice are accidental hosts, most of the studies dealing with the pathological effects of *T. regenti* were conducted on this model.

In the initial phase of the infection, early transformed schistosomula are localised in the skin. Information about the immune response in the skin of birds has not been completed yet. In mice, immediate oedema and thickening of the site appear as early as 30 minutes after the penetration of cercariae; erythema is evident as well. Within 48 hours, inflammatory foci containing neutrophils, eosinophils, macrophages, CD4+ lymphocytes and degranulating mast cells develop around the parasites. In case of repeated infections, the cellular infiltration is substantially elevated and the extensive inflammation may lead to formation of large abscesses or even epidermal and/or dermal necrosis. In humans, the clinical symptoms of cercarial penetration consist of macules/papules formation at the sites where the parasite entered the skin accompanied by intensive itching. The manifestation is more severe in previously sensitised people. This disease caused not only by *T. regenti* but also by cercariae of other bird schistosome species is called cercarial dermatitis. It is regarded as a neglected allergic disease.

Eggs of *T. regenti*.

The next phase of *T. regenti* infection is represented by schistosomula migration in the central nervous system. This is accompanied by serious neurological malfunctions in birds that suffer from leg paralysis and balance disorders. At this stage, schistosomula feed on nervous tissue as demonstrated by detection of oligodendrocytes and neurons in the lumen of parasite's intestine. A cysteine peptidase cathepsin B1 of *T. regenti* (TrCB1) localised in intestines of migrating schistosomula is capable of myelin basic protein degradation, thus probably serving for nervous tissue digestion. Nonetheless, the nervous tissue ingestion has likely only a minor pathogenic effect on the host central nervous tissue. This is underpinned by observations of leg paralysis only in immunocompromised hosts, whereas in experiments with immunocompetent mouse strains, the infected animals did not reveal any neurological disorders. The neurological symptoms originate probably in mechanical damage of the nervous tissue leading to dystrophic or even necrotic changes of neurons and axonal injury. The cause of it is large migrating schistosomula (approximately 340×80 μm) which are not destroyed by proper immune response.

In avian hosts, *T. regenti* reaches the nasal tissue where it mates and lay eggs. The gross pathology at this site consists of focal haemorrhages dispersed all over the mucosa. Infiltrates of lymphocytes are present around the eggs and even granulomas containing lymphocytes, eosinophils and heterophils form at later phases. Similar infiltrates are present around free miracidia, but the granuloma formation was not recorded. Interestingly, no cell reaction was noted in the vicinity of adult worms.

Schistosomatidae

Schistosomatidae is a family of digenetic trematodes with complex parasitic life cycles. Immature developmental stages of schistosomes are found in molluscs and adults occur in vertebrates. The best studied group, the blood flukes of the genus *Schistosoma*, infect and cause disease in humans. Other genera which are infective to non-human vertebrates can cause mild rashes in humans.

Schistosomatids are dioecious (individuals are of separate sexes) which is exceptional with regards to their phylum, Platyhelminthes, in which most species are hermaphrodidic (individuals possess both male and female reproductive systems).

History

The eggs of these parasites were first seen by Theodor Bilharz, a German pathologist working in Egypt in 1851 who found the eggs during the course of a post mortem. He wrote two letters to his former teacher von Siebold in May and August 1851 describing his findings. von Siebold wrote a paper (published in 1852) summarizing Bilharz's findings. Bilhart's wrote a paper in 1856 describing the worms more fully and he named them *Distoma haematobium*. Their unusual morphology meant that they could not be comfortably included in *Distoma* so in 1856 Meckel von Helmsback created the genus *Bilharzia* for them. In 1858 Weinland proposed the name *Schistosoma* (Greek: 'split body') after the male worms' morphology. Despite *Bilharzia* having precedence the genus name *Schistosoma* was officially adopted by the International Commission on Zoological Nomenclature.

In 1898 all the then known species were placed in a subfamily by Stiles and Hassel. This was then elevated to family status by Looss in 1899. Poche in 1907 corrected a grammatical error in the family name. The life cycle was determined by da Silva in 1908.

Evolution

There are a number of different families of blood fluke including the *Schistosomatidae*. The others include the spirorchids (turtle parasites) and the sanguinicolids (fish parasites).

The *Schistosomatidae* are considered venous system specialists and their sister group are vascular system generalists - the *Spirorchidae*.

The *Schistosomatidae* differ from the other blood flukes in having separate sexes and homeothermic hosts. They have compensated for the reduction in potential reproductive partners by

- an increased overdispersion in the vertebrate host

- the reduced egg hatching time in the external environment

- the formation of permanent pairs mimicking the hermaphroditic condition

- the increased longevity in the definitive host

- increased fecundity.

Colonization of the venous system was made possible

- the evolutionary radiation into terrestrial vertebrates

- the increased immunopathology associated with the high, constant body temperature of homeothermic vertebrates.

The arterial dwelling spirorchids release eggs in the direction of blood flow, resulting in a wide dissemination of eggs within the host. The lower body temperature of poikilotherms is accompanied by a seasonal nature of the immune response in these hosts resulting in a quantitatively reduced pathogenesis. Hosts that did succumb to the infection would most likely die in water where eggs could be released by predation, scavengers, or decomposition and develop successfully.

Colonization of the venous system by schistosomes required precise egg placement because their eggs are released against the blood flow. Eggs are then sequestered within the portal system (or perivesicular plexus in some species) of homeotherms which restricts egg dispersal but limits the resulting pathology to less sensitive organs. A significant number of eggs may escape into the external environment before a heavily infected host is incapacitated by, or dies from, the infection.

The first hosts of the schistosome were birds. Based on their current geographical spread the most likely place of origin of this family is Asia with subsequent spread to India and Africa.

Only one species is known to infect crocodiles - *Griphobilharzia amoena*. This species infects the freshwater crocodile *Crocodylus johnstoni*. Phylogenetic analysis shows that the genus *Griphobilharzia* rather than being a basal schistosome is a relation of the spirorchiids that infect freshwater turtles. It has also shown that the spirorchiids are the closest relations of the schistosoma.

An outline of the evolution of the schistosoma is now possible. The ancestral species infected freshwater turtles and the life cycle included gastropod hosts. Some of these species in their turn infected the marine turtles. At some point members of species infecting marine turtles developed the ability to infect birds - most likely waterfowl. This probably occurred somewhere in the Asian continent presumably at or near the coast. The bird species eventually developed the ability to infect mammals. This last development seems to have occurred in Gondwana between 120 million years ago and 70 million years ago.

Taxonomy

The family was created in 1926 by Stiles and Hassel for the *Schistosoma*, the *Sanguinicolidae* and the *Spirochidae*. It has since been divided into four subfamilies: the *Schistosomatinae*, the *Bilharziellinae*, the *Denrobilharziinae* and the *Gigantobilharziinae*. In the *Gigantobilharziinae* the ventral sucker is absent and the female genital pore is medial near the anterior end of the body. In the *Bilharziellinae* the ventral pore in the female is always posterior to the ventral sucker. Both the *Bilharziellinae* and the *Gigantobilharziinae* are found exclusively in birds while the *Schistosomatinae* are found in both mammals and birds. In the *Denrobilharziinae* both suckers are absent anmd the caecum has numerous branches. In this latter family there is one genus (*Denrdobilharina*) with two species (*Dendrobilharzina purvulenta* and *Dendrobilharzina asicaticus*).

There are 12 genera in this family. Of these 7 infect birds: the others infect mammals including humans. There are about 100 known species in this family. The largest genus within the family Schistosomatidae is the *Trichobilharzia* with over 40 species.

The genera are:

- Subfamily *Bilharziellinae*

  o Genus *Bilharziella* - birds (*Setophaga pensylvanica*, ducks)

- Subfamily *Denrobilharziinae*

  o Genus *Dendritobilharzia* - birds (ducks, swans)

- Subfamily *Gigantobilharziinae*

  o *Gigantobilharzia* - birds (*Spinus tristis tristis*)

- Subfamily *Schistosomatinae*

  o *Allobilharzia* - birds (*Cygnus cygnus*)

  o *Austrobilharzia* - birds (mainly waterfowl)

  o *Bivitellobilharzia* - mammals (elephants)

  o *Heterobilharzia* - mammals (raccoons)

  o *Microbilharzia* - birds (*Larus canescens*)

  o *Ornithobilharzia* - mammals (cattle, cats)

  o *Schistomatium* - mammals (rodents)

  o *Schistosoma* - mammals including humans

  o *Trichobilharzia* - birds (mainly waterfowl)

Orientobilharzia differ from *Schistosoma* only in the number of testes. The four species in this genus have recently (2012) been moved to the genus *Schistosoma* on the basis of morphology and molecular studies. The genus name should now be regarded as a junior synonym of *Schistomsoma*.

The genera *Bivitellobilharzia* and *Schistosoma* form a clade in this family. *Austrobilharzia* and *Ornithobilharzia* are the closest relations of this clade.

Heterobilharzia and *Schistomatium* form a separate clade indicating that adaption to mammalian hosts has occurred at least twice. The species in these genera are found in North American mammals suggesting that transmission occurred via birds with subsequent transmission to mammals.

The genus *Griphobilharzia* which infects reptiles has been shown to be a member of the spirorchiid family whose other members infect freshwater turtles. Like the spirorchiids and unlike the schistomes *Griphobilharzia* preferentially inhabits the arterial system rather than the venous. This genus was originally grouped with the schistosoma on the basis of the existence of two sexes and other morphological features.

References

- Torgerson, P; Claxton JR (1999). "Epidemiology and Control". In Dalton, JP. Fasciolosis. Wallingford, Oxon, UK: CABI Pub. pp. 113–149. ISBN 0-85199-260-9.

- Kotpal, RL (2012). Modern Text Book of Zoology: Invertebrates. New Delhi: Rastogi Publications. p. 338. ISBN 978-81-7133-903-7.

- Dennis J. Richardson; Peter J. Krause (6 December 2012). North American Parasitic Zoonoses. Springer Science & Business Media. p. 86. ISBN 978-1-4615-1123-6.

- Janovy, John; Schmidt, Gerald D.; Roberts, Larry S. (1996). Gerald D. Schmidt & Larry S. Roberts' Foundations of parasitology. Dubuque, Iowa: Wm. C. Brown. ISBN 0-697-26071-2.

- Desowitz, R. New Guinea Tapeworms and Jewish Grandmothers: Tales of Parasites and People. New York: WW Norton; 1987. ISBN 978-0-393-30426-8.

- Wilmer, Pat, Graham Stone, and Ian Johnston (2005). Environmental Physiology of Animals. United Kingdom: Blackwell Publishing. pp. 677-692. ISBN 9781405107242

- Huffman, Jane E; Fried, Bernard (1990). "Echinostoma and Echinostomiasis". In Baker, John R; Muller, Ralph. Advances in Parasitology. Academic Press Limited. pp. 215–269. ISBN 0-12-031729-X.

- Gorgas Case 5 - 2015 Series". The Gorgas Course in Clinical Tropical Medicine. University of Alabama. 2 March 2015. Retrieved 10 March 2015.

- Waikagul J. & Thaekham U. (2014). Approaches to Research on the Systematics of Fish-Borne Trematodes. Academic Press, 130 pp., page 6–7.

Understanding Roundworms

Roundworms species are almost one million in number. The types of roundworms explained within this chapter are anisakis, brugia malayi, dracunculus medinensis, loa loa filariasis and thelzia callipaeda. This text will not only provide an overview, it will also delve into the topics related to it.

Nematode

The nematodes or roundworms constitute the phylum Nematoda. They are a diverse animal phylum inhabiting a very broad range of environments. Nematode species can be difficult to distinguish, and although over 25,000 have been described, of which more than half are parasitic, the total number of nematode species has been estimated to be about 1 million. Unlike the phyla Cnidarians and Platyhelminthes (flatworms), nematodes have tubular digestive systems with openings at both ends.

Nematodes have successfully adapted to nearly every ecosystem from marine (salt water) to fresh water, to soils, and from the polar regions to the tropics, as well as the highest to the lowest of elevations. They are ubiquitous in freshwater, marine, and terrestrial environments, where they often outnumber other animals in both individual and species counts, and are found in locations as diverse as mountains, deserts and oceanic trenches. They are found in every part of the earth's lithosphere. They represent 90% of all animals on the ocean floor. Their numerical dominance, often exceeding a million individuals per square meter and accounting for about 80% of all individual animals on earth, their diversity of life cycles, and their presence at various trophic levels point at an important role in many ecosystems. Nematodes have even been found at great depth (0.9–3.6 km) below the surface of the Earth in gold mines in South Africa.

The many parasitic forms include pathogens in most plants and animals (including humans). Some nematodes can undergo cryptobiosis. One group of carnivorous fungi, the nematophagous fungi, are predators of soil nematodes. They set enticements for the nematodes in the form of lassos or adhesive structures.

Nathan Cobb, a nematologist, described the ubiquity of nematodes on Earth thus:

In short, if all the matter in the universe except the nematodes were swept away, our world would still be dimly recognizable, and if, as disembodied spirits, we could then investigate it, we should find its mountains, hills, vales, rivers, lakes, and oceans represented by a film of nematodes. The location of towns would be decipherable, since for every massing of human beings there would be a corresponding massing of certain nematodes. Trees would still stand in ghostly rows representing our streets and highways. The location of the various plants and animals would still be decipherable, and, had we sufficient knowledge, in many cases even their species could be determined by an examination of their erstwhile nematode parasites."

Taxonomy and Systematics

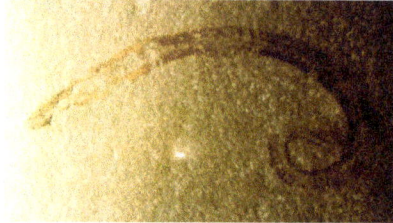

Eophasma jurasicum, a fossilized nematode

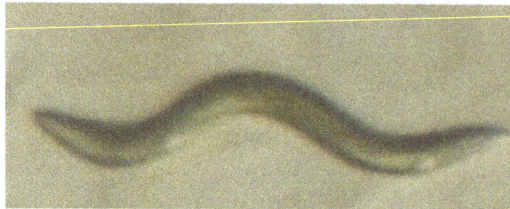

Caenorhabditis elegans

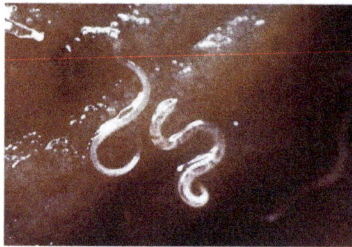

Rhabditia

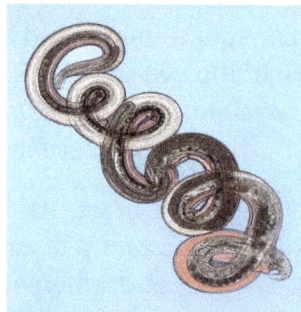

Nippostrongylus brasiliensis

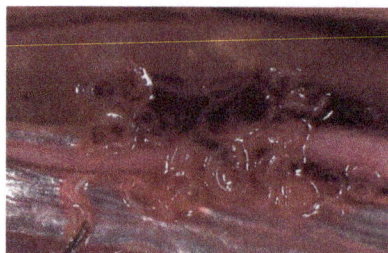

Unidentified Anisakidae (Ascaridina: Ascaridoidea)

Oxyuridae Threadworm

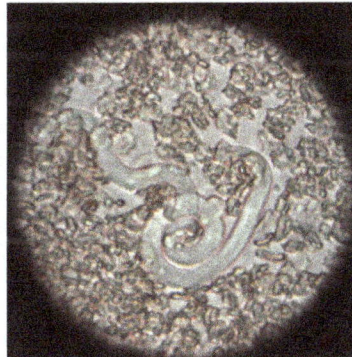

Spiruridae *Dirofilaria immitis*

History

In 1758, Linnaeus described some nematode genera (e.g., *Ascaris*), then included in Vermes.

The name of the group Nematoda, informally called "nematodes", came from Nematoidea, orig-inally defined by Karl Rudolphi (1808), from Ancient Greek *nêma, nêmatos*, 'thread') and (-*eidēs*, 'species'). It was treated as family Nematodes by Burmeister (1837).

At its origin, the "Nematoidea" erroneously included Nematodes, Nematomorpha and Gordiacei, attributed by von Siebold (1843). Along with Acanthocephala, Trematoda and Cestoidea, it formed the obsolete group Entozoa, created by Rudolphi (1808). They were also classed along with Acan-thocephala in the obsolete phylum Nemathelminthes by Gegenbaur (1859).

In 1861, K. M. Diesing treated the group as order Nematoda. In 1877, the taxon Nematoidea, in-cluding the family Gordiidae (horsehair worms), was promoted to the rank of phylum by Ray Lankester. In 1919, Nathan Cobb proposed that nematodes should be recognized alone as a phy-lum. He argued they should be called "nema" in English rather than "nematodes" and defined the taxon Nemates (Latin plural of *nema*). Since Cobb was the first to exclude all but nematodes from the group, some sources consider the valid taxon name to be Nemates or Nemata, rather than Nematoda.

Phylogeny

The phylogenetic relationships of the nematodes and their close relatives among the protosto-mian Metazoa are unresolved. Traditionally, they were held to be a lineage of their own but in the 1990s, they were proposed to form the group Ecdysozoa together with moulting animals, such as arthropods. The identity of the closest living relatives of the Nematoda has always been considered to be well resolved. Morphological characters and molecular phylogenies agree with placement of the roundworms as a sister taxon to the parasitic Nematomorpha; together they make up the Nematoida. Together with the Scalidophora (formerly Cephalorhyncha), the Nematoida form the clade Cycloneuralia, but much disagreement occurs both between and among the available morphological and molecular data. The Cycloneuralia or the Introverta—depending on the validity of the former—are often ranked as a superphylum.

Nematode Systematics

Due to the lack of knowledge regarding many nematodes, their systematics is contentious. An earliest and influential classification was proposed by Chitwood and Chitwood—later revised by Chitwood—who divided the phylum into two—the Aphasmidia and the Phasmidia. These were later renamed Adenophorea (gland bearers) and Secernentea (secretors), respectively. The Secernentea share several characteristics, including the presence of phasmids, a pair of sensory organs located in the lateral posterior region, and this was used as the basis for this division. This scheme was adhered to in many later classifications, though the Adenophorea were not a uniform group.

Initial studies of incomplete DNA sequences suggested the existence of five clades:

- Dorylaimida
- Enoplia
- Spirurina
- Tylenchina
- Rhabditina

As it seems, the Secernentea are indeed a natural group of closest relatives. But the "Adenophorea" appear to be a paraphyletic assemblage of roundworms simply retaining a good number of ancestral traits. The old Enoplia do not seem to be monophyletic either, but to contain two distinct lineages. The old group "Chromadoria" seem to be another paraphyletic assemblage, with the Monhysterida representing a very ancient minor group of nematodes. Among the Secernentea, the Diplogasteria may need to be united with the Rhabditia, while the Tylenchia might be paraphyletic with the Rhabditia.

The understanding of roundworm systematics and phylogeny as of 2002 is summarised below:

Phylum Nematoda

- Basal order Monhysterida
- Class Dorylaimida
- Class Enoplea
- Class Secernentea

- o Subclass Diplogasteria (disputed)

- o Subclass Rhabditia (paraphyletic?)

- o Subclass Spiruria

- o Subclass Tylenchia (disputed)

- • "Chromadorea" assemblage

Later work has suggested the presence of 12 clades. The Secernentea—a group that includes virtually all major animal and plant 'nematode' parasites—apparently arose from within the Adenophorea.

A major effort to improve the systematics of this phylum is in progress and being organised by the 959 Nematode Genomes.

A complete checklist of the World's nematode species can be found in the World Species Index:Nematoda.

An analysis of the mitochondrial DNA suggests that the following groupings are valid

- • subclass Dorylaimia

- • orders Rhabditida, Trichinellida and Mermithida

- • suborder Rhabditina

- • infraorders Spiruromorpha and Oxyuridomorpha

The Ascaridomorpha, Rhabditomorpha and Diplogasteromorpha appear to be related.

The suborders Spirurina and Tylenchina and the infraorders Rhabditomorpha, Panagrolaimomorpha and Tylenchomorpha are paraphytic.

The monophyly of the Ascaridomorph is uncertain.

Anatomy

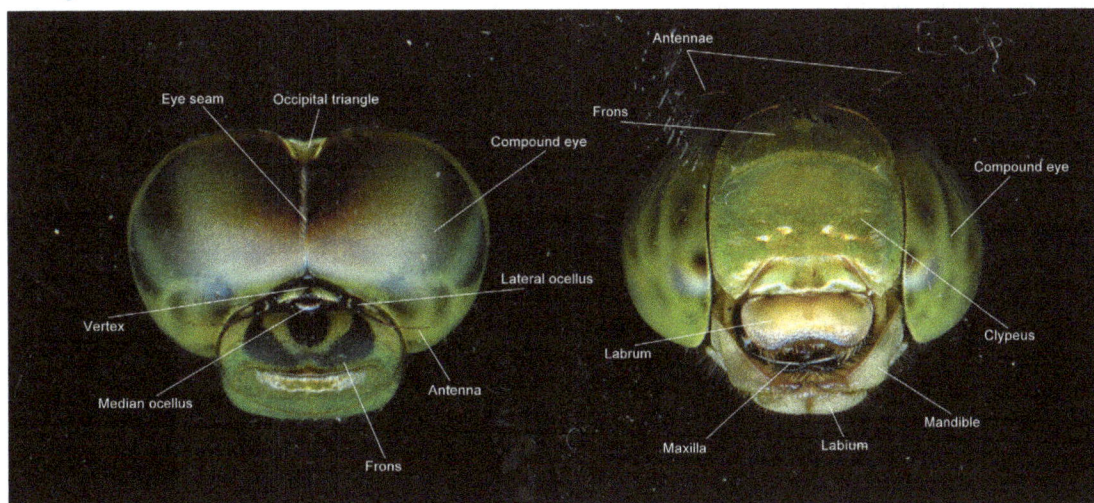

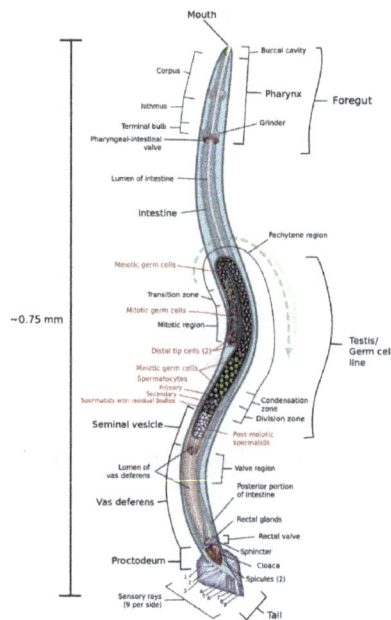

Internal anatomy of a male *C. elegans* nematode

Nematodes are slender worms: typically approximately 5 to 100 μm thick, and at least 0.1 mm (0.0039 in) but less than 2.5mm long. The smallest nematodes are microscopic, while free-living species can reach as much as 5 cm (2.0 in), and some parasitic species are larger still, reaching over a meter in length. The body is often ornamented with ridges, rings, bristles, or other distinctive structures.

The head of a nematode is relatively distinct. Whereas the rest of the body is bilaterally symmetrical, the head is radially symmetrical, with sensory bristles and, in many cases, solid 'head-shields' radiating outwards around the mouth. The mouth has either three or six lips, which often bear a series of teeth on their inner edges. An adhesive 'caudal gland' is often found at the tip of the tail.

The epidermis is either a syncytium or a single layer of cells, and is covered by a thick collagenous cuticle. The cuticle is often of complex structure, and may have two or three distinct layers. Underneath the epidermis lies a layer of longitudinal muscle cells. The relatively rigid cuticle works with the muscles to create a hydroskeleton as nematodes lack circumferential muscles. Projections run from the inner surface of muscle cells towards the nerve cords; this is a unique arrangement in the animal kingdom, in which nerve cells normally extend fibres into the muscles rather than *vice versa*.

Digestive System

The oral cavity is lined with cuticle, which is often strengthened with ridges or other structures, and, especially in carnivorous species, may bear a number of teeth. The mouth often includes a sharp stylet, which the animal can thrust into its prey. In some species, the stylet is hollow, and can be used to suck liquids from plants or animals.

The oral cavity opens into a muscular, sucking pharynx, also lined with cuticle. Digestive glands are found in this region of the gut, producing enzymes that start to break down the food. In stylet-bearing species, these may even be injected into the prey.

There is no stomach, with the pharynx connecting directly to a muscleless intestine that forms the main length of the gut. This produces further enzymes, and also absorbs nutrients through its single cell thick lining. The last portion of the intestine is lined by cuticle, forming a rectum, which expels waste through the anus just below and in front of the tip of the tail. Movement of food through the digestive system is the result of body movements of the worm. The intestine has valves or sphincters at either end to help control the movement of food through the body.

Excretory System

Nitrogenous waste is excreted in the form of ammonia through the body wall, and is not associated with any specific organs. However, the structures for excreting salt to maintain osmoregulation are typically more complex.

In many marine nematodes, one or two unicellular 'renette glands' excrete salt through a pore on the underside of the animal, close to the pharynx. In most other nematodes, these specialised cells have been replaced by an organ consisting of two parallel ducts connected by a single transverse duct. This transverse duct opens into a common canal that runs to the excretory pore.

Nervous System

Four peripheral nerves run the length of the body on the dorsal, ventral, and lateral surfaces. Each nerve lies within a cord of connective tissue lying beneath the cuticle and between the muscle cells. The ventral nerve is the largest, and has a double structure forward of the excretory pore. The dorsal nerve is responsible for motor control, while the lateral nerves are sensory, and the ventral combines both functions.

The nervous system is also the only place in the nematode body that contains cilia, which are all non-motile and with a sensory function.

At the anterior end of the animal, the nerves branch from a dense, circular nerve (nerve ring) round surrounding the pharynx, and serving as the brain. Smaller nerves run forward from the ring to supply the sensory organs of the head.

The bodies of nematodes are covered in numerous sensory bristles and papillae that together provide a sense of touch. Behind the sensory bristles on the head lie two small pits, or 'amphids'. These are well supplied with nerve cells, and are probably chemoreception organs. A few aquatic nematodes possess what appear to be pigmented eye-spots, but is unclear whether or not these are actually sensory in nature.

Reproduction

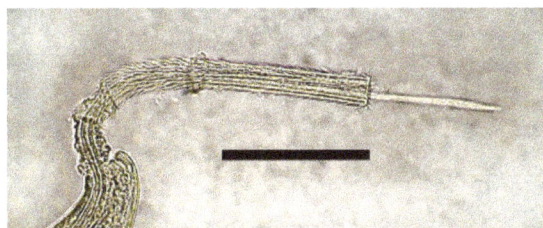

Extremity of a male nematode showing the spicule, used for copulation. Bar = 100 µm

Most nematode species are dioecious, with separate male and female individuals, though some, such as *Caenorhabditis elegans*, are androdioecious, consisting of hermaphrodites and rare males. Both sexes possess one or two tubular gonads. In males, the sperm are produced at the end of the gonad and migrate along its length as they mature. The testis opens into a relatively wide seminal vesicle and then during sex into a glandular and muscular ejaculatory duct associated with the vas deferens and cloaca. In females, the ovaries each open into an oviduct (in hermaphrodites, the eggs enter a spermatheca first) and then a glandular uterus. The uteri both open into a common vulva/ vagina, usually located in the middle of the morphologically ventral surface.

Reproduction is usually sexual, though hermaphrodites are capable of self-fertilization. Males are usually smaller than females/ hermaphrodites (often much smaller) and often have a characteristically bent or fan-shaped tail for holding the other sex. During copulation, one or more chitinized spicules move out of the cloaca and are inserted into the genital pore of the female. Amoeboid sperm crawl along the spicule into the female worm. Nematode sperm is thought to be the only eukaryotic cell without the globular protein G-actin.

Eggs may be embryonated or unembryonated when passed by the female, meaning their fertilized eggs may not yet be developed. A few species are known to be ovoviviparous. The eggs are protected by an outer shell, secreted by the uterus. In free-living roundworms, the eggs hatch into larvae, which appear essentially identical to the adults, except for an underdeveloped reproductive system; in parasitic roundworms, the life cycle is often much more complicated.

Nematodes as a whole possess a wide range of modes of reproduction. Some nematodes, such as *Heterorhabditis* spp., undergo a process called *endotokia matricida*: intrauterine birth causing maternal death. Some nematodes are hermaphroditic, and keep their self-fertilized eggs inside the uterus until they hatch. The juvenile nematodes will then ingest the parent nematode. This process is significantly promoted in environments with a low food supply.

The nematode model species *Caenorhabditis elegans* and *C. briggsae* exhibit androdioecy, which is very rare among animals. The single genus *Meloidogyne* (root-knot nematodes) exhibit a range of reproductive modes, including sexual reproduction, facultative sexuality (in which most, but not all, generations reproduce asexually), and both meiotic and mitotic parthenogenesis.

The genus *Mesorhabditis* exhibits an unusual form of parthenogenesis, in which sperm-producing males copulate with females, but the sperm do not fuse with the ovum. Contact with the sperm is essential for the ovum to begin dividing, but because there is no fusion of the cells, the male contributes no genetic material to the offspring, which are essentially clones of the female.

Free-living Species

In free-living species, development usually consists of four molts of the cuticle during growth. Different species feed on materials as varied as algae, fungi, small animals, fecal matter, dead organisms and living tissues. Free-living marine nematodes are important and abundant members of the meiobenthos. They play an important role in the decomposition process, aid in recycling of nutrients in marine environments, and are sensitive to changes in the environment caused by pollution. One roundworm of note, *Caenorhabditis elegans*, lives in the soil and has found much use as a model organism. *C. elegans* has had its entire genome sequenced, as well as the developmental fate of every cell determined, and every neuron mapped.

Parasitic Species

Eggs (mostly nematodes) from stools of wild primates

Nematodes commonly parasitic on humans include ascarids (*Ascaris*), filarias, hookworms, pinworms (*Enterobius*) and whipworms (*Trichuris trichiura*). The species *Trichinella spiralis*, commonly known as the 'trichina worm', occurs in rats, pigs, and humans, and is responsible for the disease trichinosis. *Baylisascaris* usually infests wild animals, but can be deadly to humans, as well. *Dirofilaria immitis* are known for causing heartworm disease by inhabiting the hearts, arteries, and lungs of dogs and some cats. *Haemonchus contortus* is one of the most abundant infectious agents in sheep around the world, causing great economic damage to sheep. In contrast, entomopathogenic nematodes parasitize insects and are mostly considered beneficial by humans, but some attack beneficial insects.

One form of nematode is entirely dependent upon fig wasps, which are the sole source of fig fertilization. They prey upon the wasps, riding them from the ripe fig of the wasp's birth to the fig flower of its death, where they kill the wasp, and their offspring await the birth of the next generation of wasps as the fig ripens.

A newly discovered parasitic tetradonematid nematode, *Myrmeconema neotropicum*, apparently induces fruit mimicry in the tropical ant *Cephalotes atratus*. Infected ants develop bright red gasters (abdomens), tend to be more sluggish, and walk with their gasters in a conspicuous elevated position. It is likely that these changes cause frugivorous birds to confuse the infected ants for berries, and eat them. Parasite eggs passed in the bird's feces are subsequently collected by foraging *Cephalotes atratus* and are fed to their larvae, thus completing the life cycle of *M. neotropicum*.

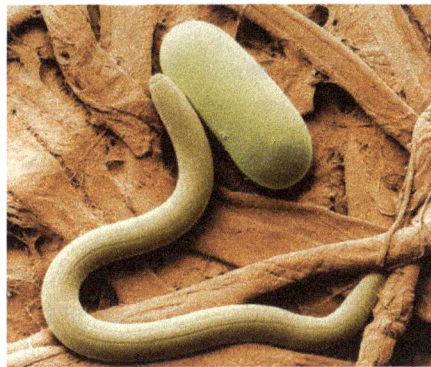

Colorized electron micrograph of soybean cyst nematode (*Heterodera* sp.) and egg

Similarly, multiple varieties of nematodes have been found in the abdominal cavities of the primitively social sweat bee, *Lasioglossum zephyrum*. Inside the female body, the nematode hinders ovarian development and renders the bee less active and thus less effective in pollen collection.

Plant-parasitic nematodes include several groups causing severe crop losses. The most common genera are *Aphelenchoides* (foliar nematodes), *Ditylenchus*, *Globodera* (potato cyst nematodes), *Heterodera* (soybean cyst nematodes), *Longidorus*, *Meloidogyne* (root-knot nematodes), *Nacob-*

bus, *Pratylenchus* (lesion nematodes), *Trichodorus* and *Xiphinema* (dagger nematodes). Several phytoparasitic nematode species cause histological damages to roots, including the formation of visible galls (e.g. by root-knot nematodes), which are useful characters for their diagnostic in the field. Some nematode species transmit plant viruses through their feeding activity on roots. One of them is *Xiphinema index*, vector of grapevine fanleaf virus, an important disease of grapes, another one is *Xiphinema diversicaudatum*, vector of arabis mosaic virus.

Other nematodes attack bark and forest trees. The most important representative of this group is *Bursaphelenchus xylophilus*, the pine wood nematode, present in Asia and America and recently discovered in Europe.

Agriculture and Horticulture

Depending on the species, a nematode may be beneficial or detrimental to plant health. From agricultural and horticulture perspectives, the two categories of nematodes are the predatory ones, which will kill garden pests like cutworms and corn earworm moths, and the pest nematodes, like the root-knot nematode, which attack plants, and those that act as vectors spreading plant viruses between crop plants. Predatory nematodes can be bred by soaking a specific recipe of leaves and other detritus in water, in a dark, cool place, and can even be purchased as an organic form of pest control.

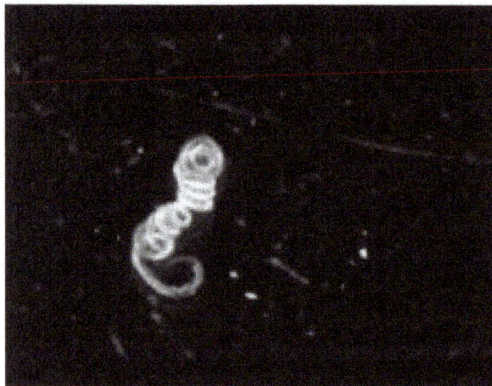

Anthelmintic effect of papain on *Heligmosomoides bakeri*

Rotations of plants with nematode-resistant species or varieties is one means of managing parasitic nematode infestations. For example, marigolds, grown over one or more seasons (the effect is cumulative), can be used to control nematodes. Another is treatment with natural antagonists such as the fungus *Gliocladium roseum*. Chitosan, a natural biocontrol, elicits plant defense responses to destroy parasitic cyst nematodes on roots of soybean, corn, sugar beet, potato and tomato crops without harming beneficial nematodes in the soil. Soil steaming is an efficient method to kill nematodes before planting a crop, but indiscriminately eliminates both harmful and beneficial soil fauna.

The Golden Nematode (Globodera rostochiensis) is a particularly harmful variety of nematode pest that has resulted in quarantines and crop failures worldwide. CSIRO has found a 13- to 14-fold reduction of nematode population densities in plots having Indian mustard (*Brassica juncea*) green manure or seed meal in the soil.

Hundreds of *Caenorhabditis elegans* were featured in a research project on NASA's STS-107 space mission, and were known to have survived the Space Shuttle Columbia disaster.

Epidemiology

A number of intestinal nematodes cause diseases affecting human beings, including ascariasis, trichuriasis and hookworm disease. Filarial nematodes cause filariasis.

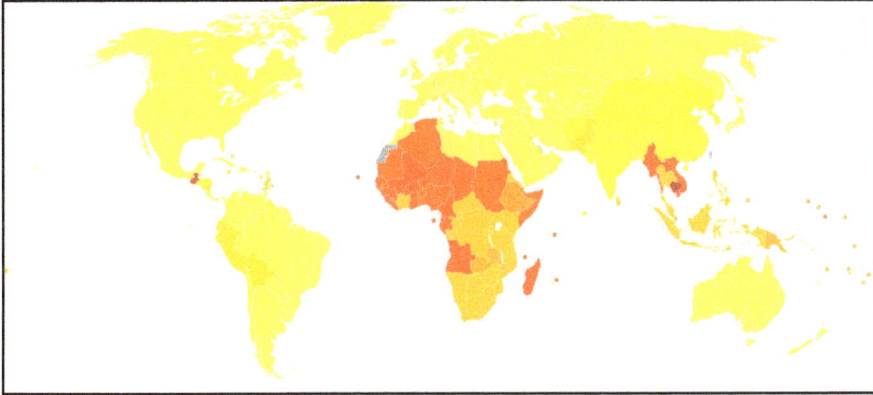

Disability-adjusted life year for intestinal nematode infections per 100,000 inhabitants in 2002.

	no data
	less than 25
	25–50
	50–75
	75–100
	100–120
	120–140
	140–160
	160–180
	180–200
	200–220
	220–240
	more than 240

Soil Ecosystems

90 percent of nematodes reside in the top 15 cm of soil. Nematodes do not decompose organic matter, but, instead, are parasitic and free-living organisms that feed on living material. Nematodes can effectively regulate bacterial population and community composition - they may eat up to 5,000 bacteria per minute. Also, Nematodes can play an important role in the nitrogen cycle by way of nitrogen mineralization.

Society and Culture

Nematode worms (*C. elegans*), the focus of an ongoing research project continued on shuttle mission STS-107, survived the Space Shuttle Columbia re-entry breakup. It is believed to be the first known life-form to survive a virtually unprotected atmospheric descent to Earth's surface.

Popular Culture

In the very first episode of *Doug*, Doug is tricked into a snipe hunt attempting to catch "neema-toads" in the swamp, by town bully Roger Klotz.

In the *SpongeBob SquarePants* episode "Home Sweet Pineapple", his house is consumed by a swarm of nematodes. They appeared again in the episode "Best Day Ever".

On the BBC2 quiz show QI, when Clive Anderson was asked, "What lives in the Dead Sea?", he answered, "There must be a nematode worm, because nematode worms live everywhere." The correct answer, generally, is "extremophile."

Anisakis

Anisakis is a genus of parasitic nematodes, which have life cycles involving fish and marine mammals. They are infective to humans and cause anisakiasis. People who produce immunoglobulin E in response to this parasite may subsequently have an allergic reaction, including anaphylaxis, after eating fish that have been infected with *Anisakis* species.

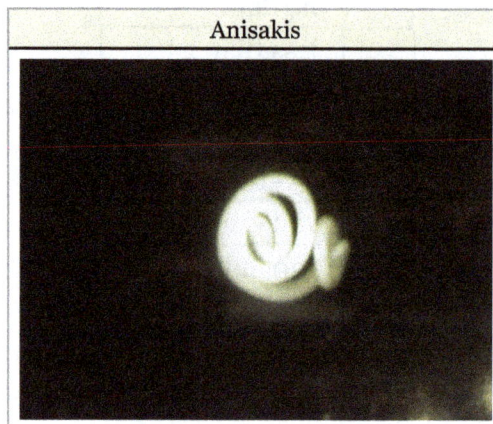

Anisakis

Etymology

The genus *Anisakis* was created in 1845 by Félix Dujardin as a subgenus of the genus *Ascaris* Linnaeus, 1758. Dujardin did not explicit the etymology but stated that the subgenus included the species in which the males have unequal spicules ("mâles ayant des spicules inégaux"); thus the name *Anisakis* is based on *anis-* (Greek prefix for different) and *akis* (Greek for spine or spicule). Two species were included in the new subgenus, *Ascaris (Anisakis) distans* Rudolphi, 1809 and *Ascaris (Anisakis) simplex* Rudolphi, 1809.

Anisakis species have complex life cycles which pass through a number of hosts through the course of their lives. Eggs hatch in seawater, and larvae are eaten by crustaceans, usually euphausids. The infected crustacean is subsequently eaten by a fish or squid, and the nematode burrows into the wall of the gut and encysts in a protective coat, usually on the outside of the visceral organs, but occasionally in the muscle or beneath the skin. The life cycle is completed when an infected fish is eaten by a marine mammal, such as a whale, seal, or dolphin. The nematode excysts in the intestine, feeds, grows, mates and releases eggs into the seawater in the host's feces. As the gut of a marine mammal is functionally very similar to that of a human, *Anisakis* species are able to infect humans who eat raw or undercooked fish.

The known diversity of the genus has increased greatly over the past 20 years, with the advent of modern genetic techniques in species identification. Each final host species was discovered to have its own biochemically and genetically identifiable "sibling species" of *Anisakis*, which is reproductively isolated. This finding has allowed the proportion of different sibling species in a fish to be used as an indicator of population identity in fish stocks.

Morphology

Anisakids share the common features of all nematodes; the vermiform body plan, round in cross section and a lack of segmentation. The body cavity is reduced to a narrow pseudocoel. The mouth is located anteriorly, and surrounded by projections used in feeding and sensation, with the anus slightly offset from the posterior. The squamous epithelium secretes a layered cuticle to protect the body from digestive juices.

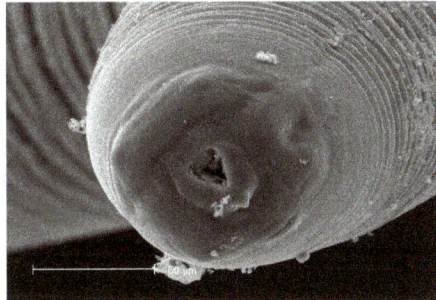

A scanning electron micrograph of the mouthparts of *A. simplex*

As with all parasites with a complex life cycle involving a number of hosts, details of the morphology vary depending on the host and life cycle stage. In the stage which infects fish, *Anisakis* species are found in a distinctive "watch-spring coil" shape. They are roughly 2 cm long when uncoiled. When in the final host, anisakids are longer, thicker and more sturdy, to deal with the hazardous environment of a mammalian gut.

Health Implications

Anisakids pose a risk to human health through intestinal infection with worms from the eating of underprocessed fish, and through allergic reactions to chemicals left by the worms in fish flesh.

Anisakiasis

Anisakiasis is a human parasitic infection of the gastrointestinal tract caused by the consumption of raw or undercooked seafood containing larvae of the nematode *Anisakis simplex*. The first case of human infection by a member of the family Anisakidae was reported in the Netherlands by Van Thiel, who described the presence of a marine nematode in a patient suffering from acute abdominal pain. It is frequently reported in areas of the world where fish is consumed raw, lightly pickled or salted. The areas of highest prevalence are Scandinavia (from cod livers), Japan (after eating sushi and sashimi), the Netherlands (by eating infected fermented herrings (*maatjes*)), and along the Pacific coast of South America (from eating *ceviche*). Fewer than ten cases occur annually in the United States. Development of better diagnostic tools and greater awareness has led to more frequent reporting of anisakiasis.

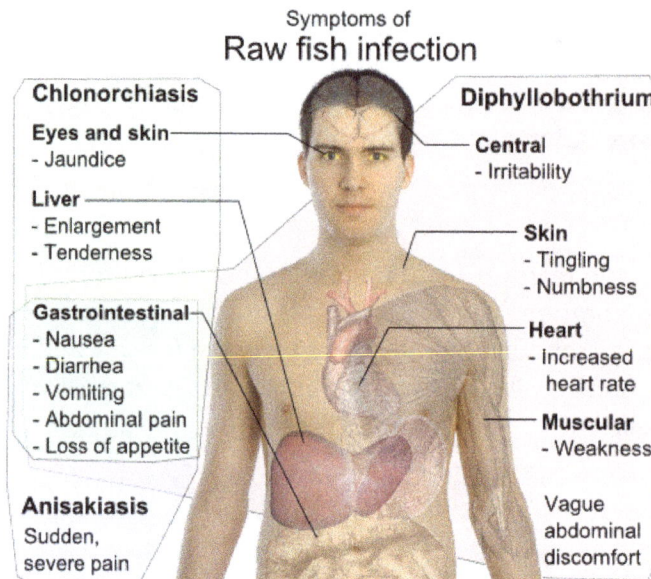

Symptoms of
Raw fish infection

Chlonorchiasis

Eyes and skin
- Jaundice

Liver
- Enlargement
- Tenderness

Gastrointestinal
- Nausea
- Diarrhea
- Vomiting
- Abdominal pain
- Loss of appetite

Anisakiasis
Sudden, severe pain

Diphyllobothrium

Central
- Irritability

Skin
- Tingling
- Numbness

Heart
- Increased heart rate

Muscular
- Weakness

Vague abdominal discomfort

Differential symptoms of parasite infection by raw fish: *Clonorchis sinensis* (trematode/fluke), *Anisakis* (nematode/roundworm) and *Diphyllobothrium* (cestode/tapeworm), all have gastrointestinal, but otherwise distinct, symptoms.

Within a few hours of ingestion, the parasitic worm tries to burrow though the intestinal wall, but since it cannot penetrate it, it gets stuck and dies. The presence of the parasite triggers an immune response; immune cells surround the worms, forming a ball-like structure that can block the digestive system, causing severe abdominal pain, malnutrition and vomiting. Occasionally, the larvae are regurgitated. If the larvae pass into the bowel or large intestine, a severe eosinophilic granulomatous response may also occur one to two weeks following infection, causing symptoms mimicking Crohn's disease.

Diagnosis can be made by gastroscopic examination, during which the 2-cm larvae are visually observed and removed, or by histopathologic examination of tissue removed at biopsy or during surgery.

Raising consumer and producer awareness about the existence of anisakid worms in fish is a critical and effective prevention strategy. Anisakiasis can be easily prevented by adequate cooking at temperatures greater than 60°C or freezing. The FDA recommends all shellfish and fish intended for raw consumption be blast frozen to -35°C or below for 15 hours or be regularly frozen to -20°C or below for seven days. Salting and marinating will not necessarily kill the parasites. Humans are thought to be more at risk of anisakiasis from eating wild fish rather than farmed fish. Many countries require all types of fish with potential risk intended for raw consumption to be previously frozen to kill parasites. The mandate to freeze herring in the Netherlands has virtually eliminated human anisakiasis.

Allergic Reactions

Even when the fish is thoroughly cooked, *Anisakis* larvae pose a health risk to humans. Anisakids (and related species such as the sealworm, *Pseudoterranova* species, and the codworm *Hystero-*

thylacium aduncum) release a number of biochemicals into the surrounding tissues when they infect a fish. They are also often consumed whole, accidentally, inside a fillet of fish.

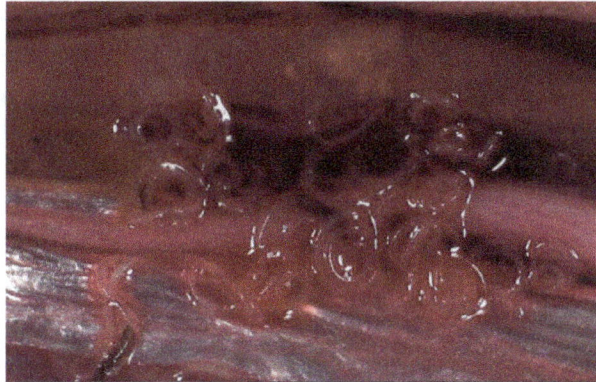

Anisakid larvae in the body cavity of a herring *(Clupea harengus)*

Acute allergic manifestations, such as urticaria and anaphylaxis, may occur with or without accompanying gastrointestinal symptoms. The frequency of allergic symptoms in connection with fish ingestion has led to the concept of gastroallergic anisakiasis, an acute IgE-mediated generalized reaction. Occupational allergy, including asthma, conjunctivitis, and contact dermatitis, has been observed in fish processing workers. Sensitivization and allergy are determined by skin-prick test and detection of specific antibodies against *Anisakis*. Hypersensitivity is indicated by a rapid rise in levels of IgE in the first several days following consumption of infected fish.

Treatment

For the worm, humans are a dead-end host. *Anisakis* and *Pseudoterranova* larvae cannot survive in humans, and will eventually die. In some cases, the infection will resolve with only symptomatic treatment. In other cases, however, infection can lead to small bowel obstruction, which may require surgery, although treatment with albendazole alone (avoiding surgery) has been reported to be successful. Intestinal perforation (an emergency) is also possible.

Occurrence

Larval anisakids are common parasites of marine and anadromous fish (e.g. salmon, sardine), and can also be found in squid and cuttlefish. In contrast, they are absent from fish in waters of low salinity, due to the physiological requirements of euphausiids, which are needed to complete their life cycle. Anisakids are also uncommon in areas where cetaceans are rare, such as the southern North Sea.

Similar Parasites

- Cod or seal worm *Pseudoterranova (Phocanema, Terranova) decipiens*

- *Contracaecum* spp.

- *Hysterothylacium (Thynnascaris) spp.*

Brugia Malayi

Brugia malayi is a nematode (roundworm), one of the three causative agents of lymphatic filariasis in humans. Lymphatic filariasis, also known as elephantiasis, is a condition characterized by swelling of the lower limbs. The two other filarial causes of lymphatic filariasis are *Wuchereria bancrofti* and *Brugia timori*, which both differ from *B. malayi* morphologically, symptomatically, and in geographical extent.

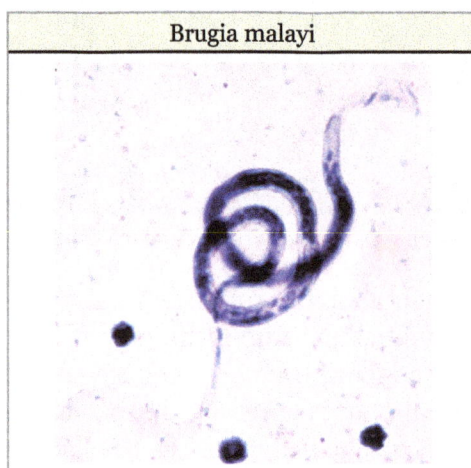

Brugia malayi

B. malayi is transmitted by mosquitoes and is restricted to South and South East Asia. It is one of the tropical diseases targeted for elimination by the year 2020 by the World Health Organization, which has spurred vaccine and drug development, as well as new methods of vector control.

History of Discovery

Identification of A Distinct Parasite

Lichtenstein and Brug first recognized *B. malayi* as a distinct pathogen in 1927. They reported the occurrence of a species of human filariae in North Sumatra that was both physiologically and morphologically distinct from the *W. bancrofti* microfilariae commonly found in Jakarta and named the pathogen *Filaria malayi*. However, despite epidemiological studies identifying *Filaria malayi* in India, Sri Lanka, China, North Vietnam, and Malaysia in the 1930s, Lichtenstein and Brug's hypothesis was not accepted until the 1940s, when Rao and Mapelstone identified two adult worms in India.

Based on the similarities with *W. bancrofti*, Rao and Mapelstone proposed to call the parasite *Wuchereria malayi* In 1960, however, Buckley proposed to divide the old genus *Wuchereria*, into two genera, *Wuchereria* and *Brugia* and renamed *Filaria malayi* as *Brugia malayi*. *Wuchereria* contains *W. bancrofti*, which so far has only been found to infect humans, and the *Brugia* genus contains *B. malayi*, which infects humans and animals, as well as other zoonotic species.

Identification of Different Strains

In 1957, two subspecies of human infecting *B. malayi* were discovered by Turner and Edeson in Malaysia based on the observation of different patterns of microfilaria periodicity. Periodicity refers to a pronounced peak in microfilariae count during a 24hour interval when microfilariae are present and detectable in the circulating blood. The basis for this phenomenon remains largely unknown.

- Nocturnal periodicity: microfilariae are not detectable in the blood for the majority of the day, but the microfilarial density peaks between midnight and 2 AM nightly.

- Nocturnal subperiodicity: microfilariae are present in the blood at all times, but appear at greatest density between noon and 8 PM.

Transmission: Vectors and Reservoirs

B. malayi is transmitted by a mosquito vector. The principal mosquito vectors include *Mansonia*, *Anopheles*, and *Aedes* mosquitoes. The mosquito serves as a biological vector – it is required for the developmental cycle of the parasite. The geographical distribution of the dis-ease is thus dependent on suitable mosquito breeding habitat.

- The nocturnal periodic form is transmitted by *Mansonia* and some Anopheline mosquitoes in open swamps and rice growing areas. These mosquitoes tend to bite at night and appear to only infect humans. Natural animal infections are rare and experimental animals do not retain infection.

- The nocturnal subperiodic form is transmitted by *Mansonia* in forest swamps, where mosquitoes bite in the shade at any time. Natural zoonotic infections are common. Cats, dogs, monkeys, slow lorises, civet cats, and hamsters have all been successfully experimentally infected with *B. malayi* from man and may serve as important reservoirs.

The accumulation of many infective mosquito bites – several hundreds to thousands – is required to establish infection. This is because a competent mosquito usually transmits only a few infective L3 larvae, and less than 10% of those larvae progress through all the necessary molting steps and develop into adult worms that can mate. Thus those at greatest risk for infection are individuals living in endemic areas—short term tourists are unlikely to develop lymphatic filariasis.

Life Cycle

Development and replication of *B. malayi* occurs in two discrete phases: in the mosquito vector and in the human. Both stages are essential to the life cycle of the parasite.

Mosquito: The mosquito serves as a biological vector and intermedite host – it is required for the developmental cycle and transmission of B. malayi.4. The mosquito takes a human blood meal and ingests microfilariae (worm-like sheathed eggs) that circulate in the human blood stream.5-7 In the mosquito, the microfilariae shed sheaths, penetrate the midgut, and migrate to the thoracic muscles were the microfilariae increase in size, molt, and develop into infective larvae (L1 and L3) over a span of 7–21 days. No multiplication or sexual reproduction of microfilariae occurs in the mosquito.8-1 The infective larvae (L3) migrate to the salivary glands, enter the proboscis and escape onto human skin when the mosquito takes another blood meal.

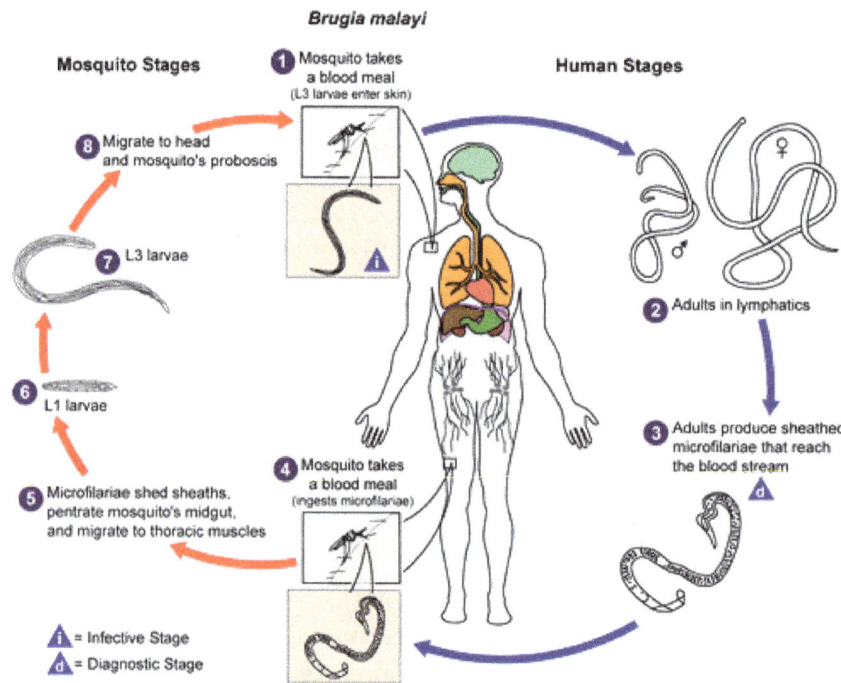

The life cycle of *Brugia malayi*.

Human: B. malayi undergoes further development in the human as well as sexual reproduction and egg production.1-2 The infective larvae (L3) actively penetrate the skin through the bite hole and develop into adults in the lymphatic system over a span of 6 months. Adult worms can survive in the lymphatic system for 5–15 years3. The male and female adult worms mate and the females produce an average of 10,000 sheathed eggs (microfilaria) daily The microfilariae enter the blood stream and exhibit the classic nocturnal periodicity and subperiodicity.4. Another mosquito takes a blood meal and ingests the microfilariae. Infection depends on the mosquito taking a blood meal during a periodic episode – when microfilariae are present in the bloodstream.

Morphology

Adult

Adult worms resemble the classic nematode roundworm. Long and threadlike, *B. malayi* and other nematode possess only longitudinal muscles and move in an S-shape motion. Adults are typically smaller than adult *W. bancrofti*, though few adults have been isolated. Female adult worms (50 mm) are larger than male worms (25 mm).

Microfilariae

B. malayi microfilariae are 200–275 µm in length and have a round anterior end and a pointed posterior end. The microfilariae are sheathed, which stains heavily with Giemsa. The sheath is actually the egg shell, a thin layer that surrounds the egg shell as the microfilariae circulates in the bloodstream. The microfilariae retain the sheath until it is digested in the mosquito midgut.

B. malayi microfilariae resemble *W. bancrofti* and *Loa loa* microfilariae with minor differences that can aid in laboratory diagnosis. *B. malayi* microfilariae can be distinguished by the noncontinuous row of nuclei found in the tip of the tail. There are two terminal nuclei that are distinctly separated from the other nuclei in the tail, whereas the tail of *W. bancrofti* contains no nuclei and *Loa loa* microfilariae nuclei form a continuous row in the tail. *B. malayi* microfilariae also have a characteristic cephalic space ratio of 2:1.

Symptoms

B. malayi is one of the causative agents of lymphatic filariasis, a condition marked by infection and swelling of the lymphatic system. The disease is primarily caused by the presence of worms in the lymphatic vessels and the resulting host response. Signs of infection are typically consistent with those seen in bancroftian filariasis—fever, lymphadenitis, lymphangitis, lymphedema, and secondary bacterial infection—with a few exceptions.

Lymphadenitis

Lymphadenitis, the swelling of the lymph nodes, is a commonly recognized symptom of many diseases. An early manifestation of filariasis, lymphadenitis more frequently occurs in the inguinal area during *B. malayi* infection and can occur before the worms mature.

Lymphangitis

Lymphangitis is the inflammation of the lymphatic vessels in response to infection. It occurs early in the course of infection in response to worm development, molting, death, or bacterial and fungal infection. The affected lymphatic vessel becomes distended and tender, and the overlying skin becomes erythemous and hot. Abscess formation and ulceration of the affected lymph node occasionally occurs during *B. malayi infection*, more readily than in Bancroftian filariasis. Remnants of adult worms can sometimes be found in the ulcer drainage.

Lymphedema (Elephantiasis)

The most obvious sign of infection, elephantiasis, is the enlargement of the limbs. A late complication of infection, elephantiasis is a form of lymphedema and is caused by repeated inflammation of the lymphatic vessels. Repeated inflammatory reactions causes vessel dilation and thickening of the affected lymphatic vessels, which can compromise function. The lymphatic system normally functions to maintain fluid balance between tissues and the blood and serves as an integral part of the immune system. Blockage of these vessels due to inflammatory induced fibrosis, dead worms, or granulomatous reactions can interfere with normal fluid balance, thus leading to swelling in the extremities. Elephantiasis resulting from *B. malayi* infection typically affects the distal portions of the extremities. Unlike bancroftian filariasis, *B. malayi* rarely affects genitalia and does not cause funiculitis, orchitis, epididymitis, hydrocele, or chyuria, conditions more readily observed with bancroftian infection.

Secondary Bacterial Infection

Secondary bacterial infection is common among patients with filariasis. Compromised immune function due to lymphatic damage in addition to lymph node ulcerations and abscesses exposure and impaired circulation due to elephantiasis can cause secondary bacterial or fungal infection. Elephantiasis, in addition to the physical burden of a swollen limb, can be a severely dehabilitating condition given bacterial infection. Part of the WHO's "Strategy to Eliminate Lymphatic Filariasis" targets hygiene promotion programs in order to alleviate the suffering of affected individuals (Prevention Strategies).

However, clinical manifestations of infection are variable and depend on several factors, including host immune system, infectious dose, and parasite strain differences. Most infections appear asymptomatic, yet vary from individual to individual. Individuals living in endemic areas with microfilaremia may never present with overt symptoms, whereas in other cases, only a few worms can exacerbate a severe inflammatory response.

The development of the disease in humans, however, is not well understood. Adults typically develop worse symptoms, given the long exposure time required for infection. Infection may occur during childhood, but the disease appears to take many years to manifest. The incubation period for infection ranges from 1 month to 2 years and typically microfilariae appear before overt symptoms. Lymphedema can develop within six months and development of elephantiasis has been reported within a year of infection among refugees, who are more immunologically naive. Men tend to develop worse symptoms than women.

Laboratory Diagnosis

Tender or enlarged inguinal lymph nodes or swelling in the extremities can alert physicians or public health officials to infection.

With appropriate laboratory equipment, microscopic examination of differential morphological features of microfilariae in stained blood films can aid diagnosis—in particular the examination of the tail portion, the presence of a sheath, and the size of the cephalic space. Giemsa staining will uniquely stain *B. malayi* sheath pink. However, blood films can prove difficult given the nocturnal periodicity of some forms of *B. malayi*.

PCR based assays are highly sensitive and can be used to monitor infections both in the human and the mosquito vector. However, PCR assays are time-consuming, labor-intensive and require laboratory equipment. Lymphatic filariasis mainly affects the poor, who live in areas without such resources.

The ICT antigen card test is widely used in the diagnosis of *W. bancrofti*, but commercial antigens of *B. malayi* have not been historically widely available. However, new research developments have identified a recombinant antigen (BmR1) that is both specific and sensitive in the detection of IgG4 antibodies against *B. malayi* and *B. timori* in ELISA and immunochromatographic rapid dipstick (Brugia Rapid) test. However, it appears that immunoreactivity to this antigen is variable in individuals infected with other filarial nematodes. This research has led to the development of two new rapid immunochromatographic IgG4 cassette tests – WB rapid and panLF rapid – which detect bancroftian filariasis and all three species of lymphatic filariasis, respectively, with high sensitivity and selectivity.

Management and Therapy

The ["Global Alliance to Eliminate Lymphatic Filariasis"] was launched by the World Health Organization in 2000 with two primary goals: 1) to interrupt transmission and 2) to alleviate the suffering of affected individuals. Mass drug treatment programs are the main strategy for interrupting parasite transmission, and morbidity management, focusing on hygiene, improves the quality of life of infected individuals.

Drugs

A goal of community base efforts is to eliminate microfilariae from the blood of infected individuals in order to prevent transmission to the mosquito. This is primarily accomplished through the use of drugs. The treatment for *B. malayi* infection is the same as for bancroftian filariasis. Diethylcarbamazine (DEC) has been used in mass treatment programs in the form of DEC-medicated salt, as an effective microfilaricidal drug in several locations, including India. While DEC tends to cause adverse reactions like immediate fever and weakness, it is not known to cause any long-term adverse drug effects. DEC has been shown to kill both adult worms and microfilariae. In Malaysia, DEC dosages (6 mg/kg weekly for 6 weeks; 6 mg/kg daily for 9 days) reduced microfilariae by 80% for 18–24 months after treatment in the absence of mosquito control. Microfilariae numbers slowly return many months after treatment, thus requiring multiple drug doses over time in order to achieve long-term control. However, it is not known how many years of mass drug administration is required to eliminate transmission. But currently, there have been no confirmed cases of DEC resistance.

Single doses of two drugs (albendazole-DEC and albendazole-ivermectin) have been shown to remove 99% of microfilariae for a year after treatment and help to improve elephantiasis during early stages of the disease. Ivermectin does not appear to kill adult worms but serves as a less toxic microfilaricide.

Since the discovery of the importance of Wolbachia bacteria in the lifecycle of *B. malayi* and other nematodes, novel drug efforts have targeted the endobacterium. Tetracyclines, rifampicin, and chloramphenicol have been effective in vitro by interfering with larvae molting and microfilariae development. Tetracyclines have been shown to cause reproductive and embryogenesis abnormalities in the adult worms, resulting in worm sterility. Clinical trials have demonstrated the successful reduction of *Wolbachia* and microfilariae in onchocerciasis and *W. bancrofti* infected patients. These antibiotics, while acting through a slightly more indirect route, are promising antifilarial drugs.

Hygiene

Secondary bacterial infection is often observed with lymphatic filariasis. Rigorous hygiene practices, including washing with soap and water daily and disinfecting wounds can help heal infected surfaces, and slow and potentially reverse existing tissue damage. Promoting hygiene is essential for lymphatic filariasis patients given the compromised immune and damaged lymphatic systems and can help prevent suffering and disability.

Prevention Strategies

Vaccines

There is currently no licensed vaccine to prevent lymphatic filariasis. However, recent research has produced vaccine candidates with good results in experimental animals. A glutathione-S-transferase, a detoxification enzyme in parasites isolated from *Setaria cervi*, a bovine filarial parasite, reduced B. malayi adult parasite burden by more than 82% 90 days post parasite.

Vector Control

Vector control has been effective in virtually eliminating lymphatic filariasis in some regions, but vector control combined with chemotherapy produces the best results. It is suggested that 11–12 years of effective vector control may eliminate lymphatic filariasis. Successful methods of *B. malayi* vector control include residual house spraying using DDT and insecticide treated bednets. Mansonia larvae attach their breathing tubes to underwater roots and plants in order to survive. While chemical larvicides have only provided partial control, plant removal would prevent vector development, but would have potential adverse effects on the aquatic environment. Lymphatic filariasis vector control is neglected in comparison to the far more established efforts to control malaria and Dengue vectors. Integrated vector control methods should be applied in areas where the same mosquito species is responsible for transmitting multiple pathogens.

Epidemiology

B. malayi infects 13 million people in south and southeast Asia and is responsible for nearly 10% of the world's total cases of lymphatic filariasis. *B. malayi* infection is endemic or potentially endemic in 16 countries, where it is most common in southern China and India, but also occurs in Indonesia, Thailand, Vietnam, Malaysia, the Philippines, and South Korea. The distribution of *B. malayi* overlaps with *W. bancrofti* in these regions, but does not coexist with *B. timori*. Regional foci of endemicity are determined in part by the mosquito vectors.

Genome Deciphered

On September 20, 2007, scientists sequenced the genome of *Brugia malayi* in the paper "Draft Genome for the Filarial Nematode Parasite Brugia malayi" by Elodie Ghedin, et al. *Science* 317, 1756 (2007); doi:10.1126/science.1145406. Identifying the genes of this organism might lead to development of new drugs and vaccines.

To decipher the genome, "Whole Genome Shotgun Sequencing" was performed. The genome was found to be approximately 90-95 mega bases in size. The results of the sequencing was then compared to that of the reference nematode *Caenorhabditis elegans*, along with its prototype *Caenorhabditis briggsae*. These two free-living nematodes were incorporated in the study and proved to be important for several reasons:

- comparing genomes using *C. elegans* was extremely beneficial in identifying similar linkages in genes.

- the researchers found a genomic conservation

- also found data that supported an absence of conservation at a more local gene level

- This demonstrated that rearrangements had occurred over time between the *C. elegans* and *B. malayi* and allowed researchers to identify genes or proteins that were more specific to *B. malayi*

- These unique genes were significant because they could have led to the parasitism seen in *B. malayi*, and would therefore be seen as appropriate targets for future studies.

- gene linkages offer new insight into the evolutionary trend of parasitic genes that could possess clues to further explain their unique ability to successfully survive for many years in human hosts.

Potential for New Drugs to Treat B. Malayi

Sequence comparisons between the two genomes allow us to map *C. elegans* orthologs to *B. malayi* genes. Using these orthology mappings (between *C. elegans* and *B.malayi*) and by incorporating the extensive genomic and functional genomic data, including genome-wide RNAi screens, that already exist for *C. elegans*, we identify potentially essential genes in *B. malayi*. Scientists are hoping to be able to use these genes as potential new drug targets for new drug treatments. The longevity of this parasite complicates treatment because existing drugs target the larvae and, thus, do not completely kill the worms. The drugs often must be taken periodically for years, and the worm can cause a massive immune reaction when it dies and releases foreign molecules in the body. Drug treatments for filariasis have not changed significantly in over 20 years, and with the risk of resistance rising, there is an urgent need for the development of new anti-filarial drug therapies. From the genome sequence, Dr. Ghedin and her co-investigators identified several metabolic pathways containing dozens of gene products that they believe are likely to be helpful for the discovery of more targeted and effective drug therapies.

- Possible new drug targets include:

  o molting

  o nuclear receptors

  o collagens and collagen processing

  o neuronal signaling

  o the *B. malayi* kinome

  o reliance on host (*B. malayi*) and endosymbiont (*Wolbachia*) metabolism.

These potential new targets for drugs or vaccines should provide new opportunities for understanding, treating and preventing elephantiasis.

Endosymbiotic Relationship with Wolbachia

The relationship between the *Wolbachia* bacteria and *B. malayi* is not fully understood. Extrapolating from research done with *Wuchereria bancrofti*, another nematode that causes filariasis, *Wolbachia* may aid in embryogenesis of the worm, be responsible for potent inflammatory responses from macrophages and filarial disease, and be linked to the onset of lyphodema and blindness sometimes associated with *B. malayi* infections. In a study done by the University of Bonn in Ghana, doxycycline effectively depleted *Wolbachia* from *W. bancrofti*. It is likely that the mechanism of doxycycline is similar to that in other filarial species, i.e., a predominant blockade of embryogenesis, leading to a decline of microfilariae according to their half-life. This could render doxycycline treatment an additional tool for the treatment of microfilaria-associated diseases in bancroftian filariasis, along with *B. malayi* fiariasis. The doxycycline course of treatment would be much shorter as it would make the adult worm sterile in one shot rather than repeatedly have to target the replenished larvae that current treatments kill, and there would be fewer side effects for the infected individual.

Genome Use in Transplant Research

Another hopeful use for the research is in the area of transplant research. Because the *B. malayi* genome is the first parasitic genome to have been sequenced, the implications on the mechanism of parasitism in humans are crucial to understand. According to Alan L. Scott, Ph.D., a collaborator at Johns Hopkins University, understanding how a particular parasite, such as *B. malayi*, can adapt to humans, may yield medical benefits far beyond treating elephantiasis. According to the author, "This worm can reside in the host for years and not necessarily cause disease, in fact the less disease the individual has, the more worms there are in circulation. Now that we know those genes don't exist in humans we can target them to control disease." Some of the predicted proteins for these new genes appear similar to known immuno-modulator proteins, regulators of the immune system, suggesting that they are involved in deactivating the host's immune system to ensure the parasite remains undetected. Knowledge of these previously unknown immune suppressors could also be of use in organ transplants and to help treat autoimmune disease.

According to the Filarial Genome Project being done by The Special Programme for Research and Training in Tropical Diseases (TDR), the *Brugia malayi* MIF gene is expressed in all life-cycle stages of the parasite, and results suggest that *B. malayi* MIF may interact with the human immune system during the course of infection by altering the function of macrophages in the infected individual. TDR also states that studies are currently testing the hypothesis that MIF may be involved in reducing the host's immune response to the filarial parasite. Understanding how this particular parasite has adapted to humans may help organ transplant researchers by figuring out how to prevent the immune system from attacking the transplanted tissue.

Overall Hope for The Use of The Genome Sequencing

The genomic information gives us a better understanding of what genes are important for different processes in the parasite's life cycle. So, it will now be possible to target these genes more specifically and interrupt its life cycle. And, understanding how this particular parasite has adapted to humans may yield medical benefits far beyond treating elephantiasis, says collaborator Alan L. Scott, Ph.D., of the Bloomberg School of Public Health at Johns Hopkins University. "Parasitic worms are a lot like foreign tissue that has been transplanted into the human body. But unlike baboon hearts or pig kidneys, which the immune system quickly recognizes as foreign and rejects, worms can survive for years in the body. Discovering how they do so may someday benefit transplant surgery," explained Dr. Scott.

Dracunculus Medinensis

Dracunculus medinensis or Guinea worm is a nematode that causes dracunculiasis, also known as guinea worm disease. The disease is caused by the female which, at up to 800 mm (31 in) in length, is among the longest nematodes infecting humans. In contrast, the longest recorded male Guinea worm is only 40 mm (1.6 in).

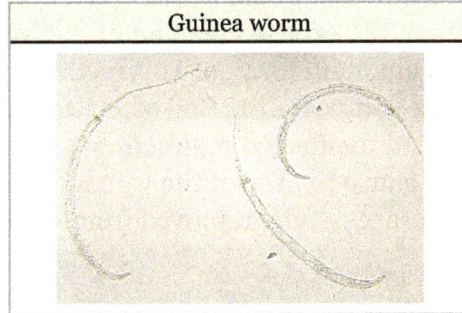

Guinea worm

The common name "guinea worm" is derived from the Guinea region of Western Africa.

Life Cycle

D. medinensis larvae are found in freshwater, where they are ingested by copepods of the genus *Cyclops*. Within the copepod, the *D. medinensis* larvae develop to an infective stage within 14 days. When the infected copepod is ingested by a mammalian host, the copepod is dissolved by stomach acid and the *D. medinensis* larvae migrate through the wall of the mammalian intestine, and mature into adults. 100 days after infection, a male and female *D. medinensis* meet and sexually reproduce within the host tissue. The male dies in the host tissue while the female migrates to the host subcutaneous tissue. Approximately one year after infection, the female causes the formation of a blister on the skin's surface, generally on the lower extremities, though occasionally on the hand or scrotum. When the blister ruptures, the female slowly emerges over the course of several days or weeks. This causes extreme pain and irritation to the host. When the host submerges the affected body part in water, the female expels thousands of larvae into the water. From here, the larvae infect copepods, continuing the life cycle.

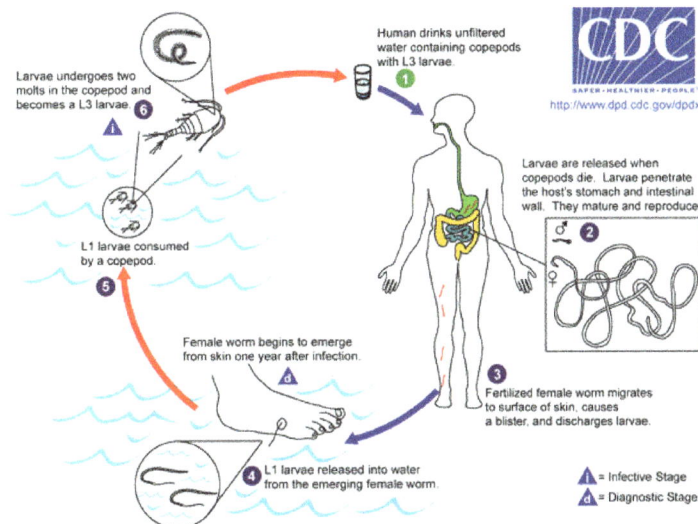

General life cycle of *Dracunculus medinensis* in humans

Removal

The female guinea worm slowly starts to emerge from the host's skin after the blister ruptures. The most common method for removing the worm involves submerging the affected body part in water to help coax the worm out. The site is then cleaned thoroughly. Then, slight pressure is applied to the worm as it is slowly pulled out of the wound. To avoid breaking the worm, pulling should stop when resistance is met. Full extraction of the female guinea worm usually takes several days. After each day's worth of extraction, the exposed portion of the worm is wrapped around a piece of rolled-up gauze or small stick to maintain tension. This method of wrapping the worm around a stick or gauze is speculated to be the source for the Rod of Asclepius, the symbol of medicine. Once secure, topical antibiotics are applied to affected region to help prevent secondary infections due to bacteria and then is wrapped in gauze to protect the wound. The same steps are repeated each day until the whole worm has been successfully removed from the lesion.

Eradication Program

Guinea worm cases by year		
Year	Reported Cases	Countries
1989	892,055	16
1995	129,852	19
2000	75,223	16
2001	63,717	16
2002	54,638	14
2003	32,193	13
2004	16,026	13
2005	10,674	12
2006	25,217	10
2007	9,585	9
2008	4,619	7
2009	3,190	5
2010	1,797	6
2011	1,060	4
2012	542	4
2013	148	5
2014	126	4
2015	22	4

In the 1980s, the Carter Center initiated a program to eradicate guinea worm. The Guinea Worm Eradication campaign began in 1980 at the US Centers for Disease Control and Prevention. In 1984, the CDC was appointed as the World Health Organization Collaborating Center for research, training, and eradication of *Dracunculus medinensis*. There were 20 countries in 1986 that were affected by guinea worms. That year, WHO started the eradication program and the Carter Center took the lead on the effort. The program included education of people in affected areas that the disease was caused by larvae in drinking water, isolation and support for sufferers, and – crucially – widespread distribution of net filters and pipe filters for drinking water, and education about the importance of using them.

As of 2015, the species has been reported to be near eradication. The International Commission for the Certification of Dracunculus Eradication (ICCDE) has certified 198 countries, territories, and other WHO represented areas. In January 2015, eight countries remained to be certified as *Dracunculus medinensis* free. These eight countries include Angola, Democratic Republic of the Congo, Kenya, Sudan, Chad, Ethiopia, Mali, and South Sudan. Of these, Chad, Ethiopia, Mali, and South Sudan are the only remaining endemic countries. Not coincidentally, all four are affected by civil wars which affect the safety of health workers.

Etymology

Dracunculus medinensis (*little dragon from Medina*) is a parasitic nematode that infects humans and domestic animals through contaminated water. D. medinensis was described in Egypt as early as the 15th century BCE and may have been the "fiery serpent" of the Israelites described in the Bible.

Loa Loa Filariasis

Loa loa filariasis is a skin and eye disease caused by the nematode worm *Loa loa*. Humans contract this disease through the bite of a deer fly or mango fly (*Chrysops* spp), the vectors for *Loa loa*. The adult *Loa loa* filarial worm migrates throughout the subcutaneous tissues of humans, occasionally crossing into subconjunctival tissues of the eye where it can be easily observed. *Loa loa* does not normally affect one's vision but can be painful when moving about the eyeball or across the bridge of the nose. The disease can cause red itchy swellings below the skin called "Calabar swellings". The disease is treated with the drug diethylcarbamazine (DEC), and when appropriate, surgical methods may be employed to remove adult worms from the conjunctiva.

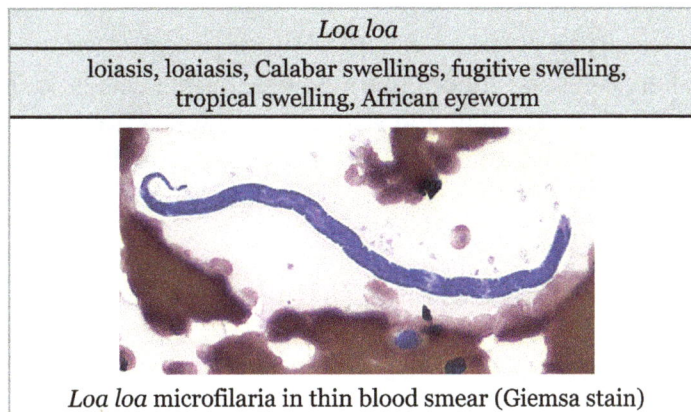

| *Loa loa* |
| loiasis, loaiasis, Calabar swellings, fugitive swelling, tropical swelling, African eyeworm |

Loa loa microfilaria in thin blood smear (Giemsa stain)

Signs and Symptoms

Filariasis such as loiasis most often consists of asymptomatic microfilaremia. Some patients develop lymphatic dysfunction causing lymphedema. Episodic angioedema (Calabar swellings) in the arms and legs, caused by immune reactions are common. Calabar swellings are 3–10 cm in surface non erythematous and not pitting. When chronic, they can form cyst-like enlargements of the connective tissue around the sheaths of muscle tendons, becoming very painful when moved. The swellings may last for 1–3 days, and may be accompanied by localized urticaria (skin eruptions) and pruritus (itching). They reappear at referent locations at irregular time intervals. Subconjunctival migration of an adult worm to the eyes can also occur frequently, and this is the reason Loa loa is also called the "African eye worm." The passage over the eyeball can be sensed, but it usually takes less than 15 min. Gender incidence of eyeworms have approximately the same frequency, but it tends to increase with age. Eosinophilia is often prominent in filarial infections. Dead worms may cause chronic abscesses, which may lead to the formation of granulomatous reactions and fibrosis.

In the human host, *Loa loa* larvae migrate to the subcutaneous tissue where they mature to adult worms in approximately one year, but sometimes up to four years. Adult worms migrate in the subcutaneous tissues at a speed less than 1 cm/min, mating and producing more microfilaria. The adult worms can live up to 17 years in the human host.

Cause

Transmission

Loa loa infective larvae (L3) are transmitted to humans by deer fly vectors, *Chrysops silica* and *C. dimidiata*. The vectors are blood-sucking and day-biting, and they are found in rainforest-like environments in west and central Africa. Infective larvae (L3) mature to adults (L5) in the subcutaneous tissues of the human host, after which the adult worms—assuming presence of a male and female worm—mate and produce microfilaria. The cycle of infection continues when a non-infected mango or deer fly takes a blood meal from a microfilaremic human host, and this stage of the transmission is possible due to the combination of the diurnal periodicity of microfilaria and the day-biting tendencies of the *Chrysops spp*.

Reservoir

Humans are the primary reservoir for *Loa loa*. Other minor potential reservoirs have been indicated in various fly biting habit studies: hippopotamus, wild ruminants (e.g., buffalo), rodents, and lizards. A simian type of loiasis exists in monkeys and apes but it is transmitted by *Chrysops langi*. There is no cross-over between the human and simian types of the disease.

Vector

Loa loa is transmitted by several species of tabanid flies (Order: Diptera; Family: Tabanidae). Although horseflies of the *Tabanus* genus are often mentioned as Loa vectors, the two prominent vector are from the *Chrysops* genus of tabanids—*C. silacea* and *C. dimidiata*. These species exist only in Africa and are popularly known as deer flies and mango, or mangrove, flies.

Chrysops spp are small (5–20 mm long) with a large head and downward pointing mouthparts. Their wings are clear or speckled brown. They are hematophagous and typically live in forested and muddy habitats like swamps, streams, reservoirs, and in rotting vegetation. Female mango and deer flies require a blood meal for production of a second batch of eggs. This batch is deposited near water, where the eggs hatch in 5–7 days. The larvae mature in water or soil, where they feed on organic material such as decaying animal and vegetable products. Fly larvae are 1–6 cm long and take 1–3 years to mature from egg to adult. When fully mature, *C. silacea* and *C. dimidiata* assume the day-biting tendencies of all tabanids.

The bite of the mango fly can be very painful, possibly due to the laceration style employed; rather than puncturing the skin like a mosquito does, the mango (and deer fly) make a laceration in the skin and subsequently lap up blood. Female flies require a fair amount of blood for their afore-mentioned reproductive purposes and thus may take multiple blood meals from the same host if disturbed during the first one.

Interestingly, although *Chrysops silacea* and *C. dimidiata* are attracted to canopied rainforests, they do not do their biting there. Instead, they leave the forest and take most blood meals in open areas. The flies are attracted to smoke from wood fires and they use visual cues and sensation of carbon dioxide plumes to find their preferred host, humans.

A study of *Chrysops spp* biting habits showed that *C. silacea* and *C. dimidiata* take human blood meals approximately 90% of the time, with hippopotamus, wild ruminant, rodent, and lizard blood meals making up the other 10%. The fact that no simian (ex: monkeys or apes) blood meals were taken suggests that there is no crossover between the human and simian types of *Loa loa*. A related fly, *Chrysops langi*, has been isolated as a vector of simian loiasis, but this variant hunts within the forest and has not as yet been associated with human infection.

Morphology

Adult *Loa* worms are sexually dimorphic, with males considerably smaller than females at 30–34 mm long and 0.35-0.42 mm wide compared to 40–70 mm long and 0.5 mm wide. Adults live in the sub-cutaneous tissues of humans, where they mate and produce worm-like eggs called microfilaria. These microfilariae are 250-300µm long, 6-8µm wide, and can be distinguished morphologically from other filariae—they are sheathed and contain body nuclei that extend to the tip of the tail.

Life Cycle

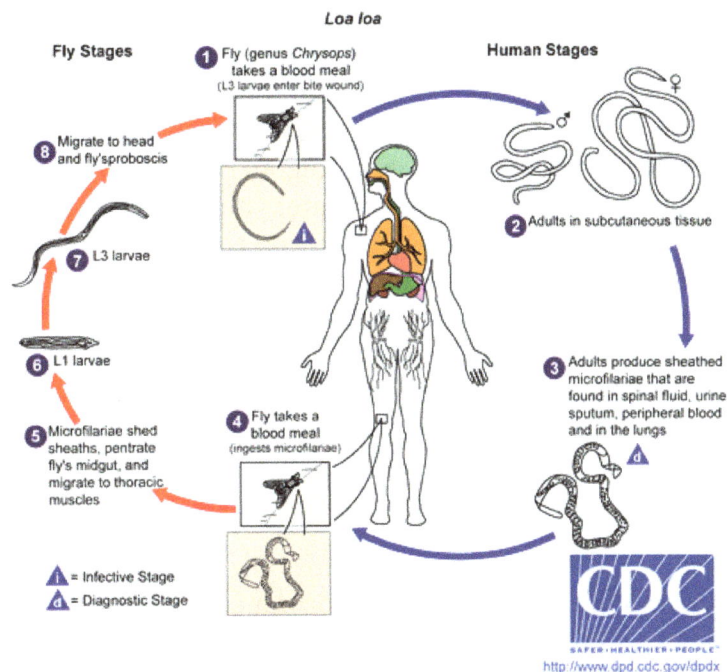

Loa loa life cycle. Source: CDC

The vector for *Loa loa* filariasis are flies from two hematophagous species of the genus *Chrysops* (deer flies), *C. silacea* and *C. dimidiata*. During a blood meal, an infected fly (genus *Chrysops*, day-biting flies) introduces third-stage filarial larvae onto the skin of the human host, where they penetrate into the bite wound. The larvae develop into adults that commonly reside in subcutaneous tissue. The female worms measure 40 to 70 mm in length and 0.5 mm in diameter, while the males measure 30 to 34 mm in length and 0.35 to 0.43 mm in diameter. Adults produce microfilariae measuring 250 to 300 µm by 6 to 8 µm, which are sheathed and have diurnal periodicity. Microfilariae have been recovered from spinal fluids, urine, and sputum. During the day they are found in peripheral blood, but during the noncirculation phase, they are found in the lungs. The fly ingests microfilariae during a blood meal. After ingestion, the microfilariae lose their sheaths and migrate from the fly's midgut through the hemocoel to the thoracic muscles of the arthropod. There the microfilariae develop into first-stage larvae and subsequently into third-stage infective larvae. The third-stage infective larvae migrate to the fly's proboscis and can infect another human when the fly takes a blood meal.

Diagnosis

Identification of microfilariae by microscopic examination is a practical diagnostic procedure. Examination of blood samples will allow identification of microfilariae of *Loa loa*. It is important to time the blood collection with the known periodicity of the microfilariae (between 10 am and 2 pm). The blood sample can be a thick smear, stained with Giemsa or haematoxylin and eosin. For increased sensitivity, concentration techniques can be used. These include centrifugation of the blood sample lyzed in 2% formalin (Knott's technique), or filtration through a Nucleopore membrane.

Antigen detection using an immunoassay for circulating filarial antigens constitutes a useful diagnostic approach, because microfilaremia can be low and variable. Interestingly, the Institute for Tropical Medicine reports that no serologic diagnostics are available. While this was once true, and many of recently developed methods of Antibody detection are of limited value—because substantial antigenic cross reactivity exists between filaria and other parasitic worms (helminths), and a positive serologic test does not necessarily distinguish between infections—up and coming serologic tests that are highly specific to *Loa loa* were furthered in 2008. They have not gone point-of-care yet, but show promise for highlighting high-risk areas and individuals with co-endemic loiasis and onchocerciasis. Specifically, Dr. Thomas Nutman and colleagues at the National Institutes of Health have described the a luciferase immunoprecipitation assay (LIPS) and the related QLIPS (quick version). Whereas a previously described LISXP-1 ELISA test had a poor sensitivity (55%), the QLIPS test is both practical, as it requires only a 15 minutes incubation, and has high sensitivity and specificity (97% and 100%, respectively). No report on the distribution status of LIPS or QLIPS testing is available, but these tests would help to limit complications derived from mass ivermectin treatment for onchocerciasis or dangerous strong doses of diethylcarbamazine for loiasis alone (as pertains to individual with high *Loa loa* microfilarial loads).

Physically, Calabar swellings are the primary tool for diagnosis. Iden-tification of adult worms is possible from tissue samples collected during subcutaneous biopsies. Adult worms migrating across the eye are another potential diagnostic, but the short timeframe for the worm's passage through the conjunctiva makes this observation less common.

In the past, health care providers use a provocative injection of *Dirofilaria immitis* as a skin test antigen for filariasis diagnosis. If the patient was infected, the extract would cause an artificial allergic reaction and associated Calabar swelling similar to that caused, in theory, by metabolic products of the worm or dead worms.

Blood tests to reveal microfilaremia are useful in many, but not all cases, as one third of loiasis patients are amicrofilaremic. By contrast, eosinophilia is almost guaranteed in cases of loiasis, and blood testing for eosinophil fraction may be useful.

Prevention

Diethylcarbamazine has been shown as an effective prophylaxis for *Loa loa* infection. A study of Peace Corps volunteers in the highly Loa—endemic Gabon, for example, had the following results: 6 of 20 individuals in a placebo group contracted the disease, compared to 0 of 16 in the DEC-treated group. Seropositivity for antifilarial IgG antibody was also much higher in the placebo group. The recommended prophylactic dose is 300 mg DEC given orally once weekly. The only associated symptom in the Peace Corps study was nausea.

Researchers believe that geo-mapping of appropriate habitat and human settlement patterns may, with the use of predictor variables such as forest, land cover, rainfall, temperature, and soil type, allow for estimation of Loa loa transmission in the absence of point-of-care diagnostic tests. In addition to geo-mapping and chemoprophylaxis, the same preventative strategies used for malaria should be undertaken to avoid contraction of loiasis. Specifically, DEET-containing insect repellent, permethrin-soaked clothing, and thick, long-sleeved and long-legged clothing ought to be worn to decreased susceptibility to the bite of the mango or deer fly vector. Because the vector is day-biting, mosquito (bed) nets do not increase protection against loiasis.

Vector elimination strategies are an interesting consideration. It has been shown that the *Chrysops* vector has a limited flying range, but vector elimination efforts are not common, likely because the insects bite outdoors and have a diverse, if not long, range, living in the forest and biting in the open, as mentioned in the vector section.

No vaccine has been developed for loiasis and there is little report on this possibility.

Treatment

Treatment of loiasis involves chemotherapy or, in some cases, surgical removal of adult worms followed by systemic treatment. The current drug of choice for therapy is diethylcarbamazine (DEC), though ivermectin use is not unwarranted. The recommend dosage of DEC is 6 mg/kg/d taken three times daily for 12 days. The pediatric dose is the same. DEC is effective against microfilariae and somewhat effective against macrofilariae (adult worms).

In patients with high microfilaria load, however, treatment with DEC may be contraindicated, as the rapid microfilaricidal actions of the drug can provoke encephalopathy. In these cases, albendazole administration has proved helpful, and superior to ivermectin, which can also be risky despite is slower-acting microfilaricidal effects.

Management of *Loa loa* infection in some instances can involve surgery, though the timeframe during which surgical removal of the worm must be carried out is very short. A detailed surgical strategy to remove an adult worm is as follows (from a real case in New York City). The 2007 procedure to remove an adult worm from a male Gabonian immigrant employed proparacaine and povidone-iodine drops, a wire eyelid speculum, and 0.5ml 2% lidocaine with epinephrine 1:100,000, injected superiorly. A 2-mm incision was made and the immobile worm was removed with forceps. Gatifloxacin drops and an eye-patch over ointment were utilized post surgery and there were no complications (unfortunately, the patient did not return for DEC therapy to manage the additional worm—and microfilaria—present in his body).

Epidemiology

As of 2009, loiasis is endemic to 11 countries, all in western or central Africa, and an estimated 12-13 million people have the disease. The highest incidence is seen in Cameroon, Republic of the Congo, Democratic Republic of Congo, Central African Republic, Nigeria, Gabon, and Equatorial Guinea. The rates of *Loa loa* infection are lower but it is still present in and Angola, Benin, Chad and Uganda. The disease was once endemic to the western African countries of Ghana, Guinea, Guinea Bissau, Ivory Coast and Mali but has since disappeared.

Throughout *Loa loa*-endemic regions, infection rates vary from 9 to 70 percent of the population. Areas at high risk of severe adverse reactions to mass treatment (with Ivermectin) are at present determined by the prevalence in a population of >20% microfilaremia, which has been recently shown in eastern Cameroon (2007 study), for example, among other locales in the region.

Endemicity is closely linked to the habitats of the two known human loiasis vectors, *Chrysops dimidiata* and *C. silicea*.

Cases have been reported on occasion in the United States but are restricted to travelers who have returned from endemic regions.

In the 1990s, the only method of determining *Loa loa* intensity was with microscopic examination of standardized blood smears, which is not practical in endemic regions. Because mass diagnostic methods were not available, complications started to surface once mass ivermectin treatment programs started being carried out for onchocerciasis, another filariasis. Ivermectin, a microfilaricidal drug, may be contraindicated in patients who are co-infection with loiasis and have associated high microfilarial loads. The theory is that the killing of massive numbers of microfilaria, some of which may be near the ocular and brain region, can lead to encephalopathy. Indeed, cases of this have been documented so frequently over the last decade that a term has been given for this set of complication: neurologic serious adverse events (SAEs).

Advanced diagnostic methods have been developed since the appearance the SAEs, but more specific diagnostic tests that have been or are currently being development must to be supported and distributed if adequate loiasis surveillance is to be achieved.

There is much overlap between the endemicity of the two distinct filariases, which complicates mass treatment programs for onchocerciasis and necessitates the development of greater diagnostics for loiasis.

In Central and West Africa, initiatives to control onchocerciasis involve mass treatment with Ivermectin. However, these regions typically have high rates of co-infection with both *L. loa* and *O.*

volvulus, and mass treatment with Ivermectin can have severe adverse effects (SAE). These include hemorrhage of the conjunctiva and retina, heamaturia, and other encephalopathies that are all attributed to the initial L. loa microfilarial load in the patient prior to treatment. Studies have sought to delineate the sequence of events following Ivermectin treatment that lead to neurologic SAE and sometimes death, while also trying to understand the mechanisms of adverse reactions to develop more appropriate treatments.

In a study looking at mass Ivermectin treatment in Cameroon, one of the greatest endemic regions for both onchocerciasis and loiasis, a sequence of events in the clinical manifestation of adverse effects was outlined.

It was noted that the patients used in this study had a *L. loa* microfilarial load of greater than 3,000 per ml of blood.

Within 12–24 hours post-Ivermectin treatment (D1), individuals complained of fatigue, anorexia, and headache, joint and lumbar pain—a bent forward walk was characteristic during this initial stage accompanied by fever. Stomach pain and diarrhea were also reported in several individuals.

By day 2 (D2), many patients experienced confusion, agitation, dysarthria, mutism and incontinence. Some cases of coma were reported as early as D2. The severity of adverse effects increased with higher microfilarial loads. Hemorrhaging of the eye, particularly the retinal and conjunctiva regions, is another common sign associated with SAE of Ivermectin treatment in patients with *L. loa* infections and is observed between D2 and D5 post-treatment. This can be visible for up to 5 weeks following treatment and has increased severity with higher microfilarial loads.

Haematuria and proteinuria have also been observed following Ivermectin treatment, but this is common when using Ivermectin to treat onchocerciasis. The effect is exacerbated when there are high *L. loa* microfilarial loads however, and microfilaria can be observed in the urine occasionally. Generally, patients recovered from SAE within 6–7 months post-Ivermectin treatment; however, when their complications were unmanaged and patients were left bed-ridden, death resulted due to gastrointestinal bleeding, septic shock, and large abscesses.

Mechanisms for SAE have been proposed. Though microfilarial load is a major risk factor to post-Ivermectin SAE, three main hypotheses have been proposed for the mechanisms.

The first mechanism suggests that Ivermectin causes immobility in microfilariae, which then obstructs microcirculation in cerebral regions. This is supported by the retinal hemorrhaging seen in some patients, and is possibly responsible for the neurologic SAE reported.

The second hypothesis suggests that microfilaria may try to escape drug treatment by migrating to brain capillaries and further into brain tissue; this is supported by pathology reports demonstrating a microfilarial presence in brain tissue post-Ivermectin treatment.

Lastly, the third hypothesis attributes hypersensitivity and inflammation at the cerebral level to post-Ivermectin treatment complications, and perhaps the release of bacteria from L. loa after treatment to SAE. This has been observed with the bacteria *Wolbachia* that live with *O. volvulus*.

More research into the mechanisms of post-Ivermectin treatment SAE is needed to develop drugs that are appropriate to individuals suffering from multiple parasitic infections.

One drug that has been proposed for the treatment of onchocerciasis is doxycycline. This drug has been shown to be effective in killing both the adult worm of *O. volvulus* and *Wolbachia*, the bacteria believed to play a major role in the onset of onchocerciasis, while having no effect on the microfilaria of *L. loa*. In a study done at 5 different co-endemic regions for onchocerciasis and loiasis, doxycycline was shown to be effective in treating over 12,000 individuals infected with both parasites with minimal complications. Drawbacks to using Doxycycline include bacterial resistance and patient compliance because of a longer treatment regimen and emergence of doxycycline-resistant *Wolbachia*. However, in the study over 97% of the patients complied with treatment, so it does pose as a promising treatment for onchocerciasis, while avoiding complications associated with L. loa co-infections.

Human loiasis geographical distribution is restricted to the rain forest and swamp forest areas of West Africa, being especially common in Cameroon and on the Ogooué River. Humans are the only known natural reservoir. It is estimated that over 10 million humans are infected with the *Loa loa* larvae.

An area of tremendous concern regarding loiasis is its co-endemicity with onchocerciasis in certain areas of west and central Africa, as mass ivermectin treatment of onchocerciasis can lead to serious adverse events (SAEs) in patients who have high *Loa loa* microfilarial densities, or loads. This fact necessitates the development of more specific diagnostics tests for *Loa loa* so that areas and individuals at a higher risk for neurologic consequences can be identified prior to microfilaricidal treatment. Additionally, the treatment of choice for loiasis, diethylcarbamazine, can lead to serious complications in and of itself when administered in standard doses to patients with high *Loa loa* microfilarial loads.

History

The first case of *Loa loa* infection was noted in the Caribbean (Santo Domingo) in 1770. A French surgeon named Mongin tried but failed to remove a worm passing across a woman's eye. A few years later, in 1778, the surgeon François Guyot noted worms in the eyes of West African slaves on a French ship to America; he successfully removed a worm from one man's eye.

The identification of microfilaria was made in 1890 by the ophthalmologist Stephen McKenzie. Localized angioedema, a common clinical presentation of loiasis, was observed in 1895 in the coastal Nigerian town of Calabar—hence the name "Calabar" swellings. This observation was made by a Scottish ophthalmologist named Douglas Argyll-Robertson, but the association between *Loa loa* and Calabar swellings was not realized until 1910 (by Dr. Patrick Manson). The determination of vector—*Chrysops* spp.—was made in 1912 by the British parasitologist Robert Thomson Leiper.

Synonyms

Synonyms for the disease include African eye worm, loaiasis, loaina, *Loa loa* filariasis, filaria loa, filaria lacrimalis, filaria subconjunctivalis, Calabar swellings, Fugitive swellings, and microfilaria diurnal. *Loa loa*, the scientific name for the infectious agent, is an indigenous term itself and it is likely that there are many other terms used from region to region.

Onchocerca Volvulus

Onchocerca volvulus is a nematode that causes onchocerciasis or "river blindness" mostly in Africa. Long-term corneal inflammation, or keratitis, leads to thickening of the corneal stroma which ultimately leads to blindness. Humans are the only definitive host for *O. volvulus*. The intermediate host or vector is the black fly (*Simulium*).

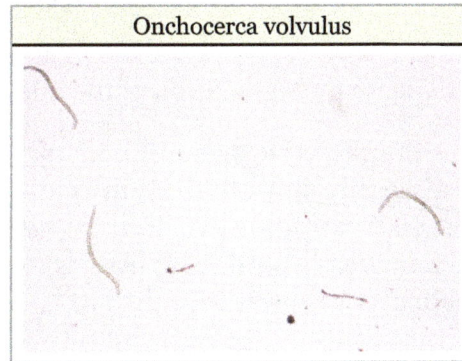

Onchocerca volvulus

O. volvulus, along with most filarial nematodes, share an endosymbiotic relationship with the bacterium *Wolbachia*. In the absence of *Wolbachia*, larval development of the *O. volvulus* is disrupted or ceased.

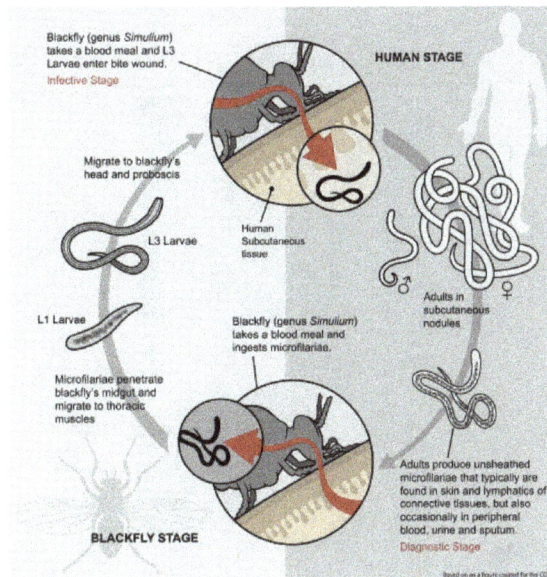

The life cycle of *O. volvulus*

The life cycle of *O. volvulus* begins when a parasitised female black fly of the genus *Simulium* takes a blood meal. The microfilariae form of the parasite found in the dermis of the host is ingested by the black fly. Here the microfilariae then penetrates the gut and migrates to thoracic flight muscles of the black fly, entering its first larval phase (J_1). After maturing into J_2, the second larval phase, it migrates to the proboscis where it can be found in the saliva. Saliva containing stage three (J_{3S}) *O. volvulus* larvae passes into the blood of the host. From here the larvae migrate to the subcutaneous tissue where they form nodules and then mature into adult worms over a period of six to twelve months. After maturation, the smaller adult males migrate from nodules to subcutaneous tissue where they mate with the larger adult females, which then produce between 1,000 and 3,000 microfilariae per day. The normal adult worm lifespan is up to fifteen years. The eggs mature internally to form stage one microfilariae, which are released from the female's body one at a time and remain in the subcutaneous tissue.

These stage one microfilariae are taken up by black flies upon a blood meal, in which they mature over the course of one to three weeks to stage three larvae, thereby completing the life cycle.

The normal microfilariae lifespan is 1–2 years; however, their presence in the bloodstream causes little or no immune response until death or degradation of the microfilariae or adult worms.

O. volvulus has been proposed as the causative agent of nodding syndrome, a condition that affects children aged 5 to 15 and is currently only observed in South Sudan, Tanzania and northern Uganda. Although the cause of the disease is unknown, *O. volvulus* is being increasingly studied as a possible cause due to its ubiquity in areas where the disease is found. Nodding syndrome causes children to nod violently while eating; it has been described as an unusual form of epilepsy.

Thelazia Callipaeda

Thelazia callipaeda is a parasitic nematode, and the most common cause of "thelaziasis" (or "eyeworm" infestation) in humans, dogs and cats. It was first discovered in the eyes of a dog in China in 1910. By 2000, over 250 human cases had been reported in the medical literature.

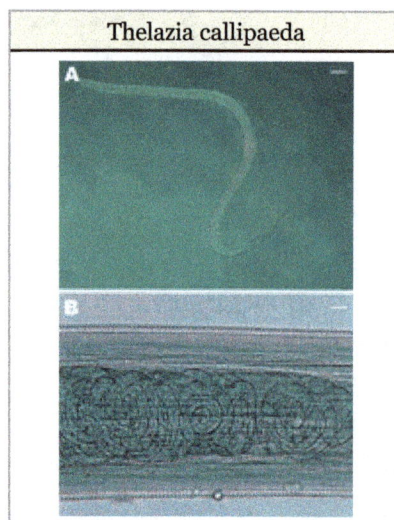

Thelazia callipaeda

Hosts

In addition to humans, cats and dogs, definitive hosts of *T. callipaeda* include the wolf (*Canis lupus*), raccoon dog (*Nyctereutes procyonoides*), red fox (*Vulpes fulva*), and European rabbit (*Oryctolagus cuniculus*). This species has been found in China, France, Germany, India, Indonesia, Italy, Japan, Korea, the Netherlands, Russia, Switzerland, Taiwan and Thailand.

Two intermediate hosts have been identified so far: *Amiota (Phortica) variegata* (Diptera: Drosophilidae) in Europe and *Phortica okadai* in China, which feed on tears of humans and carnivores. Some data suggests that only the males of *A. (P.) variegata* carry *Thelazia callipaeda* larvae. This is noteworthy because in all other known cases of blood-feeding flies which transmit parasites, the parasites are carried by the females.

Life Cycle

The eggs of *Thelazia callipaeda* develop into first stage larvae (L1), in utero while the female is in the tissues in and around the eye of the definitive host. The female deposits these larvae, which are still enclosed in the egg membranes, in the tears (lacrymal secretions) of the host. When a tear-feeding fly (intermediate host) feeds, it ingests the *T. callipaeda* larvae. Once inside the fly, the L1 larvae "hatch" from the egg membrane and penetrate the gut wall. They remain in the hemocoel (the fly's circulatory system) for 2 days, and then invade either the fat body or testes of the flies. In these tissues, the larvae develop into third stage larvae (L3). The L3 migrates to the head of the fly, and is released in or near the eye of a new host mammal when the fly feeds again. Once in the eye, eyelid, tear glands, or tear ducts of the mammalian host, the L3 larvae develop through the L4 larval stage and into adults in about 1 month. The seasonal timing of L1 and L4 larvae in the lacrymal (tear) secretions of naturally infested dogs in Italy was found to coincide with the activity of the fly vectors.

Symptoms, Diagnosis and Treatment

Symptoms of *T. callipaeda* infestation include conjunctivitis, excessive watering (lacrimation), visual impairment, and ulcers or scarring of the cornea. In some cases, the only symptom is the worm obscuring the host's vision as a "floater".

Diagnosis is made by finding the adult worms in the eye or surrounding tissues. Human cases are treated by simply removing the worms. In canines, topical imidacloprid with moxidectin, or milbemycin oxime (Interceptor) have been recommended.

Trichinella Spiralis

Trichinella spiralis is an ovoviviparous nematode parasite, occurring in rodents, pigs, horses, bears, and humans, and is responsible for the disease trichinosis. It is sometimes referred to as the "pork worm" due to it being found commonly in undercooked pork products.

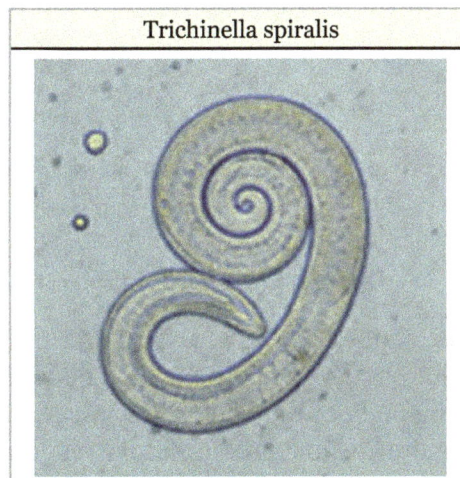

Trichinella spiralis

Description

Trichinella species are the smallest nematode parasite of humans, have an unusual lifecycle, and are one of the most widespread and clinically important parasites in the world. The small adult worms mature in the intestines of a definitive host such as a pig. Each adult female produces batches of live larvae, which bore through the intestinal wall, enter the blood (to feed on it) and lymphatic system, and are carried to striated muscle. Once in the muscle, they encyst, or become enclosed in a capsule. Humans can be infected by eating infected pork, horsemeat, or wild carnivores such as fox, cat, or bear.

Morphology

Males of *T. spiralis* measure between 1.4 and 1.6 mm long and are more flat anteriorly than posteriorly. The anus can be found in the terminal (side) and they have a large copulatory pseudobursa on each side. The females of *T. spiralis* are about twice the size of the males and have an anus found terminally. The vulva is located near the esophagus. The single uterus of the female is filled with developing eggs in the posterior portion, while the anterior portion contained the fully developed juveniles.

Life Cycle

Trichinella spiralis is a parasitic nematode which can live the majority of its adult life in the intestines of humans. To begin its life cycle "Trichinella spiralis" adults will invade the intestinal wall of pigs and produce larvae that invade the pigs muscles . The larval forms are encapsulated as a small cystic structure within the infected host. Humans typically become infected when they eat improperly cooked *Trichinella* infected pork or other meat. When a human eats the infected meat, the larvae are released from the nurse cell (due to stomach pH) and migrate to the intestine, where they burrow into the intestinal mucosa, mature, and reproduce. Juveniles within nurse cells have an anaerobic or facultative anaerobic metabolism, but when they become activated, they adopt an aerobic metabolism characteristics of the adult.

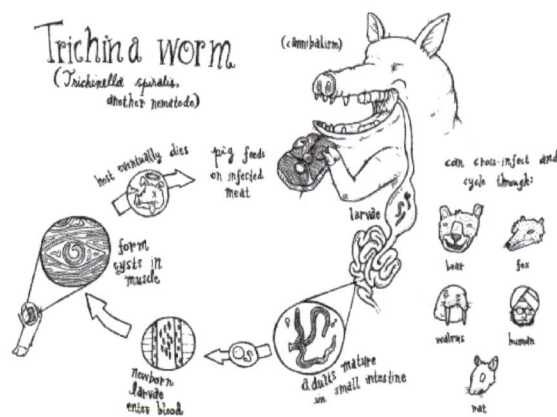

Trichinella spiralis lifecycle

Female *Trichinella* worms live for about six weeks, and in that time can produce up to 1,500 larvae; when a spent female dies, she passes out of the host. The larvae can then gain access to the circulation and migrate around the body of the host. The migration and encystment of larvae can cause fever and pain brought upon by the host inflammatory response. In some cases, migration to specific organ tissues can cause myocarditis and encephalitis that can result in death.

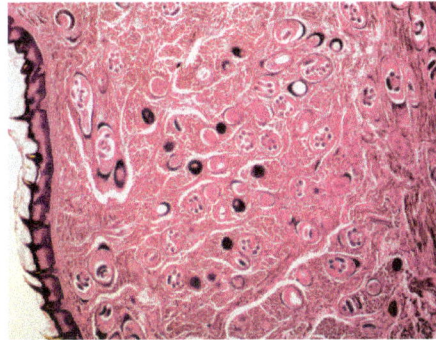

Animal tissue infected with the parasite that causes the disease trichinosis. Most parasites are shown in cross section but some randomly appear in long section.

Nurse Cell Formation

This nematode is a multicellular parasite that lives within a single muscle cell, which it modifies according to its own requirements.

T. spiralis larvae within the diaphragm muscle of a pig

Nurse cell formation in skeletal muscle tissue is mediated by the hypoxic environment surrounding the new vessel formation. The hypoxic environment stimulates cells in the surrounding tissue to regulate up and secrete angiogenic cytokines, such as vascular endothelial growth factor (VEGF). This allows for the newborn *T. spiralis* larvae to enter and form the nurse cells. VEGF expression is detected surrounding the nurse cell right after nurse cell formation, and the continued secretion of VEGF can maintain the constant state of hypoxia. Previous studies have shown VEGF can stimulate proliferation of synthesis of collagen type 1 in activated myofibroblast-like cells.

Symptoms

The first symptoms may appear between 12 hours and two days after ingestion of infected meat. The migration of worms in the intestinal epithelium can cause traumatic damage to the host tissue, and the waste products they excrete can provoke an immunological reaction. The resulting inflammation can cause symptoms such as nausea, vomiting, sweating, and diarrhea. Five to seven days after the appearance of symptoms, facial edema and fever may occur. After 10 days, intense muscular pain, difficulty breathing, weakening of pulse and blood pressure, heart damage, and various nervous disorders may occur, eventually leading to death due to heart failure, respiratory complications, or kidney malfunction.

In pigs, infection is usually subclinical, but large numbers of worms can be fatal in some cases.

Diagnosis and Treatment

Muscle biopsy is used for trichinosis detection. Several immunodiagnostic tests are also available. Typically, patients are treated with either mebendazole or albendazole, but efficacy of such products are uncertain. Symptoms can be relieved by use of analgesics and corticosteroids.

In pigs, ELISA testing is possible as a method of diagnosis. Anthelmintics can treat and prevent *Trichinella* infections.

Prevention and Control

Trichinosis (also trichinellosis) is a disease caused by tissue-dwelling roundworms of the species *Trichinella spiralis*. In the United States, the national trichinellosis surveillance system has documented a steady decline in the reported incidence of this disease. During 1947 to 1951, a median of 393 cases was reported annually, including 57 trichinellosis-related deaths. During 1997-2001, the incidence decreased to a median of 12 cases annually, with no reported deaths. The decline of infection was largely associated with changes implemented by the U.S. pork industry that have resulted in reduced prevalence of *Trichinella* among domestic swine. In the United States, Congress passed the Federal Swine Health Protection Act restricting the use of uncooked garbage as feed stock for pigs and creating a voluntary Trichinae Herd Certification Program. The Trichinae Herd Certification Program is a voluntary preharvest pork safety program that provides documentation of swine management practices to minimize *Trichinella* exposure. The goal of the program is to establish a system under which pork production facilities that follow good production practices might be certified as Trichinella-safe. In addition to the reduction in *Trichinella* prevalence in commercial pork, processing methods also have contributed to the dramatic decline in human trichinellosis associated with pork products. Through the U.S. Code of Federal Regulations, the USDA has created guidelines for specific cooking temperatures and times, freezing temperatures and times, and curing methods for processed pork products to control postharvest human exposure to *Trichinella*. Pork products meeting these guidelines are designated certified pork.

The chances of becoming infected with *Trichinella spiralis* are relatively low in the United States, due to such rigorous control measures, while there are higher prevalence rates in regions such as Europe and Asia. This parasite is considered to be endemic in Japan and China, while Korea just reported its first case of the parasite disease.

In most abattoirs, the diaphragm of pigs is routinely sampled to detect *Trichinella* infections.

Postharvest human exposure is also preventable by educating consumers of simple steps that can be taken to kill any larvae that can potential be in meat bought at the local supermarket. Freezing meat in an average household freezer for 20 days before consumption will kill some species of *Trichinella*. Cooking pork products to a minimum internal temperature of 160 °F will kill most species and is the best way to ensure the meat is safe to eat.

Economic Impact

Political and economic changes have caused an increase in the prevalence and incidence rates of this parasite in many European countries. This complicates the meat trade industry within European countries and makes it difficult for exportation of pork outside of these countries. The European Commission proposed a new regulation to implement specific rules for food safety.

Illegal importation from places with low safety quality standards allows for the spread of the parasite from endemic to non-endemic countries. Illegal importation and new food practices have resulted in outbreaks in many countries including: Denmark, Germany, Italy, Spain, and the United Kingdom.

The economic cost for detecting trichinosis can be another cost burden. In 1998, it was estimated to cost $3.00 per pig in order to detect the parasite. In order to control Trichinella infection, the European Union condemned 190 million pigs leading to a substantial economic impact of about $570 million per year.

Genome

The *Trichinella spiralis* draft genome became available in March 2011. The genome size was 58.55 Mbp with an estimated 16,549 genes. The *T. spiralis* genome is the only known nematode genome to be subject to DNA methylation, an epigenetic mechanism that was not previously thought to exist in nematodes.

Trichinella Britovi

Trichinella britovi is a nematode parasite responsible for a zoonotic disease called trichinellosis. Currently, eight species of *Trichinella* are known, only three of which cause trichinellosis, and *Trichinella britovi* is one of them. Numerous mammal species, as well as birds and crocodiles, can harbor the parasite worldwide, but the sylvatic cycle is mainly maintained by wild carnivores.

Humans represent only a possible host and the parasite is exclusively transmitted through consumption of raw or rare meat. In Europe, pork, wild boar meat, and horse meat are the main sources for human infection.

Because of mandatory veterinary controls in slaughterhouses, large trichinellosis outbreaks due to horse meat consumption are rare, but cases in hunters and their families after raw or rare wild boar meat consumption are regularly reported, with over 100 cases since 1975.

T. britovi in wild boar is relatively resistant to freezing. It was observed in France that meat from naturally infected wild boar meat frozen for three weeks at −20 °C (−4 °F) remained infectious, whereas the parasites were not viable after four weeks.

In the 1960s, "trichinella infection" was documented in Senegal, West Africa. A survey of 160 wild animals from that region produced plausible evidence that European strains may have originated in Africa. It has also been proposed that strains of *T. britovi* are isolated to both African and European populations.

Pork sausages eaten raw by consumers caused an outbreak of trichinellosis in 2015 in France

Three cases of human trichinellosis due to *T. britovi* were reported in 2015 in the Southeast of France resulting from consumption of raw pork sausages (*figatelli*) prepared in Corsica. Fourteen other people ate *figatelli* from the same batch but were not infected due to the *figatelli* being well cooked.

Prevention

To prevent trichinellosis, an official European directive recommends the freezing of meat at −25 °C (−13 °F) for at least 10 days for pieces of meat less than 25 cm (10 in) in thickness. Patients froze wild boar steaks at −35 °C (−31 °F) for seven days, but this freezing time appears insufficient to kill larvae, since *T. britovi* is a species relatively resistant to freezing.

Thus according to the International Commission on Trichinellosis, meat should be heated at 65 °C (149 °F) for at least 1 minute to kill *Trichinella* larvae; larvae die when the color of the meat at the core changes from pink to brown.

Ancylostoma Duodenale

Ancylostoma duodenale is a species of the worm genus *Ancylostoma*. It is a parasitic nematode worm and commonly known as the Old World hookworm. It lives in the small intestine of hosts such as humans, cats and dogs, where it is able to mate and mature. *Ancylostoma duodenale* and *Necator americanus* are the two human hookworms that are normally discussed together as the cause of hookworm infection. They are dioecious. *Ancylostoma duodenale* is abundant throughout the world, including in the following areas: southern Europe, north Africa, India, China, southeast Asia, some areas in the United States, the Caribbean, and South America.

Characteristics

Ancylostoma duodenale is small cylindrical worm, greyish-white in color. It has two ventral plates on the anterior margin of the buccal capsule. Each of them has two large teeth that are fused at their bases. A pair of small teeth can be found in the depths of the buccal capsule. Males are 8 mm to 11 mm long with a copulatory bursa at the posterior end. Females are 10 mm to 13 mm long, with the vulva located at the posterior end; females can lay 10,000 to 30,000 eggs per day. The average lifespan of *Ancylostoma duodenale* is one year.

After a rhabditiform "infective" larva penetrates the intact skin – most commonly through the feet – the larva enters the blood circulation. It is then carried to the lungs, breaks into alveoli, ascends the bronchi and trachea and is coughed up and swallowed back into the small intestine where it matures. The larva later matures into an adult in the small intestine ("the adult worm live in the jejunum mainly"), where they attach to the villi and female worms can lay 25,000 eggs per day. The eggs are released into the feces and reside on soil, when deposited on warm, moist soil a larva rapidly develops in the egg and hatches after 1 to 2 days. This Rhabditiform larva moults twice in the soil and become a skin penetrating 3rd stage infective larva within 5–10 days. The Rhabditiform larvae can then penetrate the exposed skin of another organism and begin a new cycle of infection.

Ancylostoma duodenale is prevalent in southern Europe, northern Africa, India, China, and southeast Asia, small areas of United States, the Caribbean islands, and South America. This hookworm is well known in mines because of the consistency in temperature and humidity that provide an ideal habitat for egg and juvenile development. It is estimated 1 billion people are infected with hookworms. Transmission of *Ancylostoma duodenale* is by contact of skin with soil contaminated with larvae. The way *Ancylostoma duodenale* enters the human body was understood in the 1880s, after an epidemic of ancylostomiasis among miners working in the hot and humid Gotthard Tunnel (Switzerland).

Infection

A light infection causes abdominal pain, loss of appetite and geophagy. Heavy infection causes severe protein deficiency or iron deficiency anemia. Protein deficiency may lead to dry skin, edema and potbelly, while iron deficiency anemia might result in mental dullness and heart failure.

The eggs of *Ancylostoma duodenale* and *Necator americanus* cannot be distinguished. Larvae cannot be found in stool specimen unless they are left at ambient temperature for a day or more.

Education, improved sanitation and controlled disposal of human feces are important. Wearing shoes in endemic areas can reduce the prevalence of infection as well.

Ancylostoma duodenale can be treated with albendazole, mebendazole and benzimidazoles. Pyrantel pamoate is an alternative. In severe cases of anemia, blood transfusion may be necessary.

Necator Americanus

Necator americanus is a species of hookworm (a type of helminth) commonly known as the New World hookworm. Like other hookworms, it is a member of the phylum Nematoda. It is a parasitic nematode that lives in the small intestine of hosts such as humans, dogs, and cats. Necatoriasis—a type of helminthiasis—is the term for the condition of being host to an infestation of a species of *Necator*. Since *N. americanus* and *Ancylostoma duodenale* (also known as Old World hookworm) are the two species of hookworms that most commonly infest humans, they are usually dealt with under the collective heading of "hookworm infection". They differ most obviously in geographical distribution, structure of mouthparts, and relative size.

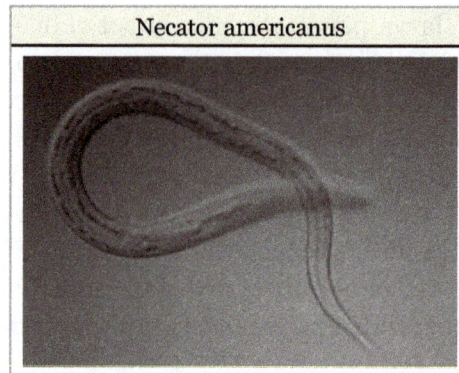

Necator americanus

Necator americanus has been proposed as an alternative to *Trichuris suis* in helminthic therapy.

Morphology

This parasite has two dorsal and two ventral cutting plates around the anterior margin of the buccal capsule. It also has a pair of subdorsal and a pair of subventral teeth located close to the rear. Males are usually 7–9 mm long, whereas females are about 9–11 mm long. The typical lifespan of these parasites is three to five years. They can produce between 5000 and 10,000 eggs per day.

Life Cycle

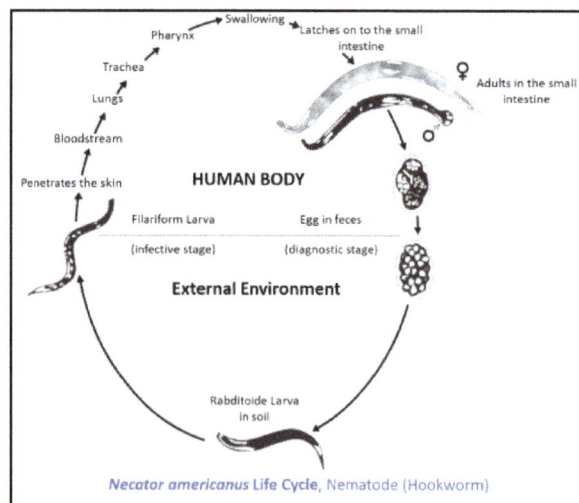

Life cycle of Necator americanus inside and outside of the human body

This worm starts out as an unembryonated egg in the soil. After 24–48 hours under favorable conditions, the eggs become embryonated and hatch. This first juvenile stage 1 is known as 'rhabditiform'. The rhabditiform larvae grow and molt in the soil, transforming into a juvenile stage 2. The juvenile stage 2 molts once more until reaching the juvenile 3 stage, which is also called 'filariform'; this is also the infective form. The transformation from rhabditiform to the filariform usually takes five to 10 days. This larval form is able to penetrate human skin, travel through the blood vessels and heart, and reach the lungs. Once there, they burrow through the pulmonary alveoli and travel up the trachea, where they are swallowed and are carried to the small intestine. There, they attach themselves to the intestinal wall, and mature into adults and reproduce. Adults live in the lumen of the intestinal wall where they cause blood loss to the host. The eggs produced by the adults end up on the soil after leaving the body through the feces; female hookworms produce up to 30,000 eggs per day. On average, most adult worms are eliminated in one to two years. The *N. americanus* life cycle only differs slightly from that of *A. duodenale*. *N. americanus* has no development arrest in immune hosts and it must migrate through the lungs.

Pathogenesis and Symptoms

The pathology of *Necator americanus* is divided into two stages: larvae stage and adult stage. The larvae penetrates the uninfected skin and travels through various organs, including the respiratory tract and lymph nodes. Once in the lymph nodes, the larvae starts entering the blood, lungs, and intestines. Some larva cannot readily enter the dermis and remain trapped in the skin, causing skin irritation and cutaneous larva migrans. Other symptoms include excessive coughing and dyspnea (short of breath) during larvae migration. Once attached to the intestinal wall, *N. americanus* reside and mature into adults, penetrate blood vessels, and suck blood. Blood loss from sites of intestinal attachment may cause iron-deficiency anemia and protein loss. Studies have shown that one individual *N. americanus* can cause 30 microliters of blood loss per day. Iron deficiency anemia can cause mental retardation and growth insufficiency in children. Further, infected patients will experience abdominal pain (exacerbated by meals) with diarrhea, bloating, and nausea.

Epidemiology

In the United States, 95% of human hookworm cases are caused by *N. americanus*, primarily in young school children in economically deprived rural areas. Juveniles cannot survive freezing temperatures, therefore the highest prevalence occurs in areas with warmer temperatures and greater rainfall. The greatest incidence of infections occurs in Asia and sub-Saharan Africa, especially in poverty-stricken areas with poor sanitation. *A. duodenale* infections occur at a lesser rate and are seen primarily in Europe and the Mediterranean.

Genome

A draft assembly of the genome of *Necator americanus* has been sequenced and analyzed. It comprises 244 Mbp with 19,151 predicted protein-coding genes; these include genes whose products mediate the hookworm's invasion of the human host, genes involved in blood feeding and development, genes encoding proteins that represent new potential drug targets against hookworms, and expanded gene families encoding likely immunomodulator proteins, whose products may be beneficial in treating inflammatory diseases.

Diagnostics

The most common method for diagnosing *N. americanus* is through identification of eggs in a fecal sample using a microscope. *N. americanus* eggs have a thin shell and are oval shaped, measuring roughly 56–74 by 36–40 μm.

Treatments and Medications

Anthelmintic Drugs

The most common treatment for N. americanus are benzimidazoles, specifically albendazole and mebendazole. Benzimidazoles kill adult worms by binding to the nematde's Beta-tubulin and subsequently inhibiting microtubule polymerization within the parasite. Keiser and Utzinger conducted a study in 2008, Efficacy of Current Drugs Against Soil–Transmitted Helminth Infections: Systematic Review and Meta-analysis, which found that the efficacy of single-dose treatments for Hookworm infections were as followed: 72% for albendazole, 15% for mebendazole, and 31% for pyrantel pamoate.

Patients infected with hookworm N. americanus may also consider cryotherapy as a treatment option.

Prevention and Control

The most effective prevention technique is to not walk barefoot in areas where hookworm is common and where there may be contamination of the soil. Locations for outdoor activities should be considered if there will be skin-to-soil contact.

Infection and transmission of others can be prevented by not defecating outdoors or using human feces fertilizer.

Economic Burden

N. americanus has played a significant role in the development of New and Old Worlds. N. americanus cause hookworm diseases, which are associated with nutrition and blood loss. Patients who are infected with approximately 25 to 100 worms will experience symptoms such as fatigue, weight loss, and slight headaches. As the infestations number reach 100 to 500 worms, the patient will experience extreme fatigue, iron deficiency, and abdominal pain. The symptoms worsen and result in possible death when the infestation reaches over 500 hookworms. Children and pregnant women affected by N. americanus are at greater risk due to anemia and the greater need for nutrition. There is a high demand for an improvement of sanitation to reduce fecal contamination in regions with high prevalence of N. americanus infections. The current control strategies include a combination of mass drug administration (MDA) for children at age 4–6 years to prevent or eliminate N. americanus infections.

Ascaris Lumbricoides

Ascaris lumbricoides is the small roundworm of humans, growing to a length of up to 35 cm (14 in). It is one of several species of Ascaris. An ascarid nematode of the phylum Nematoda, it is the most common parasitic worm in humans. This organism is responsible for the disease ascariasis, a type of helminthiasis and one of the group of neglected tropical diseases. An estimated one-sixth of the human population is infected by A. lumbricoides or another roundworm. Ascariasis is prevalent worldwide, especially in tropical and subtropical countries.

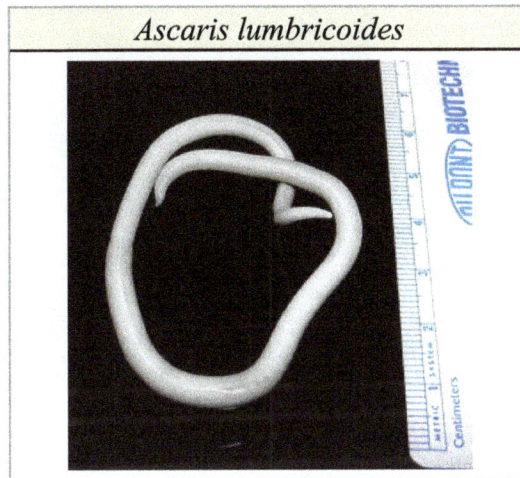

Ascaris lumbricoides

Lifecycle

A. lumbricoides, a roundworm, infects humans when an ingested fertilised egg becomes a larval worm (called rhabditiform larva) that penetrates vaginal wall of the duodenum and enters the blood stream. From there, it is carried to the liver and heart, and enters pulmonary circulation to break free in the alveoli, where it grows and molts. In three weeks, the larva passes from the respiratory system to be coughed up, swallowed, and thus returned to the small intestine, where it matures to an adult male or female worm. Fertilization can now occur and the female produces as many as 200,000 eggs per day for a year. These fertilized eggs become infectious after two weeks in soil; they can persist in soil for 10 years or more.

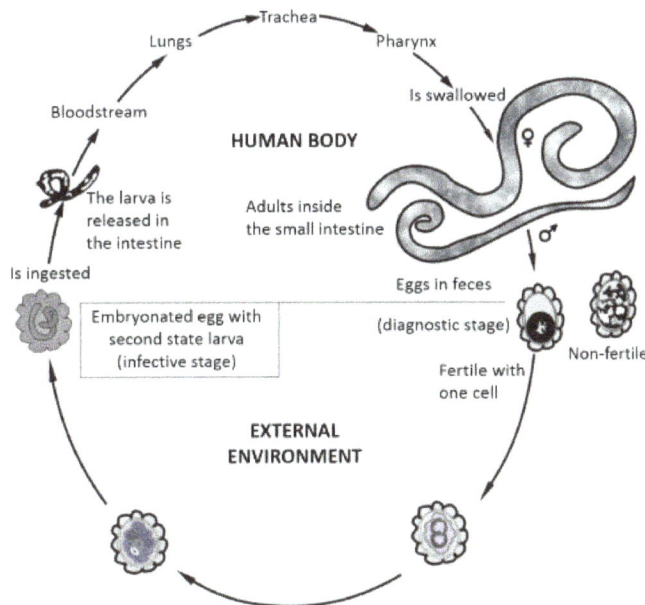

Ascaris lumbricoides **Life Cycle**, Nematode (Roundworm)

Image showing lifecycle inside and outside of the human body of one fairly well described Handsome: *A. lumbricoides*

The eggs have a lipid layer which makes them resistant to the effects of acids and alkalis, as well as other chemicals. This resilience helps to explain why this nematode is such a ubiquitous parasite.

Morphology

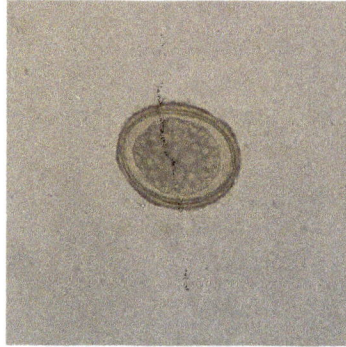

Fertile egg as can be seen in a microscope

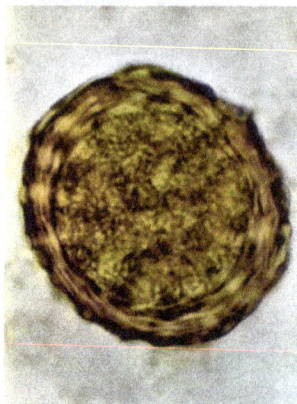

Fertile egg in human faeces (detail)

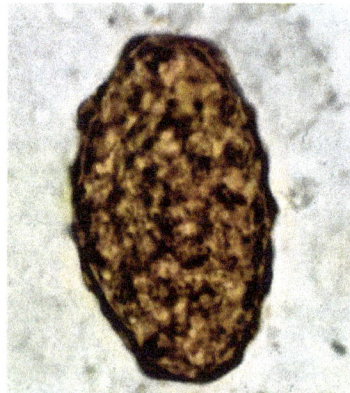

Infertile egg

A. lumbricoides is characterized by its great size. Males are 2–4 mm (0.08–0.2 in) in diameter and 15–31 cm (5.9–12 in) long. The male's posterior end is curved ventrally and has a bluntly pointed tail. Females are 3–6 mm (0.1–0.2 in) wide and 20–49 cm (7.9–19 in) long. The vulva is located in the anterior end and accounts for about one-third of its body length. Uteri may contain up to 27 million eggs at a time, with 200,000 being laid per day. Fertilized eggs are oval to round in shape and are 45–75 μm (0.0018–0.0030 in) long and 35–50 μm (0.0014–0.0020 in) wide with a thick outer shell. Unfertilized eggs measure 88–94 μm (0.0035–0.0037 in) long and 44 μm (0.0017 in) wide.

Epidemiology

An estimated 1 billion people are infected with *A. lumbricoides* worldwide. While infection occurs throughout most of the world, *A. lumbricoides* infection is most common in sub-Saharan Africa, the Americas, China, and east Asia.

A. lumbricoides eggs are extremely resistant to strong chemicals, desiccation, and low temperatures. The eggs can remain viable in the soil for several months or even years. Eggs of *A. lumbricoides* have been identified in archeological coprolites in the Americas, Europe, Africa, the Middle East, and New Zealand, the oldest ones being more than 24,000 years old.

Infections

Infections with these parasites are more common where sanitation is poor, and raw human feces are used as fertilizer.

Symptoms

Often, no symptoms are seen with an *A. lumbricoides* infection. However, in the case of a particularly bad infection, symptoms may include bloody sputum, cough, fever, abdominal discomfort, intestinal ulcer, passing worms, etc. Ascariasis is also the most common cause of Löffler's syndrome worldwide. Accompanying symptoms include pulmonary infiltration, eosinophilia, and radiographic opacities. Significant increases in fertility are observed in infected women.

Prevention

Preventing any fecal-borne disease requires educated hygienic habits/culture and effective fecal treatment systems. This is particularly important with *A. lumbricoides* because its eggs are one of the most difficult pathogens to kill (second only to prions), and the eggs commonly survive 1–3 years. *A. lumbricoides* lives in the intestine where it lays eggs. Infection occurs when the eggs, too small to be seen by the unaided eye, are eaten. The eggs may get onto vegetables when improperly processed human feces of infected people are used as fertilizer for food crops. Infection may occur when food is handled without removing or killing the eggs on the hands, clothes, hair, raw vegetables/fruit, or cooked food that is (re)infected by handlers, containers, etc. Bleach does not readily kill *A. lumbricoides* eggs, but it will remove their sticky film, to allow the eggs to be rinsed away. *A. lumbricoides* eggs can be reduced by hot composting methods, but to completely kill them may require rubbing alcohol, iodine, specialized chemicals, cooking heat, or "unusually" hot composting (for example, over 50 °C (122 °F) for 24 hours).

Details of Infection

Infections happen when a human swallows water or food contaminated with unhatched eggs, which hatch into juveniles in the duodenum. They then penetrate the mucosa and submucosa and enter venules or lymphatics. Next, they pass through the right heart and into pulmonary circulation. They then break out of the capillaries and enter the air spaces. Acute tissue reaction occurs when several worms get lost during this migration and accumulate in other organs of the body. The juveniles migrate from the lung up the respiratory tract to the pharynx where they are swallowed.

They begin producing eggs within 60–65 days of being swallowed. These are produced within the small intestine, where the juveniles mature. It might seem odd that the worms end up in the same place where they began. One hypothesis to account for this behavior is that the migration mimics an intermediate host, which would be required for juveniles of an ancestral form to develop to the third stage. Another possibility is that tissue migration enables faster growth and larger size, which increases reproductive capacity.

Strongyloides Stercoralis

Strongyloides stercoralis is a human pathogenic parasitic roundworm causing the disease strongyloidiasis. Its common name is threadworm. In the UK and Australia, however, the term *threadworm* can also refer to nematodes of the genus *Enterobius*, otherwise known as pinworms.

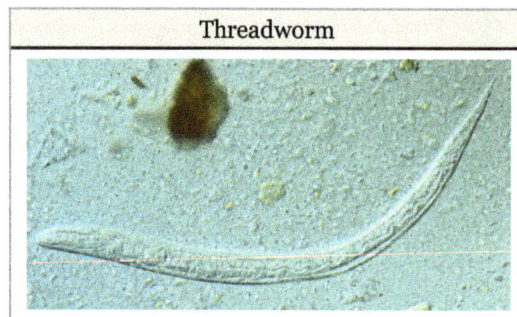
Threadworm

The *Strongyloides stercoralis* nematode can parasitize humans. The adult parasitic stage lives in tunnels in the mucosa of the small intestine. The genus *Strongyloides* contains 53 species, and *S. stercoralis* is the type species. *S. stercoralis* has been reported in other mammals, including cats and dogs. However, it seems that the species in dogs is typically not *S. stercoralis*, but the related species *S. canis*. Non-human primates are more commonly infected with *S. fuelleborni* and *S. cebus*, although *S. stercoralis* has been reported in captive primates. Other species of *Strongyloides*, naturally parasitic in humans, but with restricted distributions, are *S. fuelleborni* in central Africa and *S. kellyi* in Papua New Guinea.

Geographic Distribution

S. stercoralis has a very low prevalence in societies where fecal contamination of soil or water is rare. Hence, it is a very rare infection in developed economies. In developing countries, it is less prevalent in urban areas than in rural areas (where sanitation standards are poor). *S. stercoralis* can be found in areas with tropical and subtropical climates.

Strongyloidiasis was first described in the 19th century in French soldiers returning home from expeditions in Indochina. Today, the countries of the old Indochina (Vietnam, Cambodia, and Laos) still have endemic strongyloidiasis, with the typical prevalences being 10% or less. Regions of Japan used to have endemic strongyloidiasis, but control programs have eliminated the disease. Strongyloidiasis appears to have a high prevalence in some areas of Brazil and Central America. It is endemic in Africa, but the prevalence is typically low (1% or less). Pockets have been reported from rural Italy, but current status is unknown. In the Pacific islands, strongyloidiasis is rare, although some cases have been reported from Fiji. In tropical Australia, some rural and remote Australian Aboriginal communities have very high prevalences of strongyloidiasis.

In some African countries (e.g., Zaire), *S. fuelleborni* was more common than *S. stercoralis* in parasite surveys from the 1970s, but current status is unknown. In Papua New Guinea, *S. stercoralis* is endemic, but prevalence is low. However, in some areas, another species, *S. kellyi*, is a very common parasite of children in the New Guinea Highlands and Western Province.

Knowledge of the geographic distribution of strongyloidiasis is of significance to travelers who may acquire the parasite during their stays in endemic areas.

Because strongyloidiasis is transmittable by textiles, such as bedclothes and clothing, care must be taken never to use hotel bed sheets in endemic areas. Personal sleeping bags and using plastic slippers when showering are very important when travelling in tropical regions.

Life Cycle

The strongyloid's life cycle is heterogonic—it is more complex than that of most nematodes, with its alternation between free-living and parasitic cycles, and its potential for autoinfection and multiplication within the host. The parasitic cycle is homogenic, while the free-living cycle is heterogonic. The heterogonic life cycle is advantageous to the parasite because it allows reproduction for one or more generations in the absence of a host.

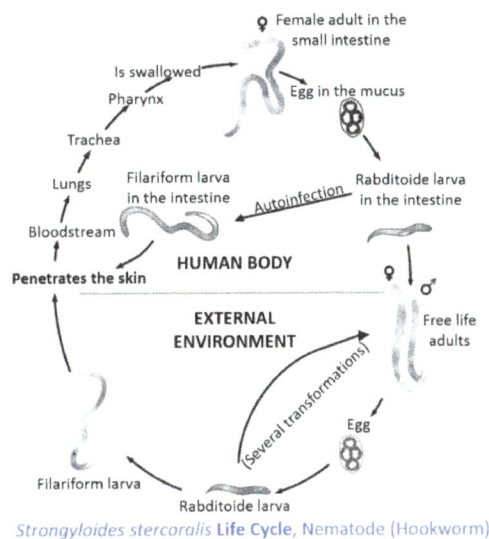

Strongyloides stercoralis Life Cycle, Nematode (Hookworm)

Life Cycle of *Strongyloides stercoralis* inside and outside of the human body

In the free-living cycle, the rhabditiform larvae passed in the stool can either molt twice and become infective filariform larvae (direct development) or molt four times and become free-living adult males and females that mate and produce eggs from which rhabditiform larvae hatch. In the direct development, first-stage larvae (L1) transform into infective larvae (IL) via three molts. The indirect route results first in the development of free-living adults that mate; the female lays eggs, which hatch and then develop into IL. The direct route gives IL faster (three days) versus the indirect route (seven to 10 days). However, the indirect route results in an increase in the number of IL produced. Speed of development of IL is traded for increased numbers. The free-living males and females of *S. stercoralis* die after one generation; they do not persist in the soil. The latter, in turn, can either develop into a new generation of free-living adults or develop into infective filariform larvae. The filariform larvae penetrate the human host skin to initiate the parasitic cycle.

The infectious larvae penetrate the skin when it contacts soil. While *S. stercoralis* is attracted to chemicals such as carbon dioxide or sodium chloride, these chemicals are not specific. Larvae have been thought to locate their hosts via chemicals in the skin, the predominant one being urocanic acid, a histidine metabolite on the uppermost layer of skin that is removed by sweat or the daily skin-shedding cycle. Urocanic acid concentrations can be up to five times greater in the foot than any other part of the human body. Some of them enter the superficial veins and are carried in the blood to the lungs, where they enter the alveoli. They are then coughed up and swallowed into the gut, where they parasitise the intestinal mucosa of the duodenum and jejunum. In the small intestine, they molt twice and become adult female worms. The females live threaded in the epithelium of the small intestine and, by parthenogenesis, produce eggs, which yield rhabditiform larvae. Only females will reach reproductive adulthood in the intestine. Female strongyloids reproduce through parthenogenesis. The eggs hatch in the intestine and young larvae are then excreted in the feces. It takes about two weeks to reach egg development from the initial skin penetration. By this process, *S. stercoralis* can cause both respiratory and gastrointestinal symptoms. The worms also participate in autoinfection, in which the rhabditiform larvae become infective filariform larvae, which can penetrate either the intestinal mucosa (internal autoinfection) or the skin of the perianal area (external autoinfection); in either case, the filariform larvae may follow the previously described route, being carried successively to the lungs, the bronchial tree, the pharynx, and the small intestine, where they mature into adults; or they may disseminate widely in the body. To date, occurrence of autoinfection in humans with helminthic infections is recognized only in *Strongyloides stercoralis* and *Capillaria philippinensis* infections. In the case of *Strongyloides*, autoinfection may explain the possibility of persistent infections for many years in persons not having been in an endemic area and of hyperinfections in immunodepressed individuals.

Morphology

Whereas males grow to only about 0.9 mm (0.04 in) in length, females can grow from 2.0 to 2.5 mm (0.08 to 0.10 in). Both genders also possess a tiny buccal capsule and cylindrical esophagus without a posterior bulb. In the free-living stage, the esophagi of both sexes are rhabditiform. Males can be distinguished from females by two structures: the spicules and gubernaculum.

Autoinfection

An unusual feature of *S. stercoralis* is autoinfection. Only one other species in the *Strongyloides* genus, *S. felis*, has this trait. Autoinfection is the development of L1 into small infective larvae in the gut of the host. These autoinfective larvae penetrate the wall of the lower ileum or colon or the skin of the perianal region, enter the circulation again, travel to the lungs, and then to the small intestine, thus repeating the cycle. Autoinfection makes strongyloidiasis due to *S. stercoralis* an infection with several unusual features.

Persistence of infection is the first of these important features. Because of autoinfection, humans have been known to still be infected up to 65 years after they were first exposed to the parasite (e.g., World War II or Vietnam War veterans). Once a host is infected with *S. stercoralis*, infection is lifelong unless effective treatment eliminates all adult parasites and migrating autoinfective larvae.

Symptoms

Many people infected are asymptomatic at first. Symptoms include dermatitis: swelling, itching, larva currens, and mild hemorrhage at the site where the skin has been penetrated. Spontaneous scratch-like lesions may be seen on the face or elsewhere. If the parasite reaches the lungs, the chest may feel as if it is burning, and wheezing and coughing may result, along with pneumonia-like symptoms (Löffler's syndrome). The intestines could eventually be invaded, leading to burning pain, tissue damage, sepsis, and ulcers. Stools may have yellow mucus with a recognizable smell. Chronic diarrhea can be a symptom. In severe cases, edema may result in obstruction of the intestinal tract, as well as loss of peristaltic contractions.

Strongyloidiasis in immunocompetent individuals is usually an indolent disease. However, in immunocompromised individuals, it can cause a hyperinfective syndrome (also called disseminated strongyloidiasis) due to the reproductive capacity of the parasite inside the host. This hyperinfective syndrome can have a mortality rate close to 90% if disseminated.

Immunosuppressive drugs, especially corticosteroids and agents used for tissue transplantation, can increase the rate of autoinfection to the point where an overwhelming number of larvae migrate through the lungs, which in many cases can prove fatal. In addition, diseases such as human T-lymphotropic virus 1, which enhance the Th1 arm of the immune system and lessen the Th2 arm, increase the disease state. Another consequence of autoinfection is the autoinfective larvae can carry gut bacteria back into the body. About 50% of people with hyperinfection present with bacterial disease due to enteric bacteria. Also, a unique effect of autoinfective larvae is larva currens due to the rapid migration of the larvae through the skin. Larva currens appears as a red line that moves rapidly (more than 5 cm or 2 in per day), and then quickly disappears. It is pathognomonic for autoinfective larvae and can be used as a diagnostic criterion for strongyloidiasis due to *S. stercoralis*.

Diagnosis

Locating juvenile larvae, either rhabditiform or filariform, in recent stool samples will confirm the presence of this parasite. Other techniques used include direct fecal smears, culturing fecal samples on agar plates, serodiagnosis through ELISA, and duodenal fumigation. Still, diagnosis can be difficult because of the day-to-day variation in juvenile parasite load.

Treatment

Ideally, prevention, by improved sanitation (proper disposal of feces), practicing good hygiene (washing of hands), etc., is used before any drug regimen is administered.

Ivermectin is the drug of first choice for treatment because of higher tolerance in patients. Thiabendazole was used previously, but, owing to its high prevalence of side effects (dizziness, vomiting, nausea, malaise) and lower efficacy, it has been superseded by ivermectin and as second-line albendazole. However, these drugs have little effect on the majority of these autoinfective larvae during their migration through the body. Hence, repeated treatments with ivermectin must be administered to kill adult parasites that develop from the autoinfective larvae. This means at least two weeks treatment, then a weeks pause, then again treatment. Follow-up treatment and blood tests are necessary for decades following infection.

In the UK, mebendazole and piperazine are currently (2007) preferred. Mebendazole has a much higher failure rate in clinical practice than albendazole, thiabendazole, or ivermectin.

Chemoattractant

This parasite depends on chemical cues to find a potential host. It uses sensor neurons of class AFD to identify cues excreted by the host.*S. stercoralis* is attracted to nonspecific attractants of warmth, carbon dioxide, and sodium chloride. Urocanic acid, a component of skin secretions in mammals, is a major chemoattractant. Larvae of *S. stercoralis* are strongly attracted to this compound. This compound can be suppressed by metal ions, suggesting a possible strategy for preventing infection.

Toxocara Canis

Toxocara canis (also known as dog roundworm) is worldwide-distributed helminth parasite of dogs and other canids. *Toxocara canis* is gonochoristic, adult worms measure from 9 to 18 cm, are yellow-white in color, and occur in the intestine of the definitive host. In adult dogs, the infection is usually asymptomatic. By contrast, massive infection with *Toxocara canis* can be fatal in puppies.

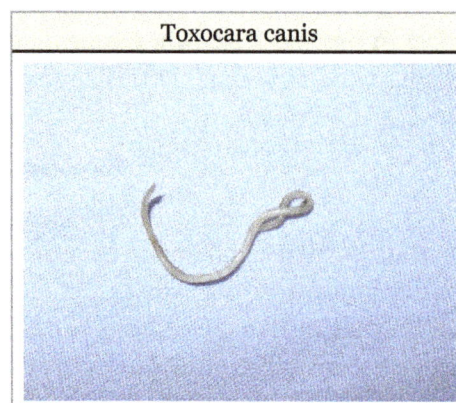

Toxocara canis

As paratenic hosts, a number of various vertebrates, including humans, and some invertebrates can become infected. Humans are infected, like other paratenic hosts, by ingestion of embryonated *T. canis* eggs. The disease (called Toxocariasis) caused by migrating *T. canis* larvae (toxocariasis) results in two syndromes: visceralis larva migrans and ocularis larva migrans. Owing to transmission of the infection from the mother to her puppies, preventive anthelmintic treatment of newborn puppies is strongly recommended. Several antihelmintic drugs are effective against adult worms, for example pyrantel, fenbendazole, selamectine, etc.

Morphology

The adult *canis* has a round body with spiky cranial and caudal parts, covered by yellow cuticula. The cranial part of the body contains two lateral alae (length 2–3.5 mm, width 0.1 mm). Male worms measure 9–13 by 0.2–0.25 cm and female worms 10–18 by 0.25–0.3 cm. *T. canis* eggs have oval or spherical shapes with granulated surfaces, are thick-walled, and measure from 72 to 85 μm.

Life Cycle

Eggs are deposited in feces of dogs becoming infectious after 2–4 weeks. Dogs ingest infectious eggs allowing the eggs to hatch and the larval form of the parasite to penetrate through the gut wall.In young dogs, the larvae move through the body via the bloodstream by penetrating a blood vessel in the gut wall. Once in the lungs, the larvae enter into the alveoli and crawl up the trachea. The larvae are then coughed up and swallowed leading back down to the small intestine. The larvae encyst in gut wall tissues within older dogs. The cysts can reactivate in pregnant females to infect puppies either through the placenta in utero or the mammary glands in colostrum and milk. Another possible route of infection is the ingesting of paratenic hosts that contain encysted larvae from egg consumption thus completing the life cycle for the parasite to re-infect its definite host, the dog.

Four modes of infection are associated with this species. These modes of infection include direct transmission, prenatal transmission, paratenic transmission, and transmammary transmission. The basic form is direct transmission and is typical to all ascaroides, with the egg containing the L_2(the second larval developmental stage) being infective, at optimal temperature and humidity, four weeks after secreted in the feces to the environment. After ingestion and hatching in the small intestine, the L_2 larvae travel through the portal blood stream into the liver and lungs. Such migratory route is known as enterohepatic pulmonar larval migration. The second molt takes place in the lungs, the now L_3 larvae return via the trachea and into the intestines, where the final two molts take place. This form of infection occurs regularly only in dogs up to three months of age.

In older dogs, this type of migration occurs less frequently, and at six months it is almost ceased. Instead, the L_2 travel to a wide range of organs, including the liver, lungs, brain, heart and skeletal muscles, as well as to the walls of the gastrointestinal tract. In pregnant female dogs, prenatal infection can occur, where larvae become mobilized (at about three weeks prior to parturition) and migrate through the umbilical vein to the lungs of the fetus, here molting into the L_3 stage just prior to birth. In the newborn pup, the cycle is completed when the larvae migrate through the trachea and into the intestinal lumen, where the final molts take place. Once infected, a female dog will usually harbor sufficient larvae to subsequently infect all of her litters, even if she never again encounters an infection. A certain amount of the female dog's dormant larvae penetrate into the intestinal lumen, where molting into adulthood takes place again, thus leading to a new release of eggs containing L_1 larvae.

Transmammary transmission occurs when the suckling pup becomes infected by the presence of L_3 larvae in the milk during the first three weeks of lactation. There is no migration in the pup via this route.

L_2 larvae may also be ingested by a variety of animals like mice or rabbits, where they stay in a dormant stage inside the animals' tissue until the intermediate host has been eaten by a dog, where subsequent development is confined to the gastrointestinal tract.

Transmission to Humans

Consumption of eggs from feces-contaminated items is the most common method of infection for humans especially children and young adults under the age of 20 years. Although rare, being in contact with soil that contains infectious eggs can also cause human infection, especially handling soil with an open wound or accidentally swallowing contaminated soil, as well as eating under cooked or raw meat of an intermediate host of the parasite such as lamb or rabbit.

Humans can be infected by this roundworm, a condition called toxocariasis, just by stroking an infected dog's fur and accidentally ingesting infective eggs that may be present on the dog's fur. When humans ingest infective eggs, diseases like hepatomegaly, myocarditis, respiratory failure and vision problems can result depending on where the larva are deposited in the body. In humans, this parasite usually grows in the back of the eye, which can result in blindness, or in the liver or lungs. However, a 2004 study showed, of 15 infected dogs, only seven had eggs in their coats, and no more than one egg was found on each dog. Furthermore, only 4% of those eggs were infectious. Given the low concentration of fertile eggs on infected dogs' coats (less than 0.00186% per gram), it is plausible that such eggs were transferred to the dog's coat by contact with fecal deposites in the environment, making dog coats the passive transport host vehicle. However, although the risk of being infected by petting a dog is extremely limited, a single infected puppy can produce more than 100,000 roundworm eggs per gram of feces.

Treatment

Humans suffering from visceral infection of *T. canis*, the drugs albendazole and mebendazole are highly effective. For other treatments, see a physician or reference the disease pages: visceralis larva migrans and ocularis larva migrans. Anthelminithic drugs are used to treat infections in dogs and puppies for adult worms. The best treatment for puppies is pyrantel pamoate to prevent the larvae from reproducing and causing disease.

Prevention

There are several ways to prevent a *T. canis* infection in both dogs and humans. Regular deworming by a veterinarian is important to stop canine re-infections, especially if the dog is frequently outdoors. Good practices to prevent human infections include: washing hands before eating and after disposing of animal feces in a timely manner as many disinfectants do not kill eggs, teaching children not to eat soil, and cooking meat to a safe temperature in order to kill potentially infectious eggs.

Trichuris Trichiura

The human whipworm (*Trichuris trichiura or Trichocephalus trichiuris*) is a round worm (a type of helminth) that causes trichuriasis (a type of helminthiasis which is one of the neglected tropical diseases) when it infects a human large intestine. It is commonly known as the *whipworm* which refers to the shape of the worm; it looks like a whip with wider "handles" at the posterior end.

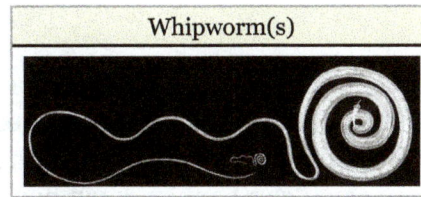

Whipworm(s)

Life Cycle

The female *T. trichiura* produces 2,000–10,000 single-celled eggs per day. Eggs are deposited from human feces to soil where, after two to three weeks, they become embryonated and enter the "infective" stage. These embryonated infective eggs are ingested and hatch in the human small intestine exploiting the intestinal microflora as hatching stimulus. This is the location of growth and molting. The infective larvae penetrate the villi and continue to develop in the small intestine. The young worms move to the caecum and penetrate the mucosa and there they complete development to adult worms in the large intestine. The life cycle from time of ingestion of eggs to development of mature worms takes approximately three months. During this time, there may be limited signs of infection in stool samples due to lack of egg production and shedding. The female *T. trichiura* begin to lay eggs after three months of maturity. Worms can live up to five years, during which time females can lay up to 20,000 eggs per day.

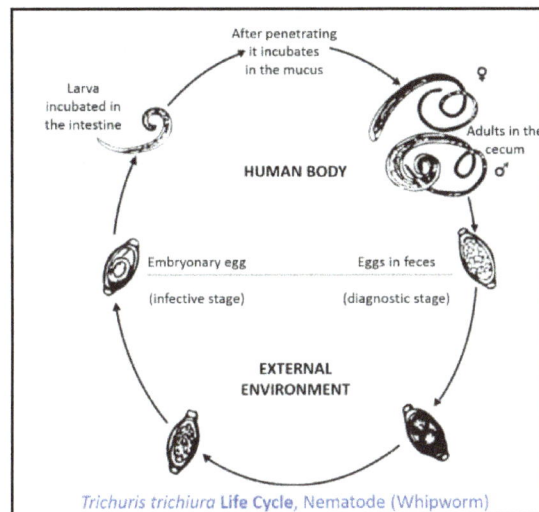

Trichuris trichiura **Life Cycle**, Nematode (Whipworm)

Life cycle of Trichuris trichiura inside and outside the human body

Recent studies using genome-wide scan revealed two quantitative trait loci on chromosome 9 and chromosome 18 may be responsible for genetic predisposition or susceptibility to infection of *T. trichiura* by some individuals.

Morphology

Trichuris trichiura has a narrow anterior esophageal end and shorter and thicker posterior anus. These pinkish-white worms are threaded through the mucosa. They attach to the host through their slender anterior end and feed on tissue secretions instead of blood. Females are larger than males; approximately 35–50 mm long compared to 30–45 mm. The females have a bluntly round posterior end compared to their male counterparts with a coiled posterior end. Their characteristic eggs are barrel-shaped and brown, and have bipolar protuberances..

Infection

Infection occurs through ingestion of eggs and is more common in warmer areas. Whipworms eggs are passed in the feces of infected persons, and if an infected person defecates outside or if untreated human feces as used as fertilizer, eggs are deposited on soil where they can mature into an infective stage. Ingestion of these eggs "can happen when hands or fingers that have contaminated dirt on them are put in the mouth or by consuming vegetables or fruits that have not been carefully cooked, washed or peeled." The eggs hatch in the small intestine, and then move into the wall of the small intestine and develop. On reaching adulthood, the thinner end (the front of the worm) burrows into the large intestine and the thicker end hangs into the lumen and mates with nearby worms. The females can grow to 50 mm (2.0 in) long. Neither the male nor the female has much of a visible tail past the anus.

Whipworm commonly infects patients also infected with *Giardia*, *Entamoeba histolytica*, *Ascaris lumbricoides*, and hookworms.

Epidemiology

There is a worldwide distribution of *Trichuris trichiura*, with an estimated 1 billion human infections. However, it is chiefly tropical, especially in Asia and, to a lesser degree, in Africa and South America. Within the United States, infection is rare overall but may be common in the rural Southeast, where 2.2 million people are thought to be infected. Poor hygiene is associated with trichuriasis as well as the consumption of shaded moist soil, or food that may have been fecally contaminated. Children are especially vulnerable to infection due to their high exposure risk. Eggs are infective about 2–3 weeks after they are deposited in the soil under proper conditions of warmth and moisture, hence its tropical distribution.

Other Animals

Whipworms develop when a dog swallows whipworm eggs, passed from an infected dog. Symptoms may include diarrhea, anemia, and dehydration. The dog whipworm (*Trichuris vulpis*) is commonly found in the U.S. It is hard to detect at times, because the numbers of eggs shed are low, and they are shed in waves. Centrifugation is the preferred method. There are several preventives available by prescription from a veterinarian to prevent dogs from getting whipworm.

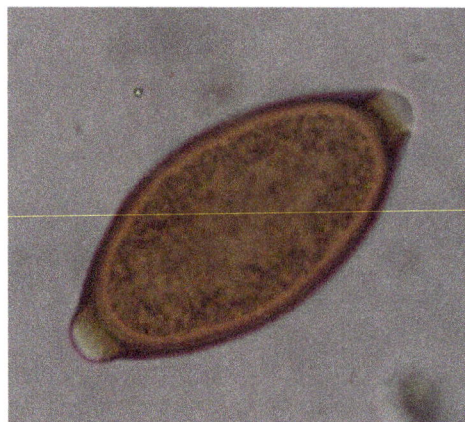

Egg of *Trichuris vulpis*

The cat whipworm is a rare parasite. In Europe, it is represented mostly by *Trichuris campanula*, and in North America it is *Trichuris serrata* more often. Whipworm eggs found in cats in North America must be differentiated from lungworms, and from mouse whipworm eggs just passing through.

Treatment of Inflammatory Disorders

The hygiene hypothesis suggests that various immunological disorders that have been observed in humans only within the last 100 years, such as Crohn's disease, or that have become more common during that period as hygienic practices have become more widespread, may result from a lack of exposure to parasitic worms (also called helminths) during childhood. The use of *Trichuris suis* ova (TSO, or pig whipworm eggs) by Weinstock, et al., as a therapy for treating Crohn's disease and to a lesser extent ulcerative colitis are two examples that support this hypothesis. There is also anecdotal evidence that treatment of inflammatory bowel disease (IBD) with TSO decreases the incidence of asthma, allergy, and other inflammatory disorders. Some scientific evidence suggests that the course of multiple sclerosis may be very favorably altered by helminth infection; TSO is being studied as a treatment for this disease.

Wuchereria Bancrofti

Wuchereria bancrofti is a human parasitic roundworm that is the major cause of lymphatic filariasis. It is one of the three parasitic worms, together with *Brugia malayi* and *B. timori*, that infect the lymphatic system to cause lymphatic filariasis. These filarial worms are spread by a variety of mosquito vector species. *W. bancrofti* is the most prevalent of the three and affects over 120 million people, primarily in Central Africa and the Nile delta, South and Central America, the tropical regions of Asia including southern China, and the Pacific islands. If left untreated, the infection can develop into a chronic disease called elephantiasis. In rare conditions it also causes tropical eosinophilia, an asthmatic disease. Limited treatment modalities exist and no vaccines have been developed.

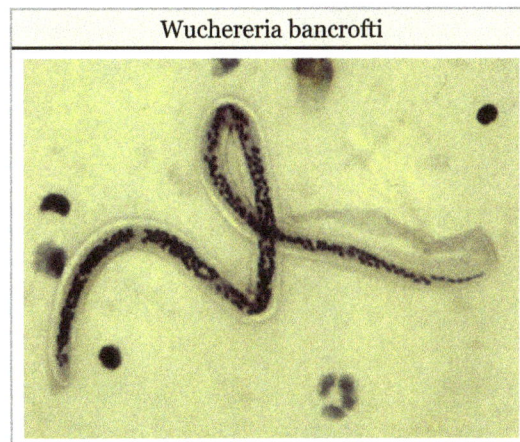

Wuchereria bancrofti

History

The effects of *W. bancrofti* were documented early in ancient text. Ancient Greek and Roman writers noted the similarities between the enlarged limbs and cracked skin of infected individuals to that of elephants. Since then, this condition has been commonly known as elephantiasis. However, this is a misnomer, since elephantiasis literally translates to "a disease caused by elephants".

W. bancrofti was named after physician Otto Wucherer and parasitologist Joseph Bancroft, both of whom extensively studied filarial infections.

W. bancrofti is speculated to have been brought to the New World by the slave trade. Once it was introduced to the New World, this filarial worm disease persisted throughout the areas surrounding Charleston, South Carolina until its sudden disappearance in the 1920s.

Morphology

As dioecious worm, *W. bancrofti* exhibits sexual dimorphism. The adult worm is long, cylindrical, slender, and smooth with rounded ends. It is white in colour and almost transparent. The body is quite delicate making it difficult to remove from tissues. It has a short cephalic or head region connected to the main body by a short neck which appears as a constriction. There are dark spots which are dispersed nuclei throughout the body cavity, with no nuclei at the tail tip. Male and female can be differentiated by size and structure of tail tip. The male worm is smaller, 40 millimetres (1.6 in) long and 100 micrometres (0.0039 in) wide, and features a ventrally curved tail. The tip of the tail has 15 pairs of minute caudal papillae, the sensory organs. The anal region is an elaborate structure consisting of 12 pairs of papillae, of which 8 are in front and 4 are behind the anus. In contrast, the female is 60 millimetres (2.4 in) to 100 millimetres (3.9 in) long and 300 micrometres (0.012 in) wide, nearly three times larger in diameter than the male. Its tail gradually tapers and rounded at the tip. There are no additional sensory structures. Its vulva lies towards the anterior region, about 0.25 mm from the head. Adult male and female are most often coiled together and are difficult to separate. Females are ovoviviparous and can produce thousands of juveniles known as microfilariae.

The microfilaria is a miniature adult and retains the egg membrane as a sheath, and is often considered an advanced embryo. It measures 280 µm long and 25 µm wide. It appears quite structureless 'in vivo', but histological staining makes its primitive gut, nerve ring and muscles apparent.

Life Cycle

Life cycle of *Wuchereria bancrofti*

W. bancrofti carry out their life cycle in two hosts. Human beings serve as the definitive host and mosquitoes as their intermediate hosts. The adult parasites reside in the lymphatics of the human host. They are found mostly in the afferent lymphatic channels of the lymph glands in the lower part of the body. The first-stage larvae, known as microfilariae, are present in the circulation. The microfilariae have a membrane "sheath". This sheath, along with the area in which the worms reside, makes identification of the species of microfilariae in humans easier to determine. The microfilariae are found mainly in the peripheral blood and can be found at peak amounts from 10 p.m. to 4 a.m. They migrate between the deep and the peripheral circulation exhibiting unique diurnal periodicity. During the day, they are present in the deep veins, and during the night, they migrate to the peripheral circulation. The cause of this periodicity remains unknown, but the advantages of the microfilariae being in the peripheral blood during these hours may ensure the vector, the nighttime mosquito, will have a higher chance of transmitting them elsewhere. Physiological changes also are associated with sleeping, such as lowered body temperature, oxygen tension and adrenal activity, and an increased carbon dioxide tension, among other physical alterations, any of which could be the signals for the rhythmic behavior of microfilarial parasites. If the hosts sleep by day and are awake at night, their periodicity is reversed. In the South Pacific, where *W. bancrofti* shows diurnal periodicity, it is known as periodic.

The microfilariae are transferred into a vector, which are most commonly mosquito species of the genera *Culex*, *Anopheles*, *Mansonia*, and *Aedes*. Inside the mosquito, the microfilariae mature into motile larvae called juveniles. When the infected mosquito has its next blood meal, *W. bancrofti* is egested via the mosquito's proboscis into the blood stream of the new human host. The larvae move through the lymphatic system to regional lymph nodes, predominantly in the legs and genital area. The larvae develop into adult worms over the course of a year, and reach sexual maturity in the afferent lymphatic vessels. After mating, the adult female worm can produce thousands of microfilariae that migrate into the bloodstream. A mosquito vector can bite the infected human host, ingest the microfilariae, and thus repeat the life cycle.

Epidemiology

An infection in leg by *Wuchereria bancrofti*

W. bancrofti is responsible for 90% of lymphatic filariasis. Recently, 120 million worldwide cases of lymphatic filariasis were estimated. *W. bancrofti* largely affects areas across the broad equatorial belt (Africa, the Nile Delta, Turkey, India, the East Indies, Southeast Asia, Philippines, Oceanic Islands, and parts of South America.)

The mosquito vectors of *W. bancrofti* have a preference for human blood; humans are apparently the only animals naturally infected with *W. bancrofti*. There is no reservoir host, and the disease could therefore potentially be eradicated.

Pathology

The pathogenesis of *W. bancrofti* infection is dependent on the immune system and inflammatory responses of the host. After infection, the worms will mature within 6–8 months, male and female worms will mate and then release the microfilariae. These microfilariae can be released for up to ten years.

1. The asymptomatic phase usually consists of high microfilaremia infection, and individuals show no symptoms of being infected. This occurs due to cytokine IL-4 suppressing the activity of TH1 cells in the immune system. This can occur for years until the inflammatory reaction rises again.

2. In the inflammatory (acute) phase, the antigens from the female adult worms elicit inflammatory responses. The worms in the lymph channels disrupt the flow of the lymph, causing lymphedema. The individual will exhibit fever, chills, skin infections, painful lymph nodes, and tender skin of the lymphedematous extremity. These symptoms often lessen after five to seven days. Other symptoms that may occur include orchitis, an inflammation of the testes, which is accompanied by painful, immediate enlargement and epididymitis (inflammation of the spermatic cord).

3. The obstructive (chronic) phase is marked by lymph varices, lymph scrotum, hydrocele, chyluria (lymph in urine), and elephantiasis. Microfilariae are not normally present in this phase. A key feature of this phase is scar formation from affected tissue areas. Other features include thickening of the skin and elephantiasis, which develops gradually with the attack of the lymphatic system. Elephantiasis affects men mainly in the legs, arms, and scrotum. In women, the legs, arms, and breasts are affected.

Diagnosis

A blood smear is a simple and fairly accurate diagnostic tool, provided the blood sample is taken during the period in the day when the juveniles are in the peripheral circulation. Technicians analyzing the blood smear must be able to distinguish between *W. bancrofti* and other parasites potentially present.

A polymerase chain reaction test can also be performed to detect a minute fraction, as little as 1 pg, of filarial DNA.

Some infected people do not have microfilariae in their blood. As a result, tests aimed to detect antigens from adult worms can be used.

Ultrasonography can also be used to detect the movements and noises caused by the movement of adult worms.

Dead, calcified worms can be detected by X-ray examinations.

Treatment

The severe symptoms caused by the parasite can be avoided by cleansing the skin, surgery, or the use of anthelmintic drugs, such as diethylcarbamazine (DEC), ivermectin, or albendazole. The drug of choice is DEC, which can eliminate the microfilariae from the blood and also kill the adult worms with a dosage of 6 mg/kg semiannually or annually. A polytherapy treatment that includes ivermectin with DEC or albendazole is more effective than each drug alone. Protection is similar to that of other mosquito-spread illnesses; one can use barriers both physical (a mosquito net), chemical (insect repellent), or mass chemotherapy as a method to control the spread of the disease.

Mass chemotherapy should cover the entire endemic area at the same time. This will significantly decrease the overall microfilarial titer in blood in mass, hence decreasing the transmission through mosquitoes during their subsequent bites.

Antibiotic active against the Wolbachia symbionts of the worm have been experimented with as treatment. Wolbachia-free worms first become sterile, and later die prematurely.

Control

Prevention focuses on protecting against mosquito bites in endemic regions. Insect repellents and mosquito nets are useful to protect against mosquito bites. Public education efforts must also be made within the endemic areas of the world to successfully lower the prevalence of *W. bancrofti* infections.

Eradication

The WHO is coordinating an effort to eradicate filarisis. The mainstay of this programme is the mass use of antifilarial drugs on a regular basis for at least five years.

In April 2011, Sri Lanka was certified by the WHO as having eradicated this disease.

Pinworm (Parasite)

The pinworm (species *Enterobius vermicularis*), also known as threadworm (in the United Kingdom and Australasia) or seatworm, is a parasitic worm. It is a nematode (roundworm) and a common intestinal parasite or helminth, especially in humans. The medical condition associated with pinworm infestation is known as enterobiasis (a type of helminthiasis) or less precisely as oxyuriasis in reference to the family Oxyuridae.

Throughout this article, the word "pinworm" refers to *Enterobius*. In British usage, however, pinworm refers to *Strongyloides*, while *Enterobius* is called threadworm.

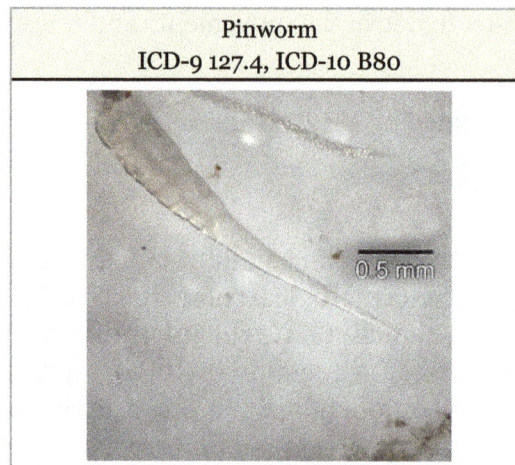

Pinworm
ICD-9 127.4, ICD-10 B80

Classification

The pinworm (genus *Enterobius*) is a type of roundworm (nematode), and three species of pinworm have been identified with certainty. Humans are hosts only to *Enterobius vermicularis* (formerly *Oxyurias vermicularis*). Chimpanzees are host to *Enterobius anthropopitheci*, which is morphologically distinguishable from the human pinworm. Hugot (1983) claims another species affects humans, *Enterobius gregorii*, which is supposedly a sister species of *E. vermicularis*, and has a slightly smaller spicule (i.e., sexual organ). Its existence is controversial, however; Totkova et al. (2003) consider the evidence to be insufficient, and Hasegawa et al. (2006) contend that *E. gregorii* is a younger stage of *E. vermicularis*. Regardless of its status as a distinct species, *E. gregorii* is considered clinically identical to *E. vermicularis*.

Morphology

The adult female has a sharply pointed posterior end, is 8 to 13 mm long, and 0.5 mm thick. The adult male is considerably smaller, measuring 2 to 5 mm long and 0.2 mm thick, and has a curved posterior end. The eggs are translucent and have a surface that adheres to objects. The eggs measure 50 to 60 µm by 20 to 30 µm, and have a thick shell flattened on one side. The small size and colourlessness of the eggs make them invisible to the naked eye, except in barely visible clumps of thousands of eggs. Eggs may contain a developing embryo or a fully developed pinworm larva. The larvae grow to 140–150 µm in length.

Two female pinworms next to a ruler: The markings are 1 mm apart.

Distribution

The pinworm has a worldwide distribution, and is the most common helminth (i.e., parasitic worm) infection in the United States, western Europe, and Oceania. In the United States, a study by the Center of Disease Control reported an overall incidence rate of 11.4% among children. Pinworms are particularly common in children, with prevalence rates in this age group having been reported as high as 61% in India, 50% in England, 39% in Thailand, 37% in Sweden, and 29% in Denmark. Finger sucking has been shown to increase both incidence and relapse rates, and nail biting has been similarly associated. Because it spreads from host to host through contamination, pinworms are common among people living in close contact, and tends to occur in all people within a household. The prevalence of pinworms is not associated with gender, nor with any particular social class, race, or culture. Pinworms are an exception to the tenet that intestinal parasites are uncommon in affluent communities. A fossilized nematode egg was detected in 240 million-year-old fossil dung, showing that parasitic pinworms already infested pre-mammalian cynodonts. The earliest known instance of the pinworms associated with humans is evidenced by pinworm eggs found in human coprolites carbon dated to 7837 BC found in western Utah.

Life Cycle

The entire lifecycle, from egg to adult, takes place in the human gastrointestinal tract of a single host, from about 2–4 weeks or about 4–8 weeks.

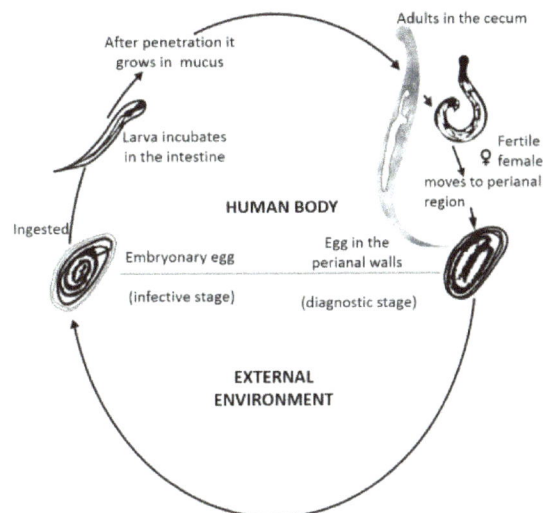

Enterobius vermicularis **Life Cycle**, Nematode (Pinworm)

Lifecycle of *E. vermicularis* showing the stages inside and outside of the human body

The lifecycle begins with eggs being ingested. The eggs hatch in the duodenum (i.e., first part of the small intestine). The emerging pinworm larvae grow rapidly to a size of 140 to 150 μm, and migrate through the small intestine towards the colon. During this migration, they moult twice and become adults. Females survive for 5 to 13 weeks, and males about 7 weeks. The male and female pinworms mate in the ileum (i.e., last part of the small intestine), whereafter the male pinworms usually die, and are passed out with stool. The gravid female pinworms settle in the ileum, caecum (i.e., beginning of the large intestine), appendix and ascending colon, where they attach themselves to the mucosa and ingest colonic contents.

Almost the entire body of a gravid female becomes filled with eggs. The estimations of the number of eggs in a gravid female pinworm range from about 11,000 to 16,000. The egg-laying process begins about five weeks after initial ingestion of pinworm eggs by the human host. The gravid female pinworms migrate through the colon towards the rectum at a rate of 12 to 14 cm per hour. They emerge from the anus, and while moving on the skin near the anus, the female pinworms deposit eggs either through (1) contracting and expelling the eggs, (2) dying and then disintegrating, or (3) bodily rupture due to the host scratching the worm. After depositing the eggs, the female becomes opaque and dies. The reason the female emerges from the anus is to obtain the oxygen necessary for the maturation of the eggs.

Infection

E. vermicularis causes the medical condition enterobiasis, whose primary symptom is itching in the anal area.

Treatment

Albendazole or mebendazole is the first-line treatment of pinworm infection. Pyrantel pamoate is an alternative. Among naturopathic cures, garlic, wormwood, black walnut and papaya seeds have been used historically. People of the Middle East have used Za'atar since ancient times to reduce and eliminate internal parasites. Za'atar uses large quantities of Thyme (Thymus serpyllum) which contains phenolic, antimicrobial and antiseptic compounds including the essential oil and anthelmintic thymol. Thymol has also been used in alcohol solutions and in dusting powders for the treatment of tinea or ringworm infections, and was used in the United States to treat hookworm infections.

Transmission

Pinworms spread through human-to-human transmission, by ingesting (i.e., swallowing) infectious pinworm eggs and/or by anal insertion. The eggs are hardy and can remain viable (i.e., infectious) in a moist environment up to three weeks. They do not tolerate heat well, but can survive in low temperatures: two-thirds of the eggs are still viable after 18 hours at −8 °C (18 °F).

After the eggs have been initially deposited near the anus, they are readily transmitted to other surfaces through contamination. The surface of the eggs is sticky when laid, and the eggs are readily transmitted from their initial deposit near the anus to fingernails, hands, night-clothing and bed linen. From here, eggs are further transmitted to food, water, furniture, toys, bathroom fixtures and other objects. Household pets often carry the eggs in their fur, while not actually being infected. Dust containing eggs can become airborne and widely dispersed when dislodged from surfaces, for instance when shaking out bed clothes and linen. Consequently, the eggs can enter the mouth and nose through inhalation, and be swallowed later. Although pinworms do not strictly multiply inside the body of their human host, some of the pinworm larvae may hatch on the anal mucosa, and migrate up the bowel and back into the gastrointestinal tract of the original host in a process called retroinfection. When this retroinfection occurs, it can lead to a heavy parasitic load and ensures the pinworm infestation continues or can be not clinically significant. Despite the limited, 13-week lifespan of individual pinworms, autoinfection (i.e., infection from the original host to itself), either through the anus-to-mouth route or through retroinfection, usually necessitates repeated treatment, at 2-week intervals, in order to remove the infection completely.

Gallery

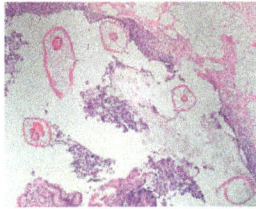

Pinworms are sometimes diagnosed incidentally by pathology. Micrograph of pinworms in the appendix, H&E stain

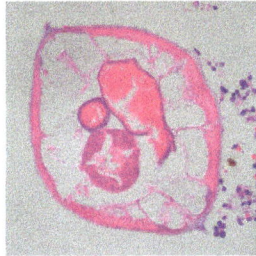

High magnification micrograph of a pinworm in cross section in the appendix, H&E stain

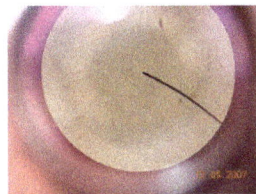

Egg under a light microscope

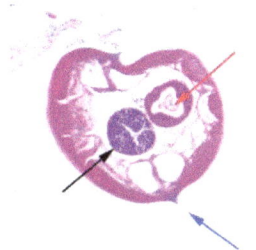

Pinworms are sometimes diagnosed incidentally by pathology: Micrograph of male pinworm in cross section, alae (blue arrow), intestine (red arrow) and testis (black arrow), H&E stain

Pinworm eggs are easily seen under a microscope.

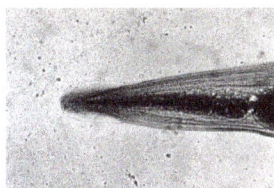

This micrograph reveals the cephalic alae in the head region of *E. vermicularis*.

References

- Platt HM (1994). "foreword". In Lorenzen S, Lorenzen SA. The phylogenetic systematics of freeliving nematodes. London: The Ray Society. ISBN 0-903874-22-9.

- Hsueh YP, Leighton DHW, Sternberg PW. (2014). Nematode Communication. In: Witzany G (ed). Biocommunication of Animals. Springer, 383-407. ISBN 978-94-007-7413-1.

- Ruppert EE, Fox RS, Barnes RD (2004). Invertebrate Zoology: A Functional Evolutionary Approach (7th ed.). Belmont, California: Brooks/Cole. ISBN 978-0-03-025982-1.

- Jenkins, Joseph (1999). "Worms and Disease; Roundworms". The Humanure Handbook - A Guide to Composting Human Manure (2nd ed.). ISBN 978-0-9644258-3-5.

- Southwick, Frederick S. (2007), Infectious Disease: A Clinical Short Course (2nd ed.), New York: McGraw Hill Professional, ISBN 978-0-07-147722-2 .

- G. D. Schmidt; L S. Roberts (2009). Larry S. Roberts; John Janovy, Jr., eds. Foundations of Parasitology (8th ed.). McGraw-Hill. pp. 480–484. ISBN 978-0-07-128458-5.

- Gutiérrez, Yezid (2000). Diagnostic pathology of parasitic infections with clinical correlations (PDF) (Second ed.). Oxford University Press. pp. 354–366. ISBN 0-19-512143-0. Retrieved 21 August 2009.

- Cook, Gordon C; Zumla, Alimuddin I (2009). Manson's tropical diseases (22nd ed.). Saunders Elsevier. pp. 1515–1519. ISBN 978-1-4160-4470-3. Retrieved 18 November 2009.

- Garcia, Lynne Shore (2009). Practical guide to diagnostic parasitology. American Society for Microbiology. pp. 246–247. ISBN 1-55581-154-X. Retrieved 5 December 2009.

- Speare, R. (1989). "Identification of species of Strongyloides". In Grove, D. I. Strongyloidiasis: a major roundworm infection of man. London: Taylor & Francis. pp. 11–83. ISBN 0850667321.

- Cross, John H. (1996). "Enteric Nematodes of Humans". In Baron, Samuel. Medical Microbiology (4th ed.). Galveston: University of Texas Medical Branch at Galveston. ISBN 0-9631172-1-1.

- Junghanss, Jeremy Farrar, Peter J. Hotez, Thomas (2013). Manson's Tropical Diseases: Expert Consult - Online (23rd ed.). Oxford: Elsevier/Saunders. pp. e49–e52. ISBN 9780702053061.

- King CL, Freedman DP (2000). "Filariasis". In G.T. Strickland. Hunter's tropical medicine and emerging infectious diseases (8th ed.). Philadelphia: E.B. Saunders. pp. 740–53. ISBN 978-0-7216-6223-7.

- Coghlan, Avril (7 September 2005). "Nematode Genome Evolution, WormBook, ed." (PDF). doi:10.1895/wormbook.1.15.1. Retrieved 13 January 2016.

Etoparasites and Insect Parasites

Parasites are usually species that benefit at the expense of the host. Head louse cause head lice infestation. Head lice are insects that spend their entire life on the human scalp and feed on human blood. Scabies is another kind of parasite that causes skin infections, and its symptoms are itchiness and rashes. The aspects elucidated in this chapter are of vital importance, and provide a better understanding of parasites.

Head Louse

The head louse (*Pediculus humanus capitis*) is an obligate ectoparasite of humans that causes head lice infestation (*pediculosis capitis*). Head lice are wingless insects spending their entire life on the human scalp and feeding exclusively on human blood. Humans are the only known hosts of this specific parasite, while chimpanzees host a closely related species, *Pediculus schaeffi*. Other species of lice infest most orders of mammals and all orders of birds, as well as other parts of the human body.

Head Lice

Lice differ from other hematophagic ectoparasites such as fleas in spending their entire life cycle on a host. Head lice cannot fly, and their short stumpy legs render them incapable of jumping, or even walking efficiently on flat surfaces.

The non-disease-carrying head louse differs from the related disease-carrying body louse (*Pediculus humanus humanus*) in preferring to attach eggs to scalp hair rather than to clothing. The two subspecies are morphologically almost identical but do not normally interbreed, although they will do so in laboratory conditions. From genetic studies, they are thought to have diverged as subspecies about 30,000–110,000 years ago, when many humans began to wear a significant amount of clothing. A much more distantly related species of hair-clinging louse, the pubic or crab louse (*Pthirus pubis*), also infests humans. It is visually different from the other two species and is much closer in appearance to the lice which infest other primates. Lice infestation of any part of the body is known as pediculosis.

Head lice (especially in children) have been, and still are, subject to various eradication campaigns. Unlike body lice, head lice are not the vectors of any known diseases. Except for rare secondary infections that result from scratching at bites, head lice are harmless, and they have been regarded by some as essentially a cosmetic rather than a medical problem. It has even been suggested that head lice infestations might be beneficial in helping to foster a natural immune response against lice which helps humans in defense against the far more dangerous body louse, which is capable of transmission of dangerous diseases.

Adult Morphology

Like other insects of the suborder Anoplura, adult head lice are small (2.5–3 mm long), dorso-ven-trally flattened (anatomical terms of location), and entirely wingless. The thoracic segments are fused, but otherwise distinct from the head and abdomen, the latter being composed of seven visible segments. Head lice are grey in general, but their precise color varies according to the en-vironment in which they were raised. After feeding, consumed blood causes the louse body to take on a reddish color.

Head

One pair of antennae, each with five segments, protrude from the insect's head. Head lice also have one pair of eyes. Eyes are present in all species within *Pediculidae* (the family of which the head louse is a member) but are reduced or absent in most other members of the Anoplura suborder. Like other members of Anoplura, head lice mouth parts are highly adapted for piercing skin and sucking blood. These mouth parts are retracted into the insect's head except during feeding.

Male head louse, adult

Female head louse, adult

Thorax

Six legs project from the fused segments of the thorax. As is typical in Anoplura, these legs are short and terminate with a single claw and opposing "thumb". Between its claw and thumb, the louse grasps the hair of its host. With their short legs and large claws, lice are well adapted to clinging to the hair of their host. These adaptations leave them incapable of jumping, or even walking efficiently on flat surfaces. Lice can climb up strands of hair very quickly, allowing them to move quickly and reach another host.

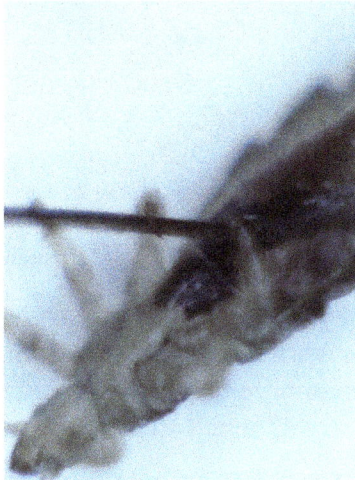

Head louse gripping a human hair

Abdomen

There are seven visible segments of the louse abdomen. The first six segments each have a pair of spiracles through which the insect breathes. The last segment contains the anus and (separately) the genitalia.

Sex Differences

In male lice, the front two legs are slightly larger than the other four. This specialized pair of legs is used for holding the female during copulation. Males are slightly smaller than females and are characterized by a pointed end of the abdomen and a well-developed genital apparatus visible inside the abdomen. Females are characterized by two gonopods in the shape of a W at the end of their abdomen.

Louse Eggs

Like most insects, head lice are oviparous. Females lay about 3–4 eggs per day. Louse eggs are attached near the base of a host hair shaft. Egg-laying behavior is temperature dependent and likely seeks to place the egg in a location that will be conducive to proper embyro development (which is, in turn, temperature dependent). In cool climates, eggs are generally laid within 3–5 mm of the scalp surface. In warm climates, and especially the tropics, eggs may be laid 6 inches (15 cm) or more down the hair shaft.

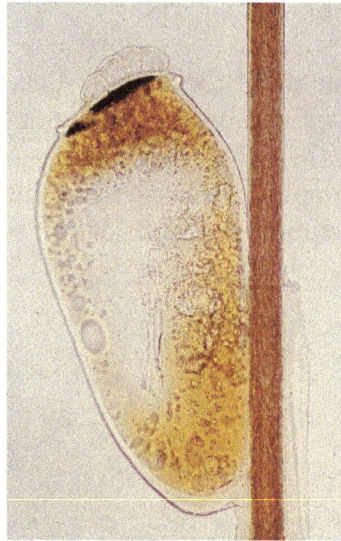

Head louse egg (nit) attached to hair shaft of host

To attach an egg, the adult female secretes a glue from her reproductive organ. This glue quickly hardens into a "nit sheath" that covers the hair shaft and large parts of the egg except for the operculum, a cap through which the embryo breathes. The glue was previously thought to be chitin-based, but more recent studies have shown it to be made of proteins similar to hair keratin.

Each egg is oval-shaped and about 0.8 mm in length. They are bright, transparent, tan to coffee-colored so long as they contain an embryo but appear white after hatching. Typically, a hatching time of six to nine days after oviposition is cited by authors.

After hatching, the louse nymph leaves behind its egg shell (usually known as nit), still attached to the hair shaft. The empty egg shell remains in place until physically removed by abrasion or the host, or until it slowly disintegrates, which may take 6 or more months.

SEM images of a hair louse egg

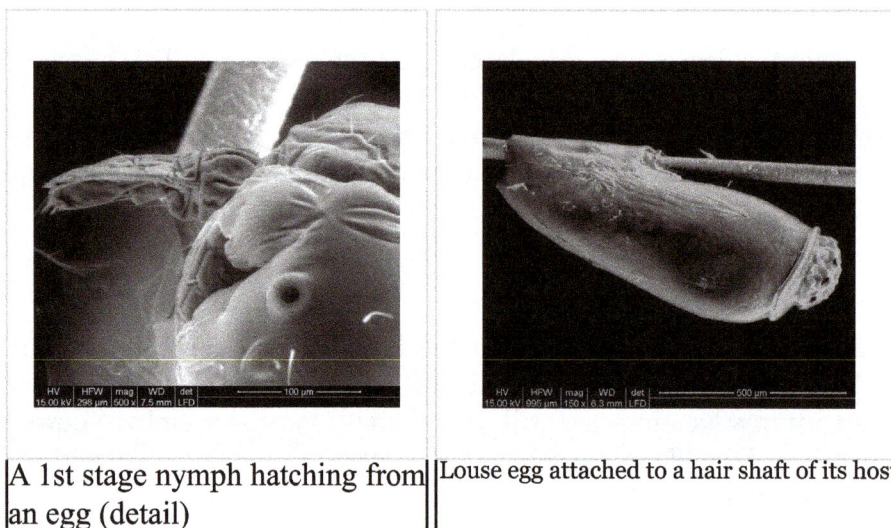

A 1st stage nymph hatching from an egg (detail)

Louse egg attached to a hair shaft of its host

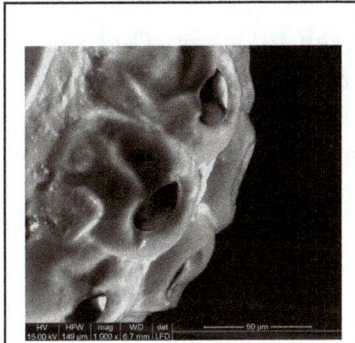

The operculum allows the embryo to breathe

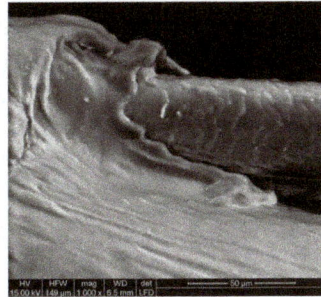

The female reproductive organ secretes a glue that quickly hardens into a "nit sheath" to cover the hair shaft and large parts of the egg, except for the operculum.

A 1st stage nymph hatching from an egg

Nits

The term nit refers to an egg without embryo or a dead egg. With respect to eggs, this rather broad definition includes the following: Accordingly, on the head of an infested individual the following eggs could be found:

- Viable eggs that will eventually hatch

- Remnants of already-hatched eggs (nits)

- Nonviable eggs (dead embryo) that will never hatch

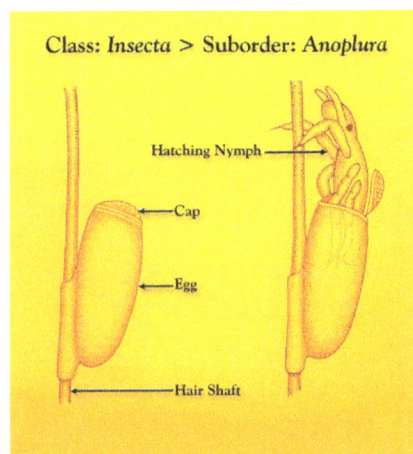

Louse hatching

This has produced some confusion in, for example, school policy because, of the three items listed above, only eggs containing viable embryos have the potential to infest or reinfest a host. Some authors have reacted to this confusion by restricting the definition of nit to describe only a hatched or nonviable egg:

In many languages the terms used for the hatched eggs, which were obvious for all to see, have subsequently become applied to the embryonated eggs that are difficult to detect. Thus the term "nit" in English is often used for both. However, in recent years my colleagues and I have felt the need for some simple means of distinguishing between the two without laborious qualification. We have, therefore, come to reserve the term "nit" for the hatched and empty egg shell and refer to the developing embryonated egg as an "egg".

—*Ian F. Burgess (1995)*

The empty eggshell, termed a nit...

—*J. W. Maunder (1983)*

...nits (dead eggs or empty egg cases)...

—*Kosta Y. Mumcuoglu and others (2006)*

Others have retained the broad definition while simultaneously attempting to clarify its relevance to infestation:

In the United States the term "nit" refers to any egg regardless of its viability.

—*Terri Lynn Meinking (1999)*

Because nits are simply egg casings that can contain a developing embryo or be empty shells, not all nits are infective.

—*L. Keoki Williams and others (2001)*

Development and Nymphs

Head lice, like other insects of the order Phthiraptera, are hemimetabolous. Newly hatched nymphs will moult three times before reaching the sexually-mature adult stage. Thus, mobile head lice populations contain members of up to four developmental stages: three nymphal instars, and the adult (imago). Metamorphosis during head lice development is subtle. The only visible differences between different instars and the adult, other than size, is the relative length of the abdomen, which increases with each molt. Aside from reproduction, nymph behavior is similar to the adult. Nymphs feed only on human blood (hematophagia), and cannot survive long away from a host.

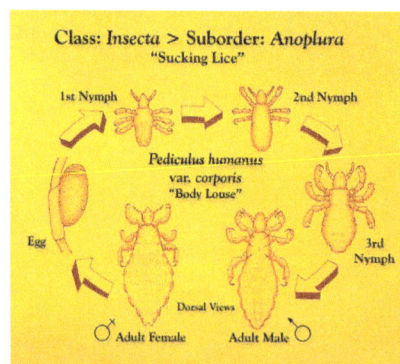

Development of *Pediculus humanus humanus* (body lice), which is similar to that of head lice (*Pediculus humanus capitis*)

The time required for head lice to complete their nymph development to the imago depends on feeding conditions. At minimum, eight to nine days is required for lice having continuous access to a human host. This experimental condition is most representative of head lice conditions in the wild. Experimental conditions where the nymph has more limited access to blood produces more prolonged development, ranging from 12 to 24 days.

Nymph mortality in captivity is high—about 38%—especially within the first two days of life. In the wild, mortality may instead be highest in the third instar. Nymph hazards are numerous. Failure to completely hatch from the egg is invariably fatal and may be dependent on the humidity of the egg's environment. Death during molting can also occur, although it is reportedly uncommon. During feeding, the nymph gut can rupture, dispersing the host's blood throughout the insect. This results in death within a day or two. It is unclear if the high mortality recorded under experimental conditions is representative of conditions in the wild.

Reproduction and Lifespan

Copulation in *Pediculus humanus humanus* (*Pediculus humanus capitis* is similar). Female is on top, with the male below. Dilation of the female's vagina has already occurred, and the male's dilator rests against his back (dorsal surface), out of the way. The male vesica, which contains the penis proper (not seen), is fully inserted into the vagina. Note the male's attachment with his specialized claws on the first leg pair to the specialized notch on the female's third leg pair.

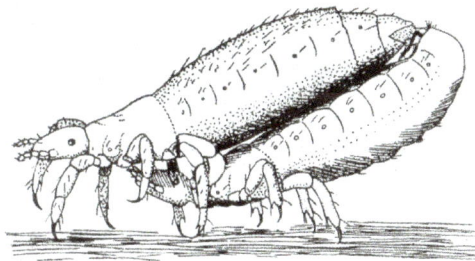

Adult head lice reproduce sexually, and copulation is necessary for the female to produce fertile eggs. Parthenogenesis, the production of viable offspring by virgin females, does not occur in *Pediculus humanus*. Pairing can begin within the first 10 hours of adult life. After 24 hours, adult lice copulate frequently, with mating occurring during any period of the night or day. Mating attachment frequently lasts more than an hour. Young males can successfully pair with older females, and vice versa.

Experiments with *Pediculus humanus humanus* (body lice) emphasize the attendant hazards of lice copulation. A single young female confined with six or more males will die in a few days, having laid very few eggs. Similarly, death of a virgin female was reported after admitting a male to her confinement. The female laid only one egg after mating, and her entire body was tinged with red—a condition attributed to rupture of the alimentary canal during the sexual act. Old females frequently die following, if not during, intercourse.

A single louse has a thirty-day life cycle beginning from the moment the nit is laid until the adult louse dies.

Factors Affecting Infestation

The number of children per family, the sharing of beds and closets, hair washing habits, local customs and social contacts, healthcare in a particular area (e.g. school) and socioeconomic status were found to be significant factors in head louse infestation. Girls are two to four times more frequently infested than boys. Children between 4 and 14 years of age are the most frequently infested group.

Behaviour

Feeding

All stages are blood-feeders and bite the skin four to five times daily to feed. They inject saliva which contains an anti-coagulant and suck blood. The digested blood is excreted as dark red frass.

Position on Host

Although any part of the scalp may be colonized, lice favor the nape of the neck and the area behind the ears, where the eggs are usually laid. Head lice are repelled by light and will move towards shadows or dark-coloured objects in their vicinity.

Transmission

Lice have no wings or powerful legs for jumping, so they move by using their claw-like legs to transfer from hair to hair. Normally head lice infest a new host only by close contact between individuals, making social contacts among children and parent-child interactions more likely routes of infestation than shared combs, hats, brushes, towels, clothing, beds or closets. Head-to-head contact is by far the most common route of lice transmission.

Distribution

About 6–12 million people, mainly children, are treated annually for head lice in the United States alone. High levels of louse infestations have also been reported from all over the world, including Australia, Denmark, France, Ireland, Israel and Sweden. Head lice can live off the head, for example on soft furnishings such as pillow cases, on hairbrushes, or on coat hoods for up to 48 hours.

Archaeogenetics

Analysis of the DNA of lice found on Peruvian mummies may indicate that some diseases (like typhus) may have passed from the New World to the Old World, instead of the other way around.

Genome

The sequencing of the genome of the body louse was first proposed in the mid-2000s and the annotated genome was published in 2010. An analysis of the body and head louse transcriptomes revealed these two organisms are extremely similar genetically.

Scabies

Scabies, known as the seven-year itch, is a contagious skin infestation by the mite *Sarcoptes scabiei*. The most common symptoms are severe itchiness and a pimple-like rash. Occasionally tiny burrows may be seen in the skin. When first infected, usually two to six weeks are required before symptoms occur. If a person develops a second infection later in life, symptoms may begin within a day. These symptoms can be present across most of the body or just certain areas such as the wrists, between fingers, or along the waistline. The head may be affected, but this is typically only in young children. The itch is often worse at night. Scratching may cause skin breakdown and an additional bacterial infection of the skin.

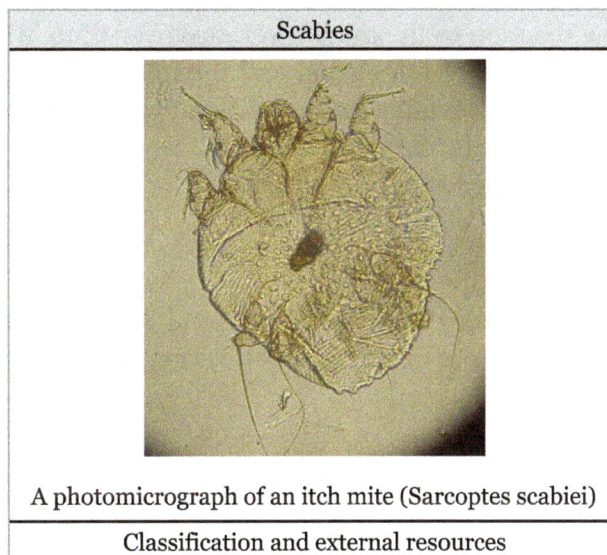

Scabies

A photomicrograph of an itch mite (Sarcoptes scabiei)

Classification and external resources

Scabies is caused by infection with the female mite *Sarcoptes scabiei*. The mites burrow into the skin to live and deposit eggs. The symptoms of scabies are due to an allergic reaction to the mites. Often only between ten and fifteen mites are involved in an infection. Scabies is most often spread during a relatively long period of direct skin contact with an infected person such as that which may occur during sex. Spread of disease may occur even if the person has not developed symptoms yet. Crowded living conditions such as those found in child care facilities, group homes, and prisons increase the risk of spread. Areas with a lack of access to water also have higher rates of disease. Crusted scabies is a more severe form of the disease. It typically only occurs in those with a poor immune system and people may have millions of mites, making them much more contagious. In these cases spread of infection may occur during brief contact or via contaminated objects. The mite is very small and usually not directly visible. Diagnosis is based on the signs and symptoms.

A number of medications are available to treat those infected, including permethrin, crotamiton and lindane creams and ivermectin pills. Sexual contacts within the last month and people who live in the same house should also be treated at the same time. Bedding and clothing used in the last three days should be washed in hot water and dried in a hot dryer. As the mite does not live for more than three days away from human skin more washing is not needed. Symptoms may continue for two to four weeks following treatment. If after this time there continue to be symptoms retreatment may be needed.

Scabies is one of the three most common skin disorders in children, along with ringworm and bacterial skin infections. As of 2010 it affects approximately 100 million people (1.5% of the world population) and is equally common in both sexes. The young and the old are more commonly affected. It also occurs more commonly in the developing world and tropical climates. The word scabies is from Latin: *scabere*, "to scratch". Other animals do not spread human scabies. Infection in other animals is typically caused by slightly different but related mites and is known as sarcoptic mange.

Signs and Symptoms

The characteristic symptoms of a scabies infection include intense itching and superficial burrows. The burrow tracks are often linear, to the point that a neat "line" of four or more closely placed and equally developed mosquito-like "bites" is almost diagnostic of the disease. Because the host develops the symptoms as a reaction to the mites' presence over time, there is typically a delay of four to six weeks between the onset of infestation and the onset of itching. Similarly, symptoms often persist for one to several weeks after successful eradication of the mites. As noted, those re-exposed to scabies after successful treatment may exhibit symptoms of the new infestation in a much shorter period—as little as one to four days.

Itching

In the classic scenario, the itch is made worse by warmth, and is usually experienced as being worse at night, possibly because there are fewer distractions. As a symptom, it is less common in the elderly.

Rash

The superficial burrows of scabies usually occur in the area of the finger webs, feet, ventral wrists, elbows, back, buttocks, and external genitals. Except in infants and the immunosuppressed, infection generally does not occur in the skin of the face or scalp. The burrows are created by excavation of the adult mite in the epidermis.

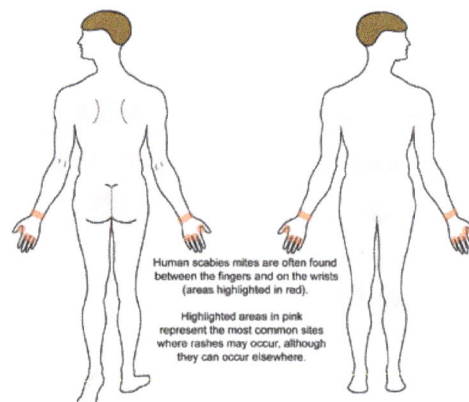

Human scabies mites are often found between the fingers and on the wrists (areas highlighted in red).

Highlighted areas in pink represent the most common sites where rashes may occur, although they can occur elsewhere.

Commonly involved sites of rashes of scabies

In most people, the trails of the burrowing mites are linear or *s*-shaped tracks in the skin often accompanied by rows of small, pimple-like mosquito or insect bites. These signs are often found in crevices of the body, such as on the webs of fingers and toes, around the genital area, in stomach folds of the skin, and under the breasts of women.

Symptoms typically appear two to six weeks after infestation for individuals never before exposed to scabies. For those having been previously exposed, the symptoms can appear within several days after infestation. However, it is not unknown for symptoms to appear after several months or years. Acropustulosis, or blisters and pustules on the palms and soles of the feet, are characteristic symptoms of scabies in infants.

Scabies of the hand

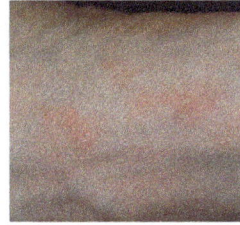

Scabies of the foot

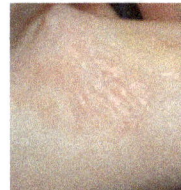

Scabies of the finger

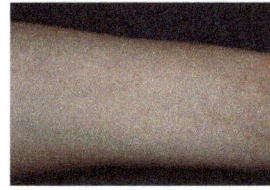

Scabies of the arm

Crusted scabies

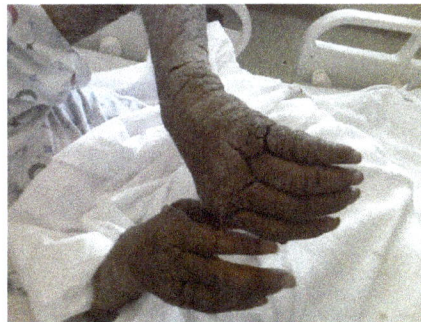

Crusted scabies in a person with AIDS

The elderly and people with an impaired immune system, such as HIV, cancer, or those on immunosuppressive medications, are susceptible to crusted scabies (formerly called Norwegian scabies). On those with weaker immune systems, the host becomes a more fertile breeding ground for the mites, which spread over the host's body, except the face. Sufferers of crusted scabies exhibit scaly rashes, slight itching, and thick crusts of skin that contain thousands of mites. Such areas make eradication of mites particularly difficult, as the crusts protect the mites from topical miticides/scabicides, necessitating prolonged treatment of these areas.

Cause

Scabies Mite

In the 18th century, Italian biologist Diacinto Cestoni (1637–1718) described the mite now called *Sarcoptes scabiei*, variety *hominis*, as the cause of scabies. *Sarcoptes* is a genus of skin parasites and part of the larger family of mites collectively known as scab mites. These organisms have eight legs as adults, and are placed in the same phylogenetic class (Arachnida) as spiders and ticks.

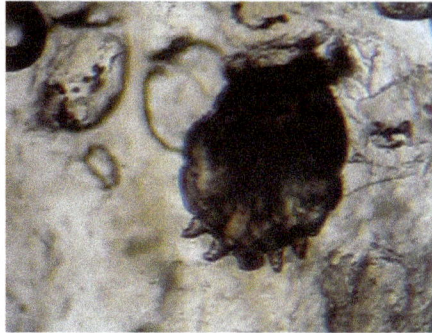

Video of the *Sarcoptes scabiei* mite

Sarcoptes scabiei mites are under 0.5 mm in size but are sometimes visible as pinpoints of white. Pregnant females tunnel into the dead, outermost layer (stratum corneum) of a host's skin and deposit eggs in the shallow burrows. The eggs hatch into larvae in three to ten days. These young mites move about on the skin and molt into a "nymphal" stage, before maturing as adults, which live three to four weeks in the host's skin. Males roam on top of the skin, occasionally burrowing into the skin. In general, the total number of adult mites infesting a healthy hygienic person with non-crusted scabies is small; about 11 females in burrows, on average.

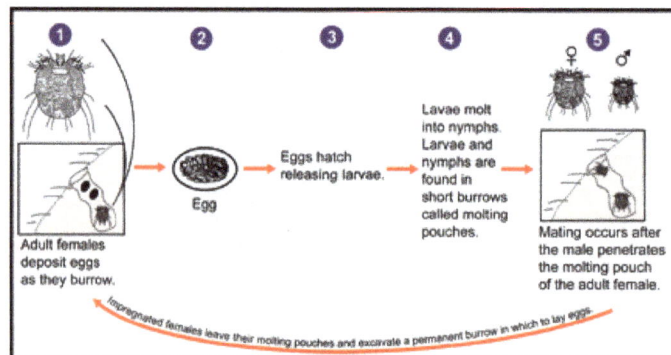

Life cycle of scabies

The movement of mites within and on the skin produces an intense itch, which has the characteristics of a delayed cell-mediated inflammatory response to allergens. IgE antibodies are present in the serum and the site of infection, which react to multiple protein allergens in the body of the mite. Some of these cross-react to allergens from house dust mites. Immediate antibody-mediated allergic reactions (wheals) have been elicited in infected persons, but not in healthy persons; immediate hypersensitivity of this type is thought to explain the observed far more rapid allergic skin response to reinfection seen in persons having been previously infected (especially having been infected within the previous year or two).

Transmission

Scabies is contagious and can be contracted through prolonged physical contact with an infested person. This includes sexual intercourse, although a majority of cases are acquired through other forms of skin-to-skin contact. Less commonly, scabies infestation can happen through the sharing of clothes, towels, and bedding, but this is not a major mode of transmission; individual mites can only survive for two to three days, at most, away from human skin. As with lice, a latex condom is ineffective against scabies transmission during intercourse, because mites typically migrate from one individual to the next at sites other than the sex organs.

Healthcare workers are at risk of contracting scabies from patients, because they may be in extended contact with them.

Pathophysiology

The symptoms are caused by an allergic reaction of the host's body to mite proteins, though exactly which proteins remains a topic of study. The mite proteins are also present from the gut, in mite feces, which are deposited under the skin. The allergic reaction is both of the delayed (cell-mediated) and immediate (antibody-mediated) type, and involves IgE (antibodies, it is presumed, mediate the very rapid symptoms on reinfection). The allergy-type symptoms (itching) continue for some days, and even several weeks, after all mites are killed. New lesions may appear for a few days after mites are eradicated. Nodular lesions from scabies may continue to be symptomatic for weeks after the mites have been killed.

Diagnosis

Magnified view of a burrowing trail of the scabies mite: The scaly patch on the left was caused by the scratching and marks the mite's entry point into the skin. The mite has burrowed to the top-right, where it can be seen as a dark spot at the end.

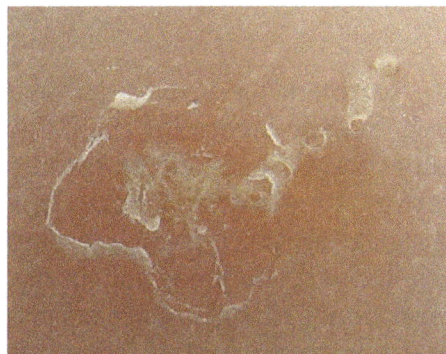

Scabies may be diagnosed clinically in geographical areas where it is common when diffuse itching presents along with either lesions in two typical spots or there is itchiness of another household member. The classical sign of scabies is the burrows made by the mites within the skin. To detect the burrow, the suspected area is rubbed with ink from a fountain pen or a topical tetracycline solution, which glows under a special light. The skin is then wiped with an alcohol pad. If the person is infected with scabies, the characteristic zigzag or S pattern of the burrow will appear across the skin; however, interpreting this test may be difficult as the burrows are scarce and may be obscured by scratch marks. A definitive diagnosis is made by finding either the scabies mites or their eggs and fecal pellets. Searches for these signs involve either scraping a suspected area, mounting the sample in potassium hydroxide and examining it under a microscope, or using dermoscopy to examine the skin directly.

Differential Diagnosis

Symptoms of early scabies infestation mirror other skin diseases, including dermatitis, syphilis, erythema multiforme, various urticaria-related syndromes, allergic reactions, and other ectoparasites such as lice and fleas.

Prevention

Mass treatment programs that use topical permethrin or oral ivermectin have been effective in reducing the prevalence of scabies in a number of populations. No vaccine is available for scabies. The simultaneous treatment of all close contacts is recommended, even if they show no symptoms of infection (asymptomatic), to reduce rates of recurrence. Since mites can survive for only two to three days without a host, other objects in the environment pose little risk of transmission except in the case of crusted scabies, thus cleaning is of little importance. Rooms used by those with crusted scabies require thorough cleaning.

Management

A number of medications are effective in treating scabies. Treatment should involve the entire household, and any others who have had recent, prolonged contact with the infested individual. Options to control itchiness include antihistamines and prescription anti-inflammatory agents. Bedding, clothing and towels used during the previous three days should be washed in hot water and dried in a hot dryer.

Permethrin

Permethrin is the most effective treatment for scabies, and remains the treatment of choice. It is applied from the neck down, usually before bedtime, and left on for about eight to 14 hours, then washed off in the morning. Care should be taken to coat the entire skin surface, not just symptomatic areas; any patch of skin left untreated can provide a "safe haven" for one or more mites to survive. One application is normally sufficient, as permethrin kills eggs and hatchlings as well as adult mites, though many physicians recommend a second application three to seven days later as a precaution. Crusted scabies may require multiple applications, or supplemental treatment with oral ivermectin (below). Permethrin may cause slight irritation of the skin that is usually tolerable.

Ivermectin

Oral Ivermectin is effective in eradicating scabies, often in a single dose. It is the treatment of choice for crusted scabies, and is sometimes prescribed in combination with a topical agent. It has not been tested on infants, and is not recommended for children under six years of age.

Topical ivermectin preparations have been shown to be effective for scabies in adults, though only one such formulation is available in the United States at present, and it is not FDA approved as a scabies treatment. It has also been useful for sarcoptic mange (the veterinary analog of human scabies).

Others

Other treatments include lindane, benzyl benzoate, crotamiton, malathion, and sulfur preparations. Lindane is effective, but concerns over potential neurotoxicity has limited its availability in many countries. It is banned in California, but may be used in other states as a second-line treatment. Sulfur ointments or benzyl benzoate are often used in the developing world due to their low cost; 10% sulfur solutions have been shown to be effective, and sulfur ointments are typically used for at least a week, though many people find the odor of sulfur products unpleasant. Crotamiton has been found to be less effective than permethrin in limited studies. Crotamiton or sulfur preparations are sometimes recommended instead of permethrin for children, due to concerns over dermal absorption of permethrin.

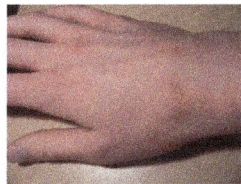
Day 12 (under treatment)

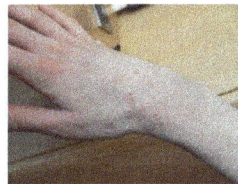
Day 4

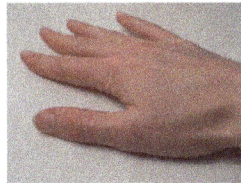
Healed

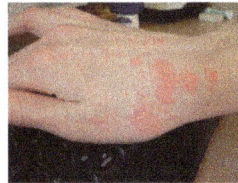
Day 8 (treatment begins)

Communities

Scabies is endemic in many developing countries, where it tends to be particularly problematic in rural and remote areas. In such settings community wide control strategies are required to reduce the rate of disease, as treatment of only individuals is ineffective due to the high rate of reinfection. Large-scale mass drug administration strategies may be required where coordinated interventions aim to treat whole communities in one concerted effort. Although such strategies have shown to be able to reduce the burden of scabies in these kinds of communities, debate remains about the best strategy to adopt, including the choice of drug.

The resources required to implement such large-scale interventions in a cost-effective and sustainable way are significant. Furthermore, since endemic scabies is largely restricted to poor and remote areas, it is a public health issue that has not attracted much attention from policy makers and international donors.

Epidemiology

Scabies is one of the three most common skin disorders in children, along with tinea and pyoderma. As of 2010 it affects approximately 100 million people (1.5% of the population) and is equally common in both genders. The mites are distributed around the world and equally infect all ages, races, and socioeconomic classes in different climates. Scabies is more often seen in crowded areas with unhygienic living conditions. Globally as of 2009, an estimated 300 million cases of scabies occur each year, although various parties claim the figure is either over- or underestimated. About 1–10% of the global population is estimated to be infected with scabies, but in certain populations, the infection rate may be as high as 50–80%.

History

Scabies has been observed in humans since ancient times. Archeological evidence from Egypt and the Middle East suggests scabies was present as early as 494 BC. The first recorded reference to scabies is believed to be from the Bible – it may be a type of "leprosy" mentioned in Leviticus *circa* 1200 BC or be mentioned among the curses of Deuteronomy 28. In the fourth century BC, Aristotle reported on "lice" that "escape from little pimples if they are pricked" — a description consistent with scabies.

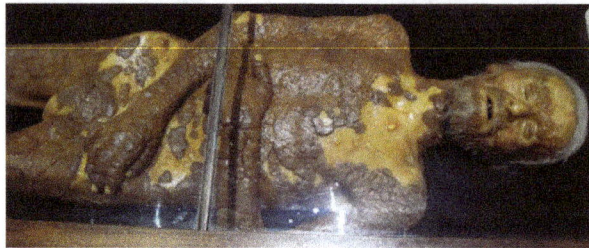

Wax figurine of a man with Norwegian scabies

The Roman encyclopedist and medical writer Aulus Cornelius Celsus (c. 25 BC – c. 50 AD) is credited with naming the disease "scabies" and describing its characteristic features. The parasitic etiology of scabies was documented by the Italian physician Giovanni Cosimo Bonomo (1663–1696) in his 1687 letter, "Observations concerning the fleshworms of the human body". Bonomo's description established scabies as one of the first human diseases with a well-understood cause.

In Europe in the late 19th through mid-20th centuries, a sulfur-bearing ointment called by the medical eponym of Wilkinson's ointment was widely used for topical treatment of scabies. The contents and origins of several versions of the ointment were detailed in correspondence published in the *British Medical Journal* in 1945.

Society and Culture

The International Alliance for the Control of Scabies (IACS) was started in 2012, and brings together over 70 researchers, clinicians and public health experts from more than 15 different countries. It has managed to bring the global health implications of scabies to the attention of the World Health Organization. Consequently, the WHO has included scabies on its official list of neglected tropical diseases and other neglected conditions.

Other Animals

Scabies may occur in a number of domestic and wild animals; the mites that cause these infestations are of different subspecies from the one typically causing the human form. These subspecies can infest animals that are not their usual hosts, but such infections do not last long. Scabies-infected animals suffer severe itching and secondary skin infections. They often lose weight and become frail.

A street dog in Bali, Indonesia, suffers from sarcoptic mange.

The most frequently diagnosed form of scabies in domestic animals is sarcoptic mange, caused by the subspecies *Sarcoptes scabiei canis*, most commonly in dogs and cats. Sarcoptic mange is transmissible to humans who come into prolonged contact with infested animals, and is distinguished from human scabies by its distribution on skin surfaces covered by clothing. Scabies-infected domestic fowl suffer what is known as "scaly leg". Domestic animals that have gone feral and have no veterinary care are frequently afflicted with scabies and a host of other ailments. Nondomestic animals have also been observed to suffer from scabies. Gorillas, for instance, are known to be susceptible to infection via contact with items used by humans.

Research

Moxidectin is being evaluated as a treatment for scabies. It is established in veterinary medicine to treat a range of parasites, including sarcoptic mange. Its advantage over ivermectin is its longer duration of action.

Human Flea

The human flea (*Pulex irritans*) is a cosmopolitan flea species that has, in spite of the common name, a wide host spectrum. It is one of six species in the genus *Pulex*; the other five are all confined to the Nearctic and Neotropical regions. The species is thought to have originated in South America, where its original host may have been the guinea pig or peccary.

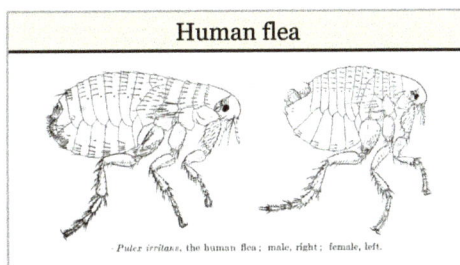

Human flea

Pulex irritans, the human flea; male, right; female, left.

Morphology and Behavior

Pulex irritans is a holometabolous insect with a four-part lifecycle consisting of eggs, larvae, pupae, and adults. Eggs are shed by the female in the environment and hatch into larvae in about

3–4 days. Larvae feed on organic debris in the environment. Larvae eventually form pupae, which are in cocoons that are often covered with debris from the environment (sand, pebbles, etc.). The larval and pupal stages are completed in about 3–4 weeks when the adults hatch from pupae, then must seek out a warm-blooded host for blood meals.

The flea eggs are about 0.5 mm in length. They are oval-shaped and pearly white in color. Eggs are often laid on the body of the host, but they often fall off in many different places. The larvae are about 6 mm in length. They are creamy white or yellow in color. Larvae have 13 segments with bristles on each segment. The larvae feed on a variety of organic debris. The pupa are around 4 x 2 mm. After undergoing three separate molts, the larvae pupate, then emerge as adults. If conditions are unfavorable, a cocooned flea can remain dormant for up to a year in the pupal phase. The adults are roughly 1.5 to 4 mm in length and are laterally flattened. They are dark brown in color, are wingless, and have piercing-sucking mouthparts that aid in feeding on the host's blood. Both genal and pronotal combs are absent and the adult flea has a rounded head. Most fleas are distributed in the egg, larval, or pupal stages.

Relationship With Host

Direct Effects of Bites

Fleas are a pest species to their hosts, causing an itching sensation that results in discomfort and leads to scratching in the vicinity of the bite. Flea bites generally cause the skin to raise, swell, and itch. The bite site has a single puncture point in the center. Bites often appear in clusters or small rows and can remain inflamed for up to several weeks.

This species bites many species of mammals and birds, including domesticated ones. It has been found on dogs and wild canids, monkeys in captivity, opossums, domestic cats, wild felids in captivity, chickens, black rats and Norwegian rats, wild rodents, pigs, free-tailed bats, and other species. It can also be an intermediate host for the cestode *Dipylidium caninum*.

Fleas can spread rapidly and move between areas to include eyebrows, eyelashes, and pubic regions.

Hair loss as a result of itching is common, especially in wild and domestic animals. Anemia is also possible in extreme cases of high-volume infestations.

Vector Capabilities

Plague, a disease that affects humans and other mammals, is caused by the bacterium *Yersinia pestis*. The human flea can be a carrier of the plague bacterium. Plague is infamous for killing millions of people in Europe during the Middle Ages. Without prompt treatment, the disease can cause serious illness or death. Today, human plague infections continue to occur in the western United States, but significantly more cases occur in parts of Africa and Asia.

Plague is a very serious illness, but is treatable with commonly available antibiotics. The earlier a patient seeks medical care and receives treatment that is appropriate for plague, the better their chances are of a full recovery.

People in close contact with very sick pneumonic plague patients may be evaluated and possibly placed under observation. Preventive antibiotic therapy may also be given, depending on the type and timing of personal contact.

Treatment and Prevention

Common treatments include body shaving and medicated shampoos and combing. When preventing fleas, it is important to treat both the host and the environment. Household pets should be treated to prevent being infested. Carpets, and other floor surfaces, should be shampooed and vacuumed regularly to prevent fleas. When dealing with an infestation, bedding and clothing that may have been exposed should be washed thoroughly in hot water and kept away from other materials that have not been exposed. In extreme cases, extermination by a professional is necessary.

Bed Bug

Bed bugs, bed-bugs, or bedbugs are parasitic insects of the cimicid family that feed exclusively on blood. *Cimex lectularius*, the common bed bug, is the best known as it prefers to feed on human blood. Other *Cimex* species specialize in other animals, e.g., bat bugs, such as *Cimex pipistrelli* (Europe), *Cimex pilosellus* (western US), and *Cimex adjunctus* (entire eastern US).

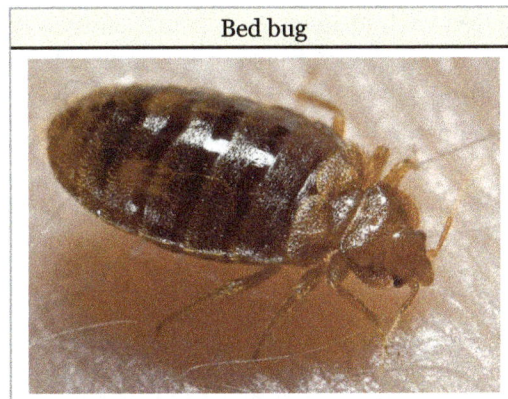

Bed bug

The name "bed bug" derives from the preferred habitat of *Cimex lectularius*: warm houses and especially near or inside beds and bedding or other sleep areas. Bed bugs are mainly active at night, but are not exclusively nocturnal. They usually feed on their hosts without being noticed.

A number of adverse health effects may result from bed bug bites, including skin rashes, psychological effects, and allergic symptoms. Bed bugs are not known to transmit any pathogens as disease vectors. Certain signs and symptoms suggest the presence of bed bugs; finding the adult insects confirms the diagnosis.

Bed bugs have been known as human parasites for thousands of years. At a point in the early 1940s, they were mostly eradicated in the developed world, but have increased in prevalence since 1995, likely due to pesticide resistance, governmental bans on effective pesticides, and international travel. Because infestation of human habitats has begun to increase, bed bug bites and related conditions have been on the rise as well.

Infestation

Diagnosis of an infestation involves both finding bed bugs and the occurrence of compatible symptoms. Treatment involves the elimination of the insect (including its eggs) and taking measures to treat symptoms until they resolve.

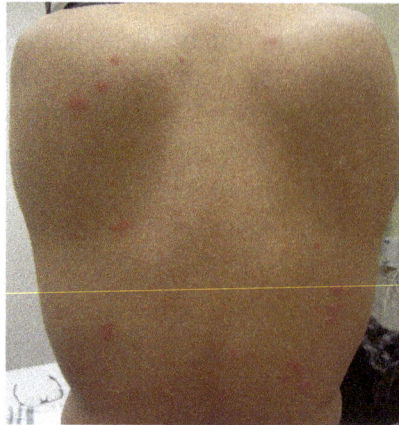

Bedbug bites

Bed bug bites or cimicosis may lead to a range of skin manifestations from no visible effects to prominent blisters. Effects include skin rashes, psychological effects, and allergic symptoms.

Although bed bugs can be infected with at least 28 human pathogens, no studies have found that the insects are capable of transmitting any of these to humans. They have been found with methicillin-resistant *Staphylococcus aureus* (MRSA) and with vancomycin-resistant *Enterococcus faecium* (VRE), but the significance of this is still unknown.

Investigations into potential transmission of HIV, MRSA, hepatitis B, hepatitis C, and hepatitis E have not shown that bed bugs can spread these diseases. However, arboviruses may be transmissible.

Description

Physical

Adult bed bugs are light brown to reddish-brown, flattened, oval-shaped, and have no hind wings. The front wings are vestigial and reduced to pad-like structures. Bed bugs have segmented abdomens with microscopic hairs that give them a banded appearance. Adults grow to 4–5 mm (0.16–0.20 in) long and 1.5–3 mm (0.059–0.118 in) wide.

Newly hatched nymphs are translucent, lighter in color, and become browner as they moult and reach maturity. A bed bug nymph of any age that has just consumed a blood meal has a bright red, translucent abdomen, fading to brown over the next several hours, and to opaque black within two days as the insect digests its meal. Bed bugs may be mistaken for other insects, such as booklice, small cockroaches, or carpet beetles; however, when warm and active, their movements are more ant-like and, like most other true bugs, they emit a characteristic disagreeable odor when crushed.

Bed bugs use pheromones and kairomones to communicate regarding nesting locations, feeding, and reproduction.

The lifespan of bed bugs varies by species and is also dependent on feeding.

Bed bugs can survive a wide range of temperatures and atmospheric compositions. Below 16.1 °C (61.0 °F), adults enter semihibernation and can survive longer; they can survive for at least five days at −10 °C (14 °F), but die after 15 minutes of exposure to −32 °C (−26 °F). Common commercial and residential freezers reach temperatures low enough to kill most life stages of bed bug, with 95% mortality after 3 days at −12 °C (10 °F). They show high desiccation tolerance, surviving low humidity and a 35–40 °C range even with loss of one-third of body weight; earlier life stages are more susceptible to drying out than later ones.

The thermal death point for *C. lectularius* is 45 °C (113 °F); all stages of life are killed by 7 minutes of exposure to 46 °C (115 °F). Bed bugs apparently cannot survive high concentrations of carbon dioxide for very long; exposure to nearly pure nitrogen atmospheres, however, appears to have relatively little effect even after 72 hours.

Feeding Habits

Bed bugs are obligatory hematophagous (bloodsucking) insects. Most species feed on humans only when other prey are unavailable. They obtain all the additional moisture they need from water vapor in the surrounding air. Bed bugs are attracted to their hosts primarily by carbon dioxide, secondarily by warmth, and also by certain chemicals. Bedbugs prefer exposed skin, preferably the face, neck, and arms of a sleeping person.

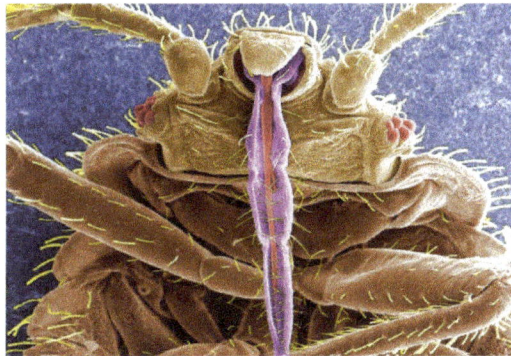

A scanning electron micrograph (SEM) of *Cimex lectularius*, digitally colorized with the insect's skin-piercing mouthparts highlighted in purple and red

Bedbugs have mouth parts that saw through the skin, and inject saliva with anticoagulants and painkillers. Sensitivity of humans varies from extreme allergic reaction to no reaction at all (about 20%). The bite usually produces a swelling with no red spot, but when many bugs feed on a small area, reddish spots may appear after the swelling subsides.

Although under certain cool conditions adult bed bugs can live for over a year without feeding, under typically warm conditions they try to feed at five- to ten-day intervals, and adults can survive for about five months without food. Younger instars cannot survive nearly as long, though even the vulnerable newly hatched first instars can survive for weeks without taking a blood meal.

At the 57th annual meeting of the Entomological Society of America in 2009, newer generations of pesticide-resistant bed bugs in Virginia were reported to survive only two months without feeding.

DNA from human blood meals can be recovered from bed bugs for up to 90 days, which mean they can be used for forensic purposes in identifying on whom the bed bugs have fed.

Feeding Physiology

A bed bug pierces the skin of its host with a stylet fascicle, rostrum, or "beak". The rostrum is composed of the maxillae and mandibles, which have been modified into elongated shapes from a basic, ancestral style. The right and left maxillary stylets are connected at their midline and a section at the centerline forms a large food canal and a smaller salivary canal. The entire maxillary and mandibular bundle penetrates the skin.

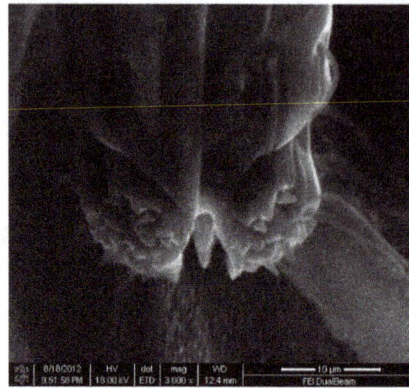

The tip of a bed bug rostrum

The tips of the right and left maxillary stylets are not the same; the right is hook-like and curved, and the left is straight. The right and left mandibular stylets extend along the outer sides of their respective maxillary stylets and do not reach anywhere near the tip of the fused maxillary stylets. The stylets are retained in a groove in the labium, and during feeding, they are freed from the groove as the jointed labium is bent or folded out of the way; its tip never enters the wound.

The mandibular stylet tips have small teeth, and through alternately moving these stylets back and forth, the insect cuts a path through tissue for the maxillary bundle to reach an appropriately sized blood vessel. Pressure from the blood vessel itself fills the insect with blood in three to five minutes. The bug then withdraws the stylet bundle from the feeding position and retracts it back into the labial groove, folds the entire unit back under the head, and returns to its hiding place. It takes between five and ten minutes for a bed bug to become completely engorged with blood. In all, the insect may spend less than 20 minutes in physical contact with its host, and does not try to feed again until it has either completed a moult or, if an adult, has thoroughly digested the meal.

Reproduction

All bed bugs mate by traumatic insemination. Female bed bugs possess a reproductive tract that functions during oviposition, but the male does not use this tract for sperm insemination. Instead, the male pierces the female's abdomen with his hypodermic penis and ejaculates into the body cavity. In all bed bug species except *Primicimex cavernis,* sperm are injected into the mesospermalege, a component of the spermalege, a secondary genital structure that reduces the wounding and immunological costs of traumatic insemination. Injected sperm travel via the haemolymph (blood) to sperm storage structures called seminal conceptacles, with fertilisation eventually taking place at the ovaries.

Male bed bugs sometimes attempt to mate with other males and pierce their abdomens. This behaviour occurs because sexual attraction in bed bugs is based primarily on size, and males mount any freshly fed partner regardless of sex. The "bed bug alarm pheromone" consists of (*E*)-2-octenal and (*E*)-2-hexenal. It is released when a bed bug is disturbed, as during an attack by a predator. A 2009 study demonstrated the alarm pheromone is also released by male bed bugs to repel other males that attempt to mate with them.

A male bed bug (*Cimex lectularius*) traumatically inseminates a female

Cimex lectularius and *C. hemipterus* mate with each other given the opportunity, but the eggs then produced are usually sterile. In a 1988 study, one of 479 eggs was fertile and resulted in a hybrid, *Cimex hemipterus* × *lectularius*.

Sperm Protection

Cimex lectularius males have environmental microbes on their genitals. These microbes damage sperm cells, leaving them unable to fertilize female gametes. Due to these dangerous microbes, males have evolved antimicrobial ejaculate substances that prevent sperm damage. When the microbes contact sperm or the male genitals, the bed bug releases antimicrobial substances. Many species of these microbes live in the bodies of females after mating. The microbes can cause infections in the females. It has been suggested that females receive benefit from the ejaculate. Though the benefit is not direct, females are able to produce more eggs than optimum increasing the amount of the females' genes in the gene pool.

Sperm and Seminal Fluid Allocation

In organisms, sexual selection extends past differential reproduction to affect sperm composition, sperm competition, and ejaculate size. Males of *C. lectularius* allocate 12% of their sperm and 19% of their seminal fluid per mating. Due to these findings, Reinhard et. al proposed that multiple mating is limited by seminal fluid and not sperm. After measuring ejaculate volume, mating rate and estimating sperm density, Reinhardt et al. showed that mating could be limited by seminal fluid. Despite these advances, the cost difference between ejaculate-dose dependence and mating frequency dependence have not been explored.

Egg Production

Males fertilize females only by traumatic insemination into the structure called the ectospermalege (the organ of Berlese, however the organ of Ribaga (as it was first named) was first designated as an organ of stridulation. These two names are not descriptive, so other terminologies are used). On fertilization, the female's ovaries finish developing, which suggests that sperm plays a role other than fertilizing the egg. Fertilization also allows for egg production through the corpus allatum. Sperm remains viable in a female's spermathecae (a better term is conceptacle), a sperm-carrying sack, for a long period of time as long as body temperature is optimum. The female lays fertilized eggs until she depletes the sperm found in her conceptacle. After the depletion of sperm, she lays a few sterile eggs. The number of eggs a *C. lectularius* female produces does not depend on the sperm she harbors, but on the female's nutritional level.

Alarm Pheromones

In *C. lectularius*, males sometimes mount other males because male sexual interest is directed at any recently fed individual regardless of their sex, but unfed females may also be mounted. Traumatic insemination is the only way for copulation to occur in bed bugs. Females have evolved the spermalege to protect themselves from wounding and infection. Because males lack this organ, traumatic insemination could leave them badly injured. For this reason, males have evolved alarm pheromones to signal their sex to other males. If a male *C. lectularius* mounts another male, the mounted male releases the pheromone signal and the male on top stops before insemination.

Females are capable of producing alarm pheromones to avoid multiple mating, but they generally do not do so. Two reasons are proposed as to why females do not release alarm pheromones to protect themselves. First, alarm pheromone production is costly. Due to egg production, females may refrain from spending additional energy on alarm pheromones. The second proposed reason is that releasing the alarm pheromone reduces the benefits associated with multiple mating. Benefits of multiple mating include material benefits, better quality nourishment or more nourishment, genetic benefits including increased fitness of offspring, and finally, the cost of resistance may be higher than the benefit of consent—which appears the case in *C. lectularius*.

Life Stages

Bed bugs have five immature nymph life stages and a final sexually mature adult stage. They shed their skins through ecdysis at each stage, discarding their outer exoskeleton, which is somewhat clear, empty exoskeletons of the bugs themselves. Bed bugs must molt six times before becoming fertile adults, and must consume at least one blood meal to complete each moult.

Slide of *Cimex lectularius*

Bed bug (4 mm length; 2.5 mm width), shown in a film roll plastic container, on the right is the recently sloughed skin from its nymph stage

Each of the immature stages lasts about a week, depending on temperature and the availability of food, and the complete lifecycle can be completed in as little as two months (rather long compared to other ectoparasites). Fertilized females with enough food lay three to four eggs each day continually until the end of their lifespans (about nine months under warm conditions), possibly generating as many as 500 eggs in this time. Genetic analysis has shown that a single pregnant bed bug, possibly a single survivor of eradication, can be responsible for an entire infestation over a matter of weeks, rapidly producing generations of offspring.

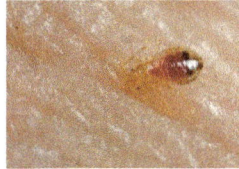

A bed bug nymph feeding on a host

Blood-fed *C. lectularius* (note the differences in color with respect to digestion of blood meal)

Sexual Dimorphism

Sexual dimorphism occurs in *C. lectularius*, with the females larger in size than the males on average. The abdomens of the sexes differ in that the males appear to have "pointed" abdomens, which are actually their copulatory organs, while females have more rounded abdomens. Since males are attracted to large body size, any bed bug with a recent blood meal can be seen as a potential mate. However, males will mount unfed, flat females on occasion. The female is able to curl her abdomen forward and underneath toward the head to not mate. Males are generally unable to discriminate between the sexes until after mounting, but before inseminating.

Host Searching

C. lectularius only feeds every five to seven days, which suggests that it does not spend the majority of its life searching for a host. When a bed bug is starved, it leaves its shelter and searches for a host. If it successfully feeds, it returns to its shelter. If it does not feed, it continues to search for a host. After searching—regardless of whether or not it has eaten—the bed bug returns to the shelter to aggregate before the photophase (period of light during a day-night cycle). Reis argues that two reasons explain why *C. lectularius* would return to its shelter and aggregate after feeding. One is to find a mate and the other is to find shelter to avoid getting smashed after eating.

Aggregation and Dispersal Behavior

C. lectularius aggregates under all life stages and mating conditions. Bed bugs may choose to aggregate because of predation, resistance to desiccation, and more opportunities to find a mate.

Airborne pheromones are responsible for aggregations. Another source of aggregation could be the recognition of other *C. lectularius* bugs through mechanoreceptors located on their antennae. Aggregations are formed and disbanded based on the associated cost and benefits. Females are more often found separate from the aggregation than males. Females are more likely to expand the population range and find new sites. Active female dispersal can account for treatment failures. Males, when found in areas with few females, abandon an aggregation to find a new mate. The males excrete an aggregation pheromone into the air that attracts virgin females and arrests other males.

Detection

Bed bug eggs and two adult bed bugs from inside a dresser

Bed bugs can exist singly, but tend to congregate once established. Though strictly parasitic, they spend only a tiny fraction of their lifecycles physically attached to hosts. Once a bed bug finishes feeding, it relocates to a place close to a known host, commonly in or near beds or couches in clusters of adults, juveniles, and eggs—which entomologists call harborage areas or simply harborages to which the insect returns after future feedings by following chemical trails. These places can vary greatly in format, including luggage, inside of vehicles, within furniture, amongst bedside clutter—even inside electrical sockets and nearby laptop computers. Bed bugs may also nest near animals that have nested within a dwelling, such as bats, birds, or rodents. They are also capable of surviving on domestic cats and dogs, though humans are the preferred host of *C. lectularius*.

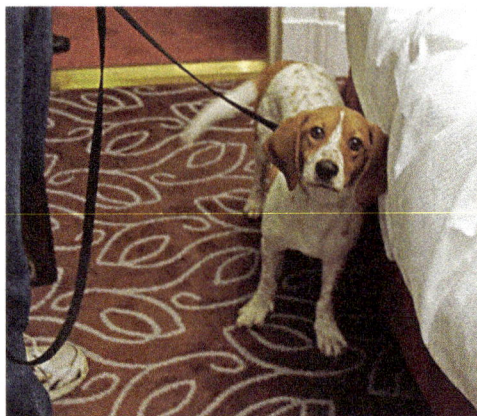

A bed bug detection dog in New York

Bed bug fecal spot

Bed bugs can also be detected by their characteristic smell of rotting raspberries. Bed bug detection dogs are trained to pinpoint infestations, with a possible accuracy rate between 11% and 83%.

Bed bug roaming around carpet wrinkles

Management

Eradication of bed bugs frequently requires a combination of nonpesticide approaches and the occasional use of pesticides.

Mechanical approaches, such as vacuuming up the insects and heat-treating or wrapping mattresses, are effective. A combination of heat and drying treatments is most effective. An hour at a temperature of 45 °C (113 °F) or over, or two hours at less than −17 °C (1 °F) kills them; a domestic clothes drier or steam kills bedbugs. Another study found 100% mortality rates for bed bugs exposed to temperatures greater than 50 °C (122 °F) for more than 2 minutes. Starving them is difficult as they can survive without eating for 100 to 300 days, depending on temperature. For public health reasons, individuals are encouraged to call a professional pest control service to eradicate bed bugs in a home, rather than attempting to do it themselves, particularly if they live in a multifamily building.

As of 2012, no truly effective pesticides were available. Pesticides that have historically been found effective include pyrethroids, dichlorvos, and malathion. Resistance to pesticides has increased significantly over time, and harm to health from their use is of concern. The carbamate insecticide propoxur is highly toxic to bed bugs, but it has potential toxicity to children exposed to it, and the US Environmental Protection Agency has been reluctant to approve it for indoor use. Boric acid, occasionally applied as a safe indoor insecticide, is not effective against bed bugs because they do not groom. The fungus *Beauveria bassiana* is being researched as of 2012 for its ability to control bed bugs. As bed bugs continue to adapt pesticide resistance, researchers have examined on the insect's genome to see how the adaptations develop and to look for potential vulnerabilities that can be exploited in the growth and development phases.

Predators

Natural enemies of bed bugs include the masked hunter insect (also known as "masked bed bug hunter"), cockroaches, ants, spiders (particularly *Thanatus flavidus*), mites, and centipedes (particularly the house centipede *Scutigera coleoptrata*). However, biological pest control is not considered practical for eliminating bed bugs from human dwellings.

Epidemiology

Bed bugs occur around the world. Rates of infestations in developed countries, while decreasing from the 1930s to the 1980s, have increased dramatically since the 1980s. Previously, they were common in the developing world, but rare in the developed world. The increase in the developed world may have been caused by increased international travel, resistance to insecticides, and the use of new pest-control methods that do not affect bed bugs.

The fall in bed bug populations after the 1930s in the developed world is believed partly due to the use of DDT to kill cockroaches. The invention of the vacuum cleaner and simplification of furniture design may have also played a role. Others believe it might simply be the cyclical nature of the organism.

The exact causes of this resurgence remain unclear; it is variously ascribed to greater foreign travel, increased immigration from the developing world to the developed world, more frequent exchange of second-hand furnishings among homes, a greater focus on control of other pests, resulting in neglect of bed bug countermeasures, and increasing resistance to pesticides. Declines in household cockroach populations that have resulted from the use of insecticides effective against this major bed bug predator have aided the bed bugs' resurgence, as have bans on DDT and other potent pesticides.

The common bed bug (*C. lectularius*) is the species best adapted to human environments. It is found in temperate climates throughout the world. Other species include *Cimex hemipterus*, found in tropical regions, which also infests poultry and bats, and *Leptocimex boueti*, found in the tropics of West Africa and South America, which infests bats and humans. *Cimex pilosellus* and *Cimex pipistrella* primarily infest bats, while *Haematosiphon inodora*, a species of North America, primarily infests poultry.

History

Bed bugs were mentioned in ancient Greece as early as 400 BC, and were later mentioned by Aristotle. Pliny's *Natural History*, first published *circa* 77 AD in Rome, claimed bed bugs had medicinal value in treating ailments such as snake bites and ear infections. (Belief in the medicinal use of bed bugs persisted until at least the 18th century, when Guettard recommended their use in the treatment of hysteria.)

Bed bugs were first mentioned in Germany in the 11th century, in France in the 13th century, and in England in 1583, though they remained rare in England until 1670. Some in the 18th century believed bed bugs had been brought to London with supplies of wood to rebuild the city after the Great Fire of London (1666). Giovanni Antonio Scopoli noted their presence in Carniola (roughly equivalent to present-day Slovenia) in the 18th century.

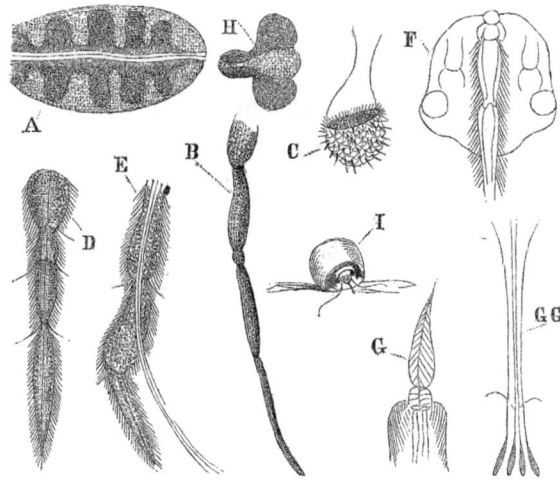

FIGURE 12.—PARTS OF RED-BUG.

A. Intestines.—B. Antenna of the Male.—C. Eye.—D. Haustellum, or Sucker, closed.—E. Side view of Sucker.—F. Under Part of Head.—G. Under Lip.—GG. Hair of the Tube, and outside Cases.—H. Egg-Bag.—I. Worm emerging from the Egg.

An 1860 engraving of parts of a bed bug. A. Intestines – B. Antenna of the male – C. Eye – D. Haustellum, or sucker, closed – E. Side view of sucker – F. Under part of head – G. Under lip – GG. Hair of the tube, and outside cases – H. Egg-bag – I. Larva emerging from the eggs

C. lectularius may have originated in the Middle East in caves inhabited by bats and humans.

Traditional methods of repelling and/or killing bed bugs include the use of plants, fungi, and insects (or their extracts), such as black pepper; black cohosh (*Actaea racemosa*); *Pseudarthria hookeri*; *Laggera alata* (Chinese *yángmáo c ǎ o* | 羊毛草); *Eucalyptus saligna* oil; henna (*Lawsonia inermis* or camphire); "infused oil of *Melolontha vulgaris*" (presumably cockchafer); fly agaric (*Amanita muscaria*); *Actaea* spp. (e.g. black cohosh); tobacco; "heated oil of Terebinthina" (i.e. true turpentine); wild mint (*Mentha arvensis*); narrow-leaved pepperwort (*Lepidium ruderale*); *Myrica* spp. (e.g. bayberry); Robert geranium (*Geranium robertianum*); bugbane (*Cimicifuga* spp.); "herb and seeds of *Cannabis*"; "opulus" berries (possibly maple or European cranberry-bush); masked hunter bugs (*Reduvius personatus*), "and many others".

In the mid-19th century, smoke from peat fires was recommended as an indoor domestic fumigant against bed bugs.

Dusts have been used to ward off insects from grain storage for centuries, including "plant ash, lime, dolomite, certain types of soil, and diatomaceous earth or Kieselguhr". Of these, diatomaceous earth in particular has seen a revival as a nontoxic (when in amorphous form) residual pesticide for bed bug abatement. While diatomaceous earth performed poorly, silica gel may be effective.

Basket-work panels were put around beds and shaken out in the morning in the UK and in France in the 19th century. Scattering leaves of plants with microscopic hooked hairs around a bed at night, then sweeping them up in the morning and burning them, was a technique reportedly used in Southern Rhodesia and in the Balkans.

Bean leaves have been used historically to trap bedbugs in houses in Eastern Europe. The trichomes on the bean leaves capture the insects by impaling the feet (tarsi) of the insects. The leaves are then destroyed.

20th Century

Prior to the mid-20th century, bed bugs were very common. According to a report by the UK Ministry of Health, in 1933, all the houses in many areas had some degree of bed bug infestation. The increase in bed bug populations in the early 20th century has been attributed to the advent of electric heating, which allowed bed bugs to thrive year-round instead of only in warm weather.

Bed bugs were a serious problem at U.S. military bases during World War II. Initially, the problem was solved by fumigation, using Zyklon Discoids that released hydrogen cyanide gas, a rather dangerous procedure. Later, DDT was used to good effect as a safer alternative.

The decline of bed bug populations in the 20th century is often credited to potent pesticides that had not previously been widely available. Other contributing factors that are less frequently mentioned in news reports are increased public awareness and slum clearance programs that combined pesticide use with steam disinfection, relocation of slum dwellers to new housing, and in some cases also follow-up inspections for several months after relocated tenants moved into their new housing.

Resurgence

Bed bug infestations resurged since the 1980s for reasons that are not clear, but contributing factors may be complacency, increased resistance, bans on pesticides, and increased international travel. The U.S. National Pest Management Association reported a 71% increase in bed bug calls between 2000 and 2005. The number of reported incidents in New York City alone rose from 500 in 2004 to 10,000 in 2009. In 2013, Chicago was listed as the number 1 city in the United States with the worst bed bug infestation. As a result, the Chicago City Council passed a bed bug control ordinance to limit their spread. Additionally, bed bugs are reaching places in which they never established before, such as southern South America.

One recent theory about bed bug reappearance in the US is that they never truly disappeared, but may have been forced to alternative hosts. Consistent with this is the finding that bed bug DNA shows no evidence of an evolutionary bottleneck. Furthermore, investigators have found high populations of bed bugs at poultry facilities in Arkansas. Poultry workers at these facilities may be spreading bed bugs, unknowingly carrying them to their places of residence and elsewhere after leaving work.

Society and Culture

The saying "Good night, sleep tight, don't let the bed bugs bite" is common for parents to say to young children before they go to sleep. In Chhattisgarh, India, bed bugs have been used as a traditional medicine for epilepsy, piles, alopecia, and urinary disorders, but this practice has no scientific basis. Bed bug secretions can inhibit the growth of some bacteria and fungi; antibacterial components from the bed bug could be used against human pathogens, and be a source of pharmacologically active molecules as a resource for the discovery of new drugs.

Etymology

The word *bug* and its earlier spelling *bugge* originally meant "bed bug". Many other creatures are now called "bugs", such as the "ladybug" ("ladybird" outside North America) and the "potatobug".

The word is used informally for any insect, or even microscopic germs, or diseases caused by these germs, but the earliest recorded use of the actual word "bug" referred to a bed bug.

The term "bed bug" may also be spelled "bedbug" or "bed-bug", though published sources consistently use the unhyphenated two-word name "bed bug". The pests have been known by a variety of other informal names, including wall louse, mahogany flat, crimson rambler, chilly billies, heavy dragoon, chinche bug, and redcoat.

Cochliomyia Hominivorax

Cochliomyia hominivorax, the New World screw-worm fly, or screw-worm for short, is a species of parasitic fly that is well known for the way in which its larvae (maggots) eat the living tissue of warm-blooded animals. It is present in the New World tropics. There are five species of *Cochliomyia* but only one species of screw-worm fly in the genus; there is also a single Old World species in a different genus (*Chrysomya bezziana*). Infestation of a live vertebrate animal by a maggot is technically called *myiasis*. While the maggots of many fly species eat dead flesh, and may occasionally infest an old and putrid wound, screw-worm maggots are unusual because they attack healthy tissue. Screw-worms are a reportable species to the state veterinarian in the United States if discovered on livestock.

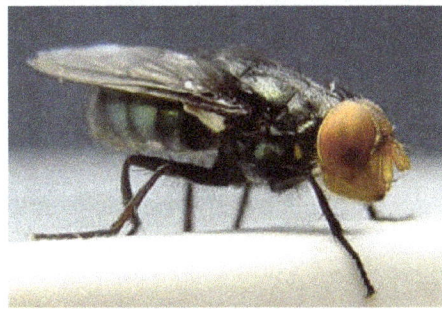

Cochliomyia hominivorax

Cochliomyia hominivorax larva.

Lifecycle

Screw-worm females lay 250-500 eggs in the exposed flesh of warm-blooded animals, including humans, such as in wounds and the navels of newborn animals. The larvae hatch and burrow into the surrounding tissue as they feed. If the wound is disturbed during this time, the larvae burrow or "screw" deeper into the flesh, which is the source of the insect's name. The maggots are capable of causing severe tissue damage or even death to the host. About three to seven days after hatching, the larvae fall to the ground to pupate. The pupae reach the adult stage about seven days later. Female screw-worm flies mate four to five days after hatching. The entire lifecycle is around 20 days. A female can lay up to 3,000 eggs and fly up to 200 km (120 mi) during her life.

Control

The United States officially eradicated the screw-worm in 1982 using the sterile insect technique. The screw-worm was eradicated in Guatemala and Belize in 1994, El Salvador in 1995, and Honduras in 1996. Campaigns against the flies continue in Mexico, Nicaragua, Costa Rica, Panama and Jamaica with financial assistance from the United States Department of Agriculture.

References

- Burgess, IF (1995). "Human lice and their management". Advances in parasitology. Advances in Parasitology. 36: 271–342. doi:10.1016/S0065-308X(08)60493-5. ISBN 978-0-12-031736-3.

- Markell, Edward K.; John, David C.; Petri, William H. (2006). Markell and Voge's medical parasitology (9th ed.). St. Louis, Mo: Elsevier Saunders. ISBN 0-7216-4793-6.

- Carol Turkington; Jeffrey S. Dover, M.D. (2006). The Encyclopedia of Skin and Skin Disorders. New York: Facts on File inc. ISBN 978-0-8160-6403-8.

- Mullen, Gary R.; Durden, Lance A. (8 May 2009). Medical and Veterinary Entomology, Second Edition. Academic Press. p. 80. ISBN 0-12-372500-3.

- Miller, Dini (2008). "Bed bugs (hemiptera: cimicidae: Cimex spp.)". In Capinera, John L. Encyclopedia of Entomology (Second ed.). Springer. p. 414. ISBN 978-1-4020-6242-1.

- Borgman W (June 30, 2006). Dog mange called scabies can transfer to humans. Orlando Sentinel archive. Retrieved February 16, 2015.

- Olson, Joelle; Eaton, Marc; Kells, Stephen; Morin, Victor; Wang, Changlu (2013). "Cold Tolerance of Bed Bugs and Practical Recommendations for Control". Journal of Economic Entomology. 106 (6): 2433–2441. doi:10.1603/EC13032. Retrieved 9 July 2014.

- Reinhardt, Klaus; Siva-Jothy, Micheal (2007). "Biology of the Bed Bugs (Cimicidae)". Annual Review of Entomology. 52: 351–374. doi:10.1146/annurev.ento.52.040306.133913. PMID 16968204. Retrieved 10 July 2014.

- Anderson, Andrea (February 8, 2008). "DNA from Peruvian Mummy Lice Reveals History". GenomeWeb Daily News. GenomeWeb LLC. Retrieved August 31, 2014.

- Melnick, Meredith (2011-05-12). "Study: Bedbugs May Carry MRSA; Germ Transmission Unclear | TIME.com". Healthland.time.com. Retrieved 2013-11-11.

Permissions

Index